Lockhart & Wiseman's
crop husbandry

Related titles from Woodhead's food science, technology and nutrition list:

Fruit and vegetable processing (ISBN 978-1-85573-548-4)

Fruit and vegetables are both major food products in their own right and key ingredients in many processed foods. This new book provides an authoritative survey of the latest research on improving the sensory, nutritional and functional qualities of fruit and vegetables whether as fresh or processed goods for the customer.

Fruit and vegetable biotechnology (ISBN 978-1-85573-467-8)

The genetic modification of food is one of the most significant and controversial developments in food processing. This important new collection reviews its application to fruit and vegetables. Part 1 looks at techniques and their applications in improving production and product quality. Part 2 discusses how genetic modification has been applied to specific crops, whilst Part 3 considers safety and consumer issues.

Kent's technology of cereals Fourth edition (ISBN 978-1-85573-361-9)

This well-established textbook provides students of food science with an authoritative and comprehensive study of cereal technology. Kent compares the merits and limitations of individual cereals as sources of food products as well as looking at the effects of processing treatments on the nutritive value of the products. The fourth edition of this classic book has been thoroughly updated with new sections including extrusion cooking and the use of cereals for animal feed.

Details of these books and a complete list of Woodhead's food science, technology and nutrition titles can be obtained by:

- visiting our web site at www.woodheadpublishing.com
- contacting Customer Services (e-mail: sales@woodheadpublishing.com; fax: +44 (0) 1223 893694; tel.: +44 (0) 1223 891358 ext. 130; address: Woodhead Publishing Limited, Abington Hall, Granta Park, Great Abington, Cambridge CB21 6AH, UK)

If you would like to receive information on forthcoming titles in this area, please send your address details to: Francis Dodds (address, tel. and fax as above; e-mail: francis.dodds@woodheadpublishing.com). Please confirm which subject areas you are interested in.

Lockhart & Wiseman's crop husbandry

including grassland

Eighth edition

H. J. S. Finch, A. M. Samuel and G. P. F. Lane

WOODHEAD PUBLISHING LIMITED

Oxford   Cambridge   New Delhi

Published by Woodhead Publishing Limited
Abington Hall, Granta Park, Great Abington
Cambridge CB21 6AH, UK
www.woodheadpublishing.com

First edition 1966 Pergamon Press Limited
Second edition 1970
Third edition 1975
Reprinted 1976
Fourth edition 1978
Reprinted with additions 1980
Fifth edition 1983
Reprinted 1984
Sixth edition 1988
Seventh edition 1993
Reprinted 2000 Woodhead Publishing Limited by agreement with Butterworth-Heinemann Limited
Eighth edition 2002 Woodhead Publishing Limited
Reprinted 2007, 2009

British Library Cataloguing in Publication Data
A catalogue record for this book is available from the British Library.

ISBN 978-1-85573-549-1 (book)
ISBN 978-1-85573-650-4 (e-book)

Printed by TJI Digital, Padstow, Cornwall, UK

Contents

x   Contents

Foreword

Since the last edition of this book was published, British agriculture has seen the start of what can almost be described as a revolution – certainly radical changes in the industry are now developing. It is probably true to say that, apart from the fact that the country is importing more of its food requirements, another important reason is a growing realisation that the countryside has a multi-purpose use – farming, conservation (wildlife and protection of the environment) and tourism. It is difficult to ignore the fact that tourism in the countryside is worth annually about £12 billion (2001) which compares with farm income in the same year of £2 billion – a lower than anticipated figure because of low cereal yields and prices, and the devastating effects of foot and mouth disease. But if agriculture is no longer the main industry in our rural areas, it remains true that prosperous farming does mean, in the main, an attractive countryside – a fact which should find favour with conservationists and those who visit the countryside for recreation.

The 1990s (particularly the second half) saw a catastrophic downturn in farm income as a result of the strong pound and the industry – through the EU's Common Agricultural Policy – starting to line itself up with world prices and, at the same time, making a definitive move towards a more environmental approach to farming. In this last context yield can no longer be the sole motive in obtaining an economic return; there are other criteria, not the least of which is one that is compatible with its effect on the environment in its widest context. More attention will certainly be paid to this in the future, and its importance is underlined in this new edition. However, yield, crop economy, conservation and other factors, important and relevant as they are in this new century, mean little in today's climate without good husbandry. As previously, particular attention has been given to the basic factors concerned with growing the crop but, as mentioned, more reference is now made to crop economy.

Steve Finch, Senior Lecturer in Crop Production at the Royal Agricultural College, and Alison Samuel, Senior Associate Lecturer from the University of Plymouth, Department of Agriculture and Food Studies, Seale-Hayne Campus, have now been joined by Gerry Lane, Principal Lecturer in Crop Production at the Royal Agricultural College as the main authors of this new edition.

The revision of Part 1 – Conditions for Crop Growth – has been the responsibility of Steve Finch for Plants, Fertilisers and Manures, Climate and Weather. This last chapter has been radically changed now that there is more awareness of the possible new pattern of climate and its effect on crop growth. Alison Samuel has written the chapters on Soils and Soil Management, Weeds, Pests and Plant Diseases. The pests and diseases tables, which in previous editions concluded the one chapter on these aspects of crop production, have now been placed at the end of the now two respective chapters. This should make for easier reading. The increasing problem of resistance to pesticides is discussed in some detail.

In Part 2 – Crop Husbandry Techniques – Steve Finch is responsible for the chapter on Cropping Techniques whilst Integrated Crop Management is written by Keith Chaney, Principal Lecturer and Head of Crops Group, Harper Adams University College. This, in some depth, discusses intensive rotational cropping, certainly a move away from continuous cereal growing which was considered an important aspect of arable cropping in earlier editions of this book.

As a practising organic farmer, Alison Samuel has enlarged on this subject which now merits a chapter on its own. This must be considered relevant. In 2000 there were 420000 hectares (2.3% of total UK farm land) in organic and in conversion land – an increase of 72% from 1999. Gerry Lane deals with the specialised subject of seed production, concluding this section of the book.

Part 3 concerns the management of the individual crops, with Alison Samuel, as in the previous edition, updating cereals, and Steve Finch root crops. The abolition of potato quotas and its implications are discussed. Gerry Lane has written the chapter on forage crops, whilst Richard Baldwin, Senior Lecturer, Pershore College is responsible for Fresh Harvested Crops (vegetable production on farms). Rhys Davies, Senior Lecturer in Crop Production, Harper Adams University College has written Combinable Break Crops; these are even more relevant with the continuing decline in returns from the cereal crop.

In Part 4 Gerry Lane has taken over all aspects of my chapters on Grassland and Grass Conservation. As well as the husbandry and management of the crop, a re-evaluation of the grass and grass/legume crops is undertaken and the central role they can play in livestock farming is discussed.

Jim Lockhart retired before the Seventh edition; I am now retired and so all the technical input for this edition is the responsibility of the aforementioned authors, and this we are more than glad to acknowledge. Colleagues of theirs who have given helpful advice are also thanked, particulary Diddie Sims assisting Alison Samuel; Sarah Cook of ADAS, Boxworth for help in compiling the section on sunflowers. John Conway, Paul Davies and Tom Overbury from the Royal Agricultural College should also be especially mentioned.

I do speak for the authors in thanking Francis Dodds, our Editor at Woodhead Publishing; his patience in dealing with them is almost beyond human comprehension! He had a trying time but was invariably polite and amazingly was able to keep the preparation up to schedule. He certainly deserves our thanks.

Tony Wiseman

Part 1

Conditions for crop growth

1

Plants

Plants are living organisms consisting of many specialised individual cells. They differ from animals in many ways and a very important difference is that they can build up valuable organic substances from simple materials. The most important part of this building process, called photosynthesis, is the production of carbohydrates such as sugars, starch and cellulose.

1.1 Plant physiology

1.1.1 Photosynthesis
In photosynthesis a special green substance called chlorophyll uses light energy (normally sunlight but sometimes artificial) to change carbon dioxide and water into sugars (carbohydrates) and oxygen in the green parts of the plant. The amount of photosynthesis per day which takes place is limited by the duration and intensity of sunlight. The amount of carbon dioxide available can also be a limiting factor. Shortage of water, low temperatures and leaf disease or damage can also reduce photosynthesis. The cells that contain chlorophyll also have yellow pigments such as carotene. Crop plants can only build up chlorophyll in the light and so any leaves that develop in the dark are yellow and cannot produce carbohydrates.

Oxygen is released back into the atmosphere during photosynthesis and the process may be set out as follows:

(a) The light stage (light dependent)
This takes place in the thylakoid membranes inside the 'chloroplast', an

organelle found inside the cells of green tissue. Light provides energy for the chlorophyll molecule that releases electrons. These split water into oxygen and hydrogen.

The chemical reaction of this stage is:

$$2H_2O \rightarrow 2H_2 + O_2 \qquad [1.1]$$

The hydrogen then moves into the next stage:

(b) The dark stage (light independent)
This takes place in the watery stroma of the chloroplast. Here the hydrogen is combined with carbon dioxide to give carbohydrates and water:

$$2H_2 + CO_2 \rightarrow CH_2O + H_2O \qquad [1.2]$$

The carbohydrates are simple sugars, which can be moved through the vascular system of the plant in solution to wherever they are needed. This process not only provides the basis for all food production but it also supplies the oxygen which animals and plants need for respiration. The simple carbohydrates, such as glucose, may be built up to form starch for storage purposes or as cellulose for building cell walls. Fats and oils are formed from carbohydrates. Protein material, which is an essential part of all living cells, is made from carbohydrates and nitrogen compounds and also frequently contains sulphur.

Most plants consist of roots, stems, leaves and reproductive parts and need a medium in which to grow. These media could be soil, compost or even water where plants are grown hydroponically. In soil the roots spread through the spaces between the particles and anchor the plant. The amount of root growth can be phenomenal. For example, in a single plant of wheat the root system may extend to many miles.

The leaves, with their broad surfaces, are the main parts of the plant where photosynthesis occurs (Fig. 1.1). A very important feature of the leaf structure is the presence of large numbers of tiny pores (stomata) on the surface of the leaf (Fig. 1.2). There are usually thousands of stomata per square centimetre of leaf surface. Each pore (stoma) is oval-shaped and surrounded by two guard cells. The carbon dioxide used in photosynthesis diffuses into the leaf through the stomata. Most of the water vapour leaving the plant, as well as the oxygen from photosynthesis, diffuses out through the stomata.

1.1.2 Transpiration
The evaporation of water from plants is called transpiration. It mainly occurs through the stomata and has a cooling effect on the leaf cells. Water in the cells of the leaf can pass into the pore spaces in the leaf and then out through the stomata as water vapour (Fig. 1.3).

The rate of transpiration varies considerably. It is greatest when the plant is well supplied with water and the air outside the leaf is warm and dry. When the guard cells are turgid (full of water) the stomata are open. When the plant

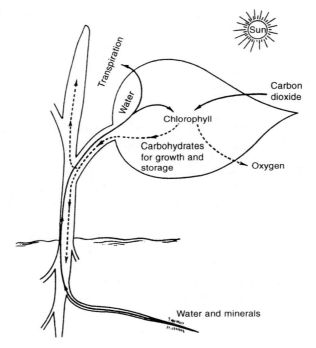

Fig. 1.1   Photosynthesis.

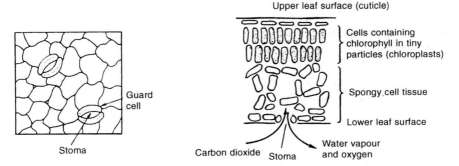

Fig. 1.2   Stomata on leaf surface.

Fig. 1.3   Cross-section of green leaf showing gaseous movements during daylight.

is under drought stress the guard cells lose water and the stoma closes, slowing down the loss of water vapour (transpiration) from the plant. It also slows down the rate of photosynthesis. The stomata also close in very cold weather, e.g. 0 °C. Transpiration is also retarded if the humidity of the atmosphere is high because there is only a very small water vapour gradient between the inside of the leaf and the outside atmosphere. The stomata guard cells close (and so transpiration ceases) during darkness. This is because photosynthesis ceases and water is lost from the guard cells when some of the sugars present change to starch.

1.1.3   Respiration

Plants breathe, like animals, i.e. they take in oxygen that combines with organic foodstuffs and releases energy, carbon dioxide and water. Plants are likely to be checked in growth if the roots are deprived of oxygen for respiration, which might occur in a waterlogged soil. Respiration appears, superficially, to be the reverse of photosynthesis, with carbohydrates combining with oxygen to give carbon dioxide and water with a release of energy. However, it is much more complicated with a very different metabolic process taking place.

There are two main processes. The first is glycolysis where simple sugars are split to release energy and to form pyruvic acid, water and a carrier molecule. The second stage is the 'Citric Acid' or 'Krebs' cycle where the pyruvic acid is converted to citric acid, which cycles within the system through intermediate molecules releasing energy in the form of ATP (adenosine triphosphate), carbon dioxide and water. The energy-rich ATP can then be moved around the plant to provide it with the energy it requires for metabolism, growth and development.

1.1.4   Conduction

The conductive flow of water through the plant takes place in the xylem tissue that runs in bundles along the length of the root and stem and into the organs of the plant. The xylem or wood vessels, which carry the water and mineral salts from the roots to the leaves, are tubes made from dead cells called tracheids and vessel elements. The cross walls of these cells are no longer present and the longitudinal walls are thickened with lignum to form wood. These tubes help to strengthen the stem.

1.1.5   Translocation

The movement of food materials through the plant is known as translocation. The phloem tubes carry organic material through the plant, e.g. sugars and amino acids from the leaves to storage parts or growing points. These vessels are chains of living cells, not lignified, and with cross walls which are perforated. They are sometimes referred to as sieve tubes. These form a branched system throughout the plant and transport assimilates from sites of production (sometimes called sources) to sites of demand or consumption (sinks). The actual mechanism of transport is still not clear.

In the stem of dicotyledons, the xylem and phloem tubes are usually found in a ring near the outside of the stem with phloem on the outside and xylem on the inside, separated by a cambium layer. In monocotyledons the vascular bundles are more randomly arranged within the stem. In the root, the xylem and phloem tubes form separate bundles and are found near the centre.

1.1.6 Uptake of water

Water is taken into the plant from the soil. This occurs mainly through the root hairs near the root tip. As the root grows the hairs are constantly replaced. A few centimetres back from the root tip the hairs disappear. Their function is to increase the surface area available for absorption of water. There are thousands of root tips on a single healthy crop plant (Fig. 1.4).

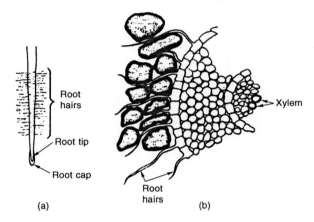

Fig. 1.4 (a) Section of root tip and root hair region. (b) Cross-section of root showing the root hairs as tube-like elongations of the surface cells in contact with soil particles.

The absorption of water into the plant in this way is due to suction pull, which starts in the leaves. As water transpires (evaporates) from the cells in the leaf, more water is drawn from the xylem tubes which extend from the leaves to the root tips. In these tubes the water is stretched like a taut wire. This is possible because the tiny particles of water hold together very firmly when in narrow tubes. The pull of this water in the xylem tubes of the root is transferred through the root cells to the root hairs and so water is absorbed into the roots and up to the leaves. In general, the greater the rate of transpiration, the greater is the amount of water taken into the plant.

However, this 'transpiration pull' cannot completely explain water movement within plants. There appears to be a positive pressure exerted by roots which forces water up into the stem, and the very narrow vascular tubes act as a kind of wick (capillarity) which helps to hold water inside them.

The rate of absorption is slowed down by:

1 Shortage of water in the soil.
2 Lack of oxygen for root respiration (e.g. in waterlogged soils).
3 A high concentration of salts in the soil water near the roots.

Normally, the concentration of the soil solution does not interfere with water absorption. High soil water concentration can occur in salty soils and near bands of fertiliser. Too much fertiliser near developing seedlings may damage germination and subsequent emergence by restricting the uptake of water.

1.1.7   Osmosis

Much of the water movement into and from cell to cell in plants is due to osmosis. This is a process in which a solvent, such as water, will flow through a semi-permeable membrane (e.g. a cell wall) from a weak solution to a more concentrated one. The cell wall only allows the water to pass through, as the molecules in solution are too big. The force exerted by such a flow is called the osmotic pressure. In plants the normal movement of the water is from the soil solution into the cell. However, if the concentration of a solution outside the cell is greater than that inside, there is a loss of water from the cell, and its content contracts; this is called plasmolysis.

1.1.8   Uptake of nutrients

The absorption of chemical substances (nutrients) into the root cells is partly due to a diffusion process but it is mainly due to the ability of the cells near the root tips to accumulate such nutrients. Nutrients are taken into the root in the form of charged ions through the root hairs, along with water. The water and solutes move through the cells into the inner ring of xylem. They are prevented from leaking back into the soil between the cell walls by a waxy layer of cells called the Casparian Strip. In this way an electrochemical gradient is produced which allows the flow of nutrient ions into the plant from the soil solution.

1.2   Plant groups

There are many ways of classifying plant groups but, from an agricultural and horticultural point of view, a useful way is to divide them into annuals, biennials and perennials according to their total length of life.

1.2.1   Annuals

Typical examples are wheat, barley and oats that complete their life history in one growing season, i.e. starting from the seed, in one year they develop roots, stem and leaves and then produce flowers and seed before dying.

1.2.2 Biennials

These plants grow for two years. They spend the first year in producing roots, stem and leaves, and the following year in producing the flowering stem and seeds, after which they die. Sugar beet, swedes and turnips are typical biennials, although the grower treats these crops as annuals, harvesting them at the end of the first year when all of the foodstuff is stored up in the root and before the plant moves on to produce the seed head. Those plants that do behave as annuals and throw a seed head in their first year are called 'bolters'.

1.2.3 Perennials

They live for more than two years and, once fully developed, they usually produce seeds each year. Many of the grasses and forage legumes are perennials, as are many of the horticultural fruit crops such as raspberries or apples.

1.3 Structure of the seed

Plants are also classified as dicotyledons and monocotyledons according to the structure of the seed.

1.3.1 Dicotyledon

A good example of a dicotyledon seed is the field bean. If its pod is opened when nearly ripe it will be seen that each seed is attached to the inside of the pod by a short stalk called the funicle. All the nourishment that the developing seed requires passes through the funicle from the bean plant. When the seed is ripe and has separated from the pod, a black scar, known as the hilum, can be seen where the funicle was attached. Near one end of the hilum is a minute hole called the micropyle (Fig. 1.5).

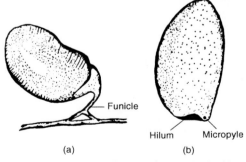

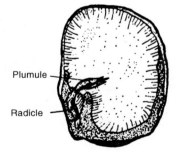

Fig. 1.5   (a) Bean seed attached to the inside of the pod by the funicle. (b) Bean seed showing the hilum and micropyle.

Fig. 1.6   Bean seed with one cotyledon removed.

If a bean is soaked in water the seed coat can be removed easily and all that is left is largely made up of the embryo (germ). This consists of two seed leaves (cotyledons) which contain the food for the young seedling. Lying between the two cotyledons is the radicle (which eventually forms the primary root) and a continuation of the radicle at the other end, the plumule (Fig. 1.6). This develops into the young shoot and is the first bud of the plant.

1.3.2   Monocotyledon

This important class includes all the cereals and grasses. The wheat grain is a typical example. It is not a true seed (it should be called a single-seeded fruit). The seed completely fills the whole grain, being almost united with the inside wall of the grain or fruit. This fruit wall is made up of many different layers which are separated on milling into varying degrees of fineness, e.g. bran and pollards which are valuable livestock feed.

Most of the interior of the grain is taken up by the floury endosperm. The embryo occupies the small raised area at the base. The scutellum, a shield-like structure, separates the embryo from the endosperm. Attached to the base of the scutellum are the five roots of the embryo, one primary and two pairs of secondary rootlets. The roots are enclosed by a sheath called the coleorhiza while the shoot is enclosed by the coleoptile. The position of the radicle and the plumule can be seen in the diagram (Fig. 1.7). The scutellum can be regarded as the cotyledon of the seed. There is only one cotyledon present and so wheat is a monocotyledon.

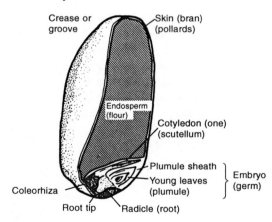

Fig. 1.7   Wheat seed cut in half at the crease.

1.3.3   Germination of the bean – the dicotyledon

Given suitable conditions for germination, i.e. water, heat and air, the seed coat of the dormant, but living, seed splits near the micropyle and the radicle begins to grow downwards through this split to form the main, or primary, root from which lateral branches will soon develop (Fig. 1.8). When the root is firmly held in the soil, the plumule starts to grow by pushing its way out of the same opening in the seed coat. As it grows upwards its tip is bent to protect it from injury in passing through the soil, but it straightens out on reaching the surface, and leaves develop very quickly from the plumular shoot.

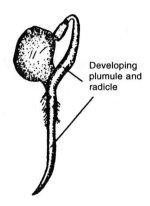

Developing plumule and radicle

Fig. 1.8   Germination of bean seed.

In the field bean the cotyledons remain underground, gradually giving up their stored food materials to the developing plant. This is called hypogeal germination. However, in the French bean and in many other dicotyledon seeds, the cotyledons are brought above ground with the plumule. This is epigeal germination.

1.3.4   Germination of wheat – the monocotyledon

When the grain germinates the coleorhiza expands and splits open the seed coat and, at the same time, the roots break through the coleorhiza and enter the soil (Fig. 1.9). The primary root is soon formed, supported by the two pairs of secondary rootlets, but this root system (the seminal roots) is only temporary and is soon replaced by adventitious roots (Figs 1.10 and 1.11). As the first root system is being formed at the base of the stem, so the plumule starts to grow upwards, and its first leaf, the coleoptile, appears above the ground as a single pale tube-like structure. From a slit in the top of the coleoptile there appears the first true leaf followed by others, the younger leaves growing from the older leaves (Fig. 1.12).

As the wheat embryo grows the floury endosperm is used up by the developing roots and plumule. The scutellum has the important function of changing the endosperm into digestible food for the growing parts, converting starch into simple, mobile sugars. In the field bean, the cotyledons provide the food for the early nutrition of the plant, whilst the wheat grain is dependent upon the endosperm and scutellum. In both cases it is not until the plumule has reached

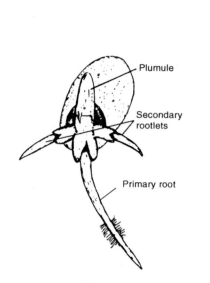

Plumule

Secondary
rootlets

Primary root

Fig. 1.9   Germination of the wheat grain.

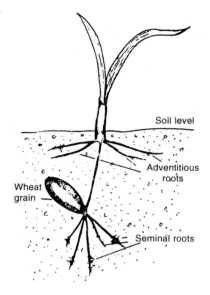

Soil level

Adventitious
roots

Wheat
grain

Seminal roots

Fig. 1.10   Developing wheat plant.

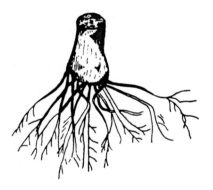

Fig. 1.11   Adventitious root system.

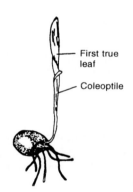

First true
leaf

Coleoptile

Fig. 1.12   Seedling wheat plant.

the light, and turned green, that the plant can begin to be independent. This point is important in relation to the depth at which seeds should be sown. Small seeds such as brassicas, clovers and many of the grasses must be, as far as possible, sown shallow. Their food reserves will be exhausted before the shoot reaches the surface if sown too deep. Larger seeds such as beans and peas can and should be sown deeper.

When the leaves of the plant begin to manufacture food by photosynthesis, and when the primary root has established itself sufficiently well to absorb nutrients from the soil, the plant can develop independently, provided there is sufficient moisture and air present.

The main differences between the two groups of plants can be summarised as follows:

Dicotyledons	Monocotyledons
The embryo has two seed leaves.	The embryo has one seed leaf.
A primary root system is developed and persists.	A primary root system is developed, but is replaced by an adventitious root system.
Usually broad-leaved plants, e.g. legumes and sugar beet.	Usually narrow-leaved plants, e.g. cereals and grasses.

These two great groups of flowering plants can be further divided in the following way:

Families or orders, e.g.	Legumes
Genus	Clovers of the legume family
Species	Red clover
Cultivar or Variety	Late red clover

1.4   Plant structure

The plant can be divided into four parts: the root system, stem, leaf and flower.

1.4.1   The root system

The root system is concerned with the parts of the plant growing in the soil; there are two main types:

(1) The tap root or primary system
This is made up of the primary root called the tap root with lateral secondary roots branching out from it and, from these, tertiary roots possibly developing obliquely to form, in some cases, a very extensive system of roots (Fig. 1.13). The root of the bean plant is a good example of a tap root system. If this is split it will be seen that there is a slightly darker central woody core, the skeleton of the root, which helps to anchor the plant and transport foodstuffs. The lateral secondary roots arise from this central core (Fig. 1.14). Carrots and other true root crops such as sugar beet and mangels have very well developed tap roots. These biennials store food in their roots during the first year of growth to be used in the following year for the production of the flowering shoot and seeds. However, they are normally harvested after one season and the roots are used as food for man and stock.

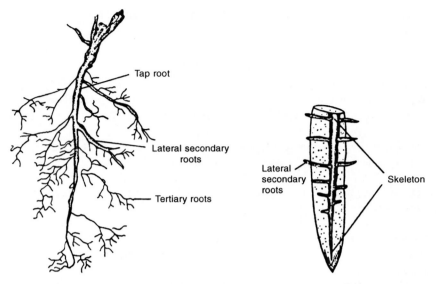

Fig. 1.13 Tap root or primary root system.   **Fig. 1.14** Tap root of the bean plant.

(2) The adventitious root system
This is found on all grasses and cereals, and it is the main root system of most monocotyledons. The primary root is soon replaced by adventitious roots, which arise from the base of the stem (Fig. 1.11). These roots can, in fact, develop from any part of the stem, and they are found on some dicotyledons as well (but not as the main root system), e.g. underground stems of the potato.

Root hairs (Fig. 1.4) are very small white, hair-like structures that are found near the tips of all roots. As the root grows, the hairs on the older parts die off and others develop on the younger parts of the root. They play a very important part in the nutrition and water uptake of a plant.

1.4.2   The stem
The second part of the flowering plant is the shoot that normally grows upright above the ground. It is made up of a main stem and branches. Stems are either soft (herbaceous) or hard (woody) and in UK agriculture it is only the soft and green herbaceous stems which are of any importance. These usually die back every year. All stems start life as buds and the increase in length takes place at the tip of the shoot called the terminal bud. If a Brussels sprout is cut longitudinally and examined it will be seen that the young leaves arise from the bud axis. This axis is made up of different types of cell tissue, which is continually making new cells and thus growing (Fig. 1.15).

Stems are usually jointed, each joint forming a node, the part between two nodes called the internode. At the nodes the stem is usually solid and thicker, and this swelling is caused by the storing up of material at the base of the leaf

(Fig. 1.16). The bud consists of closely packed leaves arising from a number of nodes. It is, in fact, a condensed portion of the stem that develops by a lengthening of the internodes.

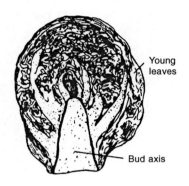

Fig. 1.15   Longitudinal section
    of a Brussels sprout.

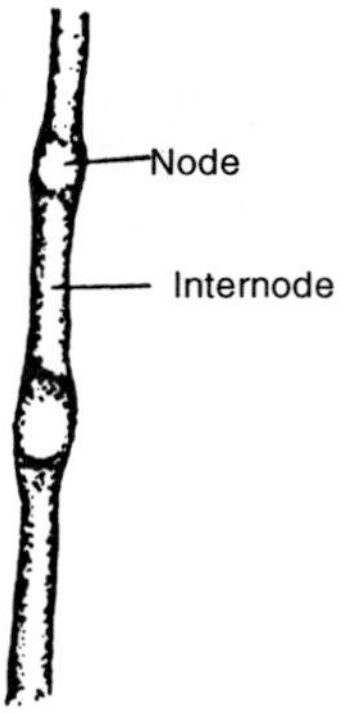

Fig. 1.16   Jointed stem.

Axillary buds are formed in the angle between the stem and leaf stalk. These buds, which are similar to the terminal bud, develop to form lateral branches, leaves and flowers.

1.4.2.1 Modified stems

1   A stolon is a stem that grows along the ground surface.
    Adventitious roots are produced at the nodes, and buds on the runner can develop into upright shoots, and separate plants can be formed, e.g. strawberry plants (Fig. 1.17).

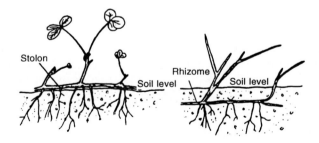

Fig. 1.17   Modified stems.

2   A rhizome is similar to a stolon but grows under the surface of the ground, e.g. common couch (Fig. 1.17).
3   A tuber is really a modified rhizome. The ends of the rhizomes swell to form tubers. The tuber is therefore a swollen stem. The potato is a well-

known example with 'eyes' (buds) which develop shoots when the potato tuber is planted. Potato tubers will also turn green by producing chlorophyll when exposed to light; this is another characteristic of stems.

4   A tendril is found on certain legumes, such as the pea. The terminal leaflet is modified as in the diagram (Fig. 1.18). This is useful for climbing purposes to support the plant.

Tendril

Fig. 1.18   Modified stems.

Corms and suckers are other examples of modified stems.

1.4.3   The leaf

Leaves in all cases arise from buds. They are extremely important organs, being not only responsible for the manufacture of sugar and starch from the atmosphere for the growing parts of the plant, but they are also the organs through which transpiration of water takes place.

A typical leaf of a dicotyledon consists of three main parts:

1   The blade.
2   The stalk or petiole.
3   The basal sheath connecting the leaf to the stem. This may be modified (as with legumes) into a pair of wing-like stipules (Fig. 1.19a).

The blade is the most obvious part of the leaf and it is made up of a network of veins.

There are two main types of dicotyledonous leaves:

1   Those with a prominent central midrib, from which lateral veins branch off on either side. These side veins branch into smaller and smaller ones (Fig. 1.19a).
2   Those with no single midrib, but several main ribs spread out from the top of the leaf stalk; between these the finer veins spread out as before, e.g. horse-chestnut leaf (Fig. 1.19b).

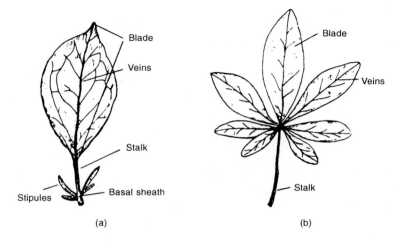

Fig. 1.19   (a) Simple leaf. (b) Compound leaf.

The veins are the essential supply lines for the process of photosynthesis. They consist of two main parts: the xylem for bringing the required raw material up the leaf and the phloem that carries the finished product away from the leaf.

Leaves can show great variation in shape and type of margin, as in Fig. 1.19. They can also be divided into two broad classes as follows:

1   Simple leaves. The blade consists of one continuous piece (Fig. 1.19a).
2   Compound leaves. Simple leaves may become deeply lobed and when the division between the lobes reaches the midrib it becomes a compound leaf, and the separate parts of the blade are called the leaflets (Fig. 1.19b).

The blade surface may be smooth (glabrous) or hairy, according to variety. This is important in legumes because it can affect their palatability. Surfaces of leaves can also be glaucous, i.e. covered with a bloom of wax. This wax coating gives the leaf some protection from disease or pest attack.

Monocotyledonous leaves are dealt with in Part 4, Chapters 17 and 18 (Grassland).

1.4.3.1   Modified leaves

1   Cotyledons or seed leaves are usually of a very simple form.
2   Scales are normally rather thin, yellowish to brown membranous leaf structures, very variable in size and form. On woody stems they are present as bud scales which protect the bud; they are also found on rhizomes such as common couch.
3   Leaf tendrils. The terminal leaflet like the stem can be modified into thin thread-like structures, e.g. the pea plant.

Other examples of modified leaves are leafspines and bracts.

1.4.4   The flower

In the centre of the flower is the axis that is simply the continuation of the flower stalk. It is known as the receptacle and on it are arranged four kinds of organ:

1   The lowermost is a ring of green leaves called the calyx, made up of individual sepals. The sepals offer the flower protection while still at the bud stage, when they are still closed.

2   Immediately above the calyx is a ring of petals known as the corolla. The number of petals on a flower can differ widely between species. Petals are usually brightly coloured and their function is to attract insects for pollination. At the base of the petal are modified structures called nectaries. These, as their name suggests, produce the sweet nectar that acts as a reward for a visiting insect.

3   Above the corolla are the stamens, again arranged in a ring. They are similar in appearance to an ordinary match, the swollen tip called the anther sitting on top of the filament. When ripe the anther bursts open to release the pollen grains. The filaments can be of varying sizes, either keeping the anther inside the flower or allowing it to hang outside (useful in wind pollinated plants).

4   The highest position on the receptacle is occupied by the pistil which consists of one or more small green bottle-shaped bodies called carpels. These are made up of three parts: the stigma, style and the ovary (containing ovules). It is within the ovary that the future seeds are produced (Figs 1.20 and 1.21).

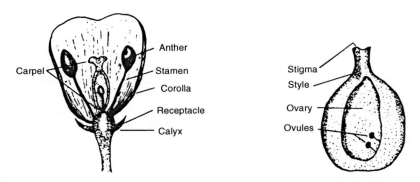

Fig. 1.20   Longitudinal section of a simple flower.   **Fig. 1.21**   Carpel.

Most flowers are more complicated in appearance than the above, but basically they consist of these four main parts.

1.4.5   The formation of seeds

Pollination precedes fertilisation, which is the union of the male and female reproductive cells. When pollination takes place the pollen grain is transferred from the anther to the stigma. This may be self-pollination where the pollen is transferred from the anther to the stigma of the same flower, or cross-pollination when it is carried to a different flower (Fig. 1.22). The vectors of pollen transfer can also vary. Some flowers are insect pollinated (entomophilous) and are usually scented and brightly coloured. The pollen itself is usually sticky or oily. Other flowers are wind pollinated (anemophilous) and do not need to be brightly coloured. They produce huge amounts of pollen (most is lost) and the pollen grains are smooth, light and small. The flowers are often unisexual with a predominance of males. Stamens and stigmas often hang outside the flowers, and the stigmas are often feathery to give them a better chance of trapping a pollen grain as it blows past.

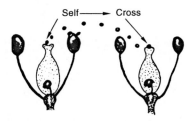

Fig. 1.22   Self- and cross-pollination.

With fertilisation the pollen grain grows a pollen tube down the style of the carpel. This takes place very quickly and the tube grows around the ovary sac and enters through the micropyle. Three nucleii travel down the tube, one tube nucleus and two male nucleii. The tube nucleus disintegrates once the tube has reached the embryo sac, one male nucleus fuses with the egg cell in the ovule and the other joins with a second nucleus to form the primary endosperm nucleus. This double fertilisation is unique to flowering plants. The ovule itself goes on to form the seed. The ovary also changes after fertilisation to form the fruit, as distinct from the seed.

With the grasses and cereals there is only one seed formed in the fruit and, being so closely united with the inside wall of the ovary, it cannot easily be separated from it. The one-seeded fruit is called a grain.

1.4.6   The inflorescence
Special branches of the plant are modified to bear the flowers, and they form the inflorescence. There are two main types of inflorescence:

1   Where the branches bearing the flowers continue to grow, so that the youngest flowers are nearest the apex and the oldest farthest away – an indeterminate inflorescence (Fig. 1.23a). A well-known example of this inflorescence is the spike found in many species of grasses.
2   Where the main stem is terminated by a single flower and ceases to grow

in length; any further growth takes place by lateral branches, and they
eventually terminate in a single flower and growth is stopped – a determinate
inflorescence (Fig. 1.23b). e.g. linseed.

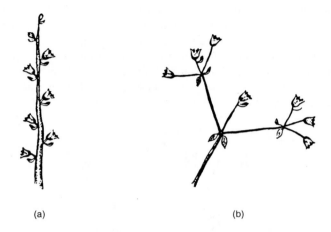

(a)   (b)

Fig. 1.23   (a) Indeterminate inflorescence. (b) Determinate inflorescence.

There are many variations of these two main types of inflorescence.

1.5   Plant requirements

To grow satisfactorily a plant needs warmth, light, water, carbon dioxide and
other chemical elements which it can obtain from the soil.

1.5.1   Warmth

Most crop plants in this country start growing when the average daily temperature
reaches 6 °C. Growth is best between 16 °C and 27 °C. These temperatures apply
to thermometer readings taken in the shade about 1.5 m above ground. Crops
grown in hotter countries usually have higher temperature requirements. Several
plants require specific soil temperatures before they can successfully germinate,
e.g. maize. Plants require a certain number of 'heat units' before they can successfully
complete their development. This explains the difficulty of trying to grow sub-
tropical or Mediterranean plants in the UK, e.g. durum wheat, grain maize or
soya.

Cold frosty conditions may seriously damage plant growth. Crop plants differ
in their ability to withstand extreme cold. For example, winter rye and wheat
can stand colder conditions better than winter oats. Potato plants and stored
tubers are easily damaged by frost. Sugar beet plants may go to seed (bolt) if

there are frosts after germination. Frost in December and January may destroy crops left in the ground. Cold may, however, be beneficial. A period of cold temperature is necessary to allow some autumn-sown crops to change from vegetative to sexual development. This is called 'vernalisation'. Some seeds may have to undergo cold conditions before they will germinate. This is known as 'stratification'. Cold temperatures can also help to reduce the incidence of weeds, pests and disease in crops.

1.5.2 Light
Without light, flowering plants cannot produce carbohydrates and will soon die. The amount of photosynthesis that takes place daily in a plant is partly due to the length of daylight, and partly to the intensity of the sunlight. Bright sunlight is very important where there is dense plant growth. The periods of daylight and darkness will vary according to the distance from the equator and also from season to season. This affects the flowering and seeding of crop plants and is another of the limiting factors in introducing new crops into a country.

1.5.3 Water
Water is an essential part of all plant cells and it is also required in extravagant amounts for the process of transpiration. Water carries nutrients from the soil into and through the plant; it also carries the products of photosynthesis from the leaves to wherever they are needed. It is also used to give the plant rigidity (turgor); is used in many of the biochemical pathways inside the plant and is also used to regulate the temperature of the plant. Plants take up about 200 tonnes of water for every tonne of dry matter produced.

1.5.4 Carbon dioxide
Plants need carbon dioxide for photosynthesis. This is taken into the leaves through the stomata, and the amount which can go in is affected by the rate of transpiration. Another limiting factor is the small amount (0.03 %) of carbon dioxide in the atmosphere. The percentage can increase just above the surface of soils rich in organic matter where soil bacteria are active and releasing carbon dioxide. This is possibly one of the reasons why crops grow better on such soils. In protected crops, in greenhouses and polytunnels, the levels of carbon dioxide can be artificially raised by the grower to improve production.

1.5.5 Chemical elements required by plants
Many chemical elements are needed by the plant in order that it may live and flourish. Most soils supply the majority of these nutrients and in farming

practice it is only with regard to nitrogen, phosphorus, potassium and magnesium (the major elements) that there is any widespread necessity to supplement the natural supplies from the soil. Other major elements may be required in certain situations. Calcium is an essential plant food but, as lime, it is regarded more as a soil conditioner. Sodium is highly desirable for maritime crops such as sugar beet and fodder beet, when it can replace some or all of the potassium requirements. Sulphur, in areas away from industry, could be needed for the grass crop when it is cut more than once in the year, and for certain brassicas.

The main plant foods are discussed more fully in Chapter 3 (Fertilisers and Manures) and see also Table 1.1.

Table 1.1

The major nutrients	Use	Source
Carbon (C) Hydrogen (H) Oxygen (O)	Used in making carbohydrates	The air and water
Nitrogen (N)	Very important for building proteins	Organic matter (including farmyard manure); rainfall Nitrogen-fixing soil micro-organisms Nitrogen fertilizers such as ammonium and nitrate compounds and urea
Phosphorus (P) (phosphate)	Essential for cell division and many chemical reactions	Small amounts from the mineral and organic matter in the soil Mainly from phosphate fertilizers, e.g. superphosphate, ground rock phosphate, basic slag and compounds, and residues of previous fertilizer applications
Potassium (K) (potash)	Helps with formation of carbohydrates and proteins Regulates water in and through the plant	Small amounts from mineral and organic matter in the soil. Potash fertilizers, e.g. muriate and sulphate of potash
Calcium (Ca)	Essential for development of growth tissues, e.g. root tips	Usually enough in the soil. Applied as chalk or limestone to neutralize acidity
Magnesium (Mg)	A necessary part of chlorophyll	If soil is deficient, may be added as magnesium limestone or magnesium sulphate, also FYM
Sulphur (S)	Part of many proteins and some oils	Usually sufficient in the soil. Atmospheric sulphur absorbed by the soil and plant. Added in some fertilizers (e.g. sulphate of ammonia and superphosphate)

Those elements required only in small amounts by the plant are known as the minor or trace elements. They are, nevertheless, essential and a shortage or lock-up in the soil, especially of boron and manganese (often as a result of liming or high organic matter), will cause deficiency diseases in particular crops. Other trace elements include magnesium, chlorine, iron, molybdenum and zinc, but these rarely cause trouble on most farm soils (see pages 57–8).

1.6   Legumes and nitrogen fixation

Legumes are plants that have a number of interesting features such as:

1   A special type of fruit called a legume, which splits along both sides to release its seeds, e.g. pea pod.
2   Nodules on the roots containing special types of bacteria (rhizobia) which can 'fix' (convert) nitrogen from the air into nitrogen compounds. These bacteria enter the plant through the root hairs from the surrounding soil.

This fixation of nitrogen is of considerable agricultural importance. Many of our farm crops are legumes, for example, peas, beans, vetches, lupins, clovers, lucerne (alfalfa), sainfoin and trefoil. The bacteria obtain carbohydrates from the plant and in return they supply nitrogen as ammonium compounds which are released into the soil or directly into the host plant in the form of soluble nitrates. In the soil the ammonium part of the compound is changed to nitrate and taken up by neighbouring plants, e.g. by grasses in a grass and clover sward, or by the following crop, e.g. wheat after clover or beans. The amount of nitrogen that can be fixed by legume bacteria varies widely – estimates of 50–450 kg/ha of nitrogen have been made. Some of the reasons for variations are:

- *The type of plant.* Some crop plants fix more nitrogen than others, e.g. lucerne and clovers (especially if grazed) are usually better than peas and beans.
- *The conditions in the soil.* The bacteria usually work best in soils that favour the growth of the plant on which they live. A good supply of calcium and phosphate in the soil is usually beneficial, although lupins grow well on acid soils.
- *The strains of bacteria present.* The majority of soils in this country contain the strains of bacteria required for most of the leguminous crops which are grown. Lucerne is an exception and, unless it has been grown in the field within the previous three years, it is necessary to inoculate the seed, i.e. treat it with an inoculum containing *Rhizobium meliloti* before sowing to encourage effective nodulation.

1.7   The control of plant growth and development

1.7.1   Growth substances

In order for plants to grow and develop in an orderly manner there needs to be some kind of regulation of cell division, elongation and differentiation. As the plant has no nervous system, this regulation is entirely chemical in nature. The chemicals involved are called 'growth substances' and there are five main types:

1   Auxins – these substances mostly control cell enlargement and differentiation. Most auxins are made at the apex of stems and in young leaves.
2   Gibberellins – also involved in cell enlargement, but also play a key role in the breaking of dormancy in seeds and the mobilisation of seed reserves by the production of the enzyme, α-amylase.
3   Cytokinins – associated with cell division in the presence of auxins. Found in large amounts in fruits and seeds where they are associated with embryo growth.
4   Abscisic acid – a growth inhibitor associated with dormant buds and some seeds.
5   Ethene (ethylene) – a gas which controls fruit ripening and senescence. Controls fruit and leaf drop from some trees.

One of the ways that plants react to external stimuli is by movement. Roots, shoots, leaves and flowers can all move either towards or away from these stimuli. This movement is called 'tropism'.

1.7.2   Phototropism

This is the movement response to light. Shoots will be positively phototropic and move towards light. Auxins play a major part in this movement by moving to the dark side of the shoot, encouraging cell growth and thus bending the shoot towards the light source. This is a vital response if maximum sunlight interception is to take place by the plant so that this can be converted to yield.

1.7.3   Geotropism

This is the response to gravity. Shoots are negatively geotropic and therefore grow upwards while roots are positively geotropic and grow downwards. This is particularly important in the germinating and emerging seed and means that the grower does not have to ensure that the seed is planted the right way up! Other organs of the plant have an intermediate geotropism, e.g. leaves and side branches.

1.7.4   Hydrotropism

Roots are positively hydrotropic and will grow towards water.

1.7.5 Thigmotropism

This is a very specialised type related to touch. Tendrils of peas are positively thigmotropic allowing them to wrap around each other and any supporting objects. Similarly, hop stems will climb up strings in the hop garden by wrapping themselves around for support.

1.7.6 The effect of light

As well as movement caused by differentiation in cell enlargement and division, the plant's development is also controlled in part by external stimuli. By far the most important one is light, although temperature does have a part to play. Light affects plants in several ways. Without light plants become 'etiolated' – they lack chlorophyll and are therefore yellow rather than green and are fragile and collapse easily. The intensity and spectrum of the light has an effect on production, on breaking dormancy in some seeds and in the regulation of flowering.

Probably the most important aspect of light is its duration. The response of a plant to day length is called 'photoperiodism'. In fact it is the length of the dark period that is critical. The biggest effect is on flowering, although fruit and seed production, dormancy and leaf fall are also affected. We can categorise plants into three groups when we look at the differences in photoperiod requirements for flowering:

1.7.6.1 Short day (long night) plants
Flowering is induced by dark periods **longer** than a critical length. The length can vary according to species.

1.7.6.2 Long day (short night) plants
Flowering is induced by dark periods **shorter** than a critical length. Many spring crops, e.g. spring wheat and spring barley, are long day plants with the move into summer, and the subsequent shortening of the nights, stimulating them to move from vegetative growth to sexual development.

1.7.6.3 Day neutral plants
Flowering is independent of photoperiod. There are a large number of plants in this category. All this becomes much more important the further away you are from the equator, with the seasons very strongly associated with varying day lengths.

1.8 Further reading

Barnes C and Poole N, *Plant Science in Action*, Hodder, 1994.
Rost N Y *et al.*, *Botany: A Brief Introduction to Plant Biology*, John Wiley, 1984.
Soper R (ed.), *Biological Science 1 and 2*, Cambridge, 1991.

2

Soils and soil management

Soils are very complex and most have developed over a very long period of time. They provide a suitable medium for plants to obtain water, nutrients and oxygen for growth and development. Most soils also have enough depth to allow plant roots to provide a firm anchorage. Mineral soils are formed initially by the weathering of parent rock, often accompanied by deposition of material by ice, water and/or wind. Organic material is added by the growth and decay of living organisms. If a farmer is to provide the best possible conditions for crop growth it is necessary for him to understand what soils are, how they were formed and how they should be managed.

The topsoil is a layer up to 30 cm deep which may be taken as the greatest depth which a farmer can plough or cultivate and in which most of the plant roots are found. The subsoil, which lies underneath, is an intermediate stage in the formation of soil from the rock below. Some deep-rooting plants such as cereals and oilseed rape can grow in the subsoil down to depths of 1.5–2 m.

A soil profile is a section taken vertically through the soil. In some cases this may consist only of a shallow surface soil of 10–15 cm on top of rock such as chalk or limestone. In deeper well-developed soils there are usually three or more definite layers (or horizons) which vary in colour, texture and structure (Fig. 2.1). The soil profile can be examined by digging a trench or by taking out cores of soil from various depths using a soil auger. A careful examination of the layers can be useful in deciding how the soil was formed and the cropping potential. The colour of the soil in the various horizons will indicate whether the soil is well or poorly drained.

There are a number of ways of classifying soil for crop production. Soils have been grouped into *soil associations*. Each association consists of a number

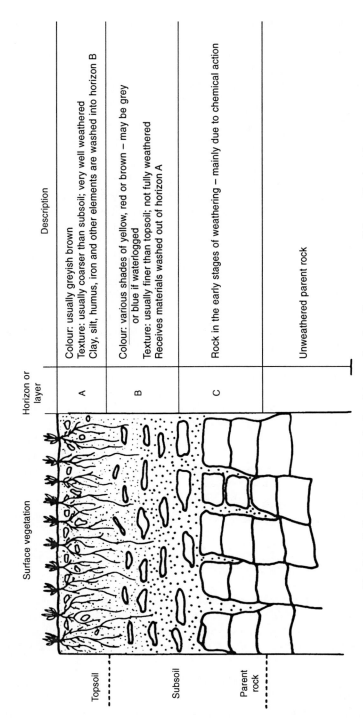

Fig. 2.1 Soil profile showing the breakdown of rock to form various soil layers (horizons).

Surface vegetation

Horizon or layer

Description

A

Colour: usually greyish brown
Texture: usually coarser than subsoil; very well weathered
Clay, silt, humus, iron and other elements are washed into horizon B

B

Colour: various shades of yellow, red or brown – may be grey or blue if waterlogged
Texture: usually finer than topsoil; not fully weathered
Receives materials washed out of horizon A

C

Rock in the early stages of weathering – mainly due to chemical action

Unweathered parent rock

Topsoil

Subsoil

Parent rock

of *soil series* each of which has distinct characteristics, both of parent material, soil profile and topsoil. The soil series is usually named after the place where the soil was first described. The same soil series can occur in different regions. Soil characteristics together with relief and climate and cropping potential have also been used to classify land for farming (see Appendix 8).

2.1   Soil formation

There are very many different types of topsoils and subsoils. The differences are partly due to the kind of material from which they are formed. However, other factors such as climate, topography, plant and animal life, the age of the developing soil material and farming operations affect the type of soil which develops.

2.1.1   The more important rock formations

- *Igneous rocks*, e.g. granite (coarse crystals) and basalt (fine crystals), were formed from the very hot molten material. The minerals (chemical compounds) in these rocks are mostly in the form of crystals. Igneous rocks are very hard and usually weather very slowly.
- *Sedimentary* or *transported rocks* have been formed from weathered material (e.g. clay, silt and sand) carried and deposited by water and wind. The sediments later became compressed by more material on top and cemented to form new rocks such as sandstone and shale.
- *The chalks* and *limestones* were formed from the shells and skeletons of sea animals of various sizes. These rocks are mainly calcium carbonate but in some cases also contain magnesium carbonate. The calcareous soils are formed from them (pages 43–4).
- *Metamorphic rocks*, e.g. marble (from limestone) and slate (from shale), are rocks which have been changed in various ways such as by heat or pressure.

2.1.2   Some other deposits

2.1.2.1   Organic soils

Deep deposits of raw organic matter or peat are found in places where waterlogged soil conditions did not allow the breakdown of dead plant material by micro-organisms and oxidation (due to the anaerobic conditions). There are different sorts of peat depending on the type of vegetation and area where it developed.

2.1.2.2   Glacial drift

Many soils in northern Europe are not derived from the rocks underneath but from material deposited by glaciers, often known as boulder clays. This makes the study of such soils very complicated.

2.1.2.3 Alluvium
This is material which has been deposited recently, for example, by river flooding. It has a very variable composition. The texture depends on the speed of river flow (e.g. fast rivers–stones and sand, slow rivers–silt and clay).

2.1.3 Weathering of rocks
The breakdown of rocks is mainly caused by physical and chemical action.

Physical weathering, due to frost action, causes the mineral crystals in rocks to expand and contract by different amounts, resulting in the occurrence of cracking and shattering. Water can cause pieces of rock surfaces to split off when it freezes and expands in cracks and crevices. The pieces of rock broken off are usually sharp-edged, but if they are carried and knocked about by glaciers, rivers or wind, they become more rounded in shape, e.g. sand and stones in a river bed.

Chemical weathering is the breakdown of the mineral matter in a developing soil brought about by the action of water, oxygen, carbon dioxide and nitric acid from the atmosphere, and by carbonic and organic acids from the biological activity in the soil. The soil water, which is a weak acid, dissolves some minerals and allows chemical reactions to take place.

Clay is produced by chemical weathering of some primary minerals. In the case of rocks such as granite, when the clay-producing parts are weathered away, the more resistant quartz crystals are left as sand or silt. In the later stages of chemical weathering the soil minerals are broken down to release plant nutrients. This is a continuing process in most soils.

In poorly drained soils, which become waterlogged from time to time, various complex chemical reactions (including a reduction process) occur, referred to as *gleying*. This process, which is very important in the formation of some soils, results in ferrous iron, manganese and some other trace elements moving around more freely and producing colour changes in the soil. Gleyed soils are generally greyish in colour (but may also be greenish or blueish). Rusty-coloured deposits of ferric iron (oxidised iron) also occur in root and other channels, and along the boundaries between the waterlogged and aerated soil, so producing a mottled appearance. Glazing or coating of the soil structure units with fine clay is also associated with gleying.

2.2 Other factors in soil formation

2.2.1 Climate
The rate of weathering partly depends on the climate. For example, the wide

variations of temperature and the high rainfall of the tropics make for much faster soil development than would be possible in the colder and drier climatic regions.

2.2.2   Topography
The slope of the ground can considerably affect the depth of soil. Weathered soil tends to erode from steep slopes and build up on the flatter land at the bottom. Level land is more likely to produce uniform weathering.

2.2.3   Biological activity
Plants, animals and micro-organisms, during their life cycles, leave many organic substances in the soil. Some of these substances may dissolve some components of the mineral material; dead material may partially decompose to give humus. The roots of plants may open up cracks in the soil. Vegetation such as mosses and lichens can attack and break down the surface of rocks. Holes made in the soil by burrowing animals such as earthworms, moles and rabbits help to break down soft and partly weathered rocks. Biological activity usually increases with higher temperatures and decreases under waterlogged and/or acid conditions.

2.2.4   Farming operations
Deep ploughing and cultivation, artificial drainage and liming can speed up the soil formation processes very considerably (Chapter 14).

2.3   The physical make-up of soil and its effect on plant growth

The farmer considers the soil from the point of view of its ability to grow crops. To produce good crops the soil must provide suitable conditions in which plant roots can grow. It should supply nutrients, water and air. The temperature must also be suitable for the growth of the crop.

The soil is composed of:

- *Solids.* Mineral matter (stones, gravel, sand, silt and clay) and organic matter (remains of plants and animals).
- *Liquids.* Soil water (a weak acid).
- *Gases.* Soil air which occupies a variable amount of the pore spaces.
- *Living organisms.* Micro-organisms (bacteria, fungi, small soil animals, earthworms, etc.).

2.3.1   Mineral matter and soil texture

The relative proportion of the mineral material, clay, silt and sand, in a soil is called the *soil texture*. There is a large variation in the texture of soils found on farms. Soil texture can be assessed accurately in the laboratory using mechanical analysis or in the field using hand texturing.

In mechanical analysis the organic matter, stones and gravel are first removed. The rest of the mineral fraction is separated using sieving and sedimentation techniques. This gives an accurate measurement of the amount of clay, silt and sand particles present. The particle size for each mineral fraction is shown in Table 2.1. Clays are the smallest particles and sands and gravel the largest. If a known volume of sand and clay are compared, pore size will be greatest for the sand and surface area greatest for the clay. Because these mineral components have very different properties it is important to know the relative amount of each, hence the importance of knowing the soil texture.

Table 2.1   Particle size and surface area of soils

Material	Diameter of particles (mm)	Surface area
Clay	Less than 0.002	$100\,000 \times a$
Silt	0.002–0.06	$1\,000 \times a$
Fine sand	0.06–0.02	$100 \times a$
Medium sand	0.2–0.6	$10 \times a$
Coarse sand	0.6–2.0	a
Gravel	More than 2.0	

Soil texture is classified into 11 textural classes in the UK (see Appendix 1).

Hand texturing is a very quick and simple technique; with practice it can be a reasonably accurate method. It uses the difference in feel and binding properties of moist soil to differentiate between the textural classes.

The main characteristics of the three mineral fractions are as follows (see Appendix 1 for a full summary):

- *Sands.* Feel gritty and are moulded with difficulty.
- *Silts.* Feel smooth and silky. The soil ball is easily deformed.
- *Clays.* Feel sticky and bind to form a strong soil ball.

The texture of the soil has a major influence on many properties including:

1 *Soil structure.* Soil structure is the arrangement of the individual particles into larger units or aggregates. It is mainly the clay and the organic matter particles that bind the aggregates together.
2 *Air and water supply.* Clay soils have the largest number of small pores which tend to be filled with water rather than air. Some of this water is unavailable for root uptake. The sandy soils have the greatest number of large pores that tend to be filled with air rather than water. Because of these properties clay soils usually need draining for arable cropping and sandy soils are drought prone. It is the medium-textured silty loams that hold the most water available for crop growth.

3   *Cultivation system.* The texture of the soil will affect the cultivation system used to establish crops (pages 195–6). Sandy soils are very easy and require little power to cultivate. Clays, on the other hand, can be very difficult with limited periods when the land can be worked without damaging the structure and with higher power requirements.

4   *Lime and fertiliser requirements.* Some soil types have different abilities both to hold and release nutrients. Most clay minerals have a negative charge which attracts positively charged ions such as potassium, magnesium and calcium. This is called the *cation exchange capacity.* The presence of these charged sites affects the availability of nutrients. Sand and silt are relatively inert. Different soil textures with the same level of acidity have different recommendations for the application of lime. This is due to differences in the cation exchange capacity of the soil.

5   *Erosion.* Some of the more poorly structured soils such as the fine sandy loams can suffer from both wind and water erosion. The peat soils can also be affected by wind erosion.

6   *Herbicide use.* Some soil textures are more prone to herbicide leaching, particularly of some soil applied residuals. Recommended application rates of these chemicals are affected by the soil texture in order to avoid any risk of crop damage. On clay soils rates may be higher as some chemicals are adsorbed onto the soil particles and are then unavailable for weed uptake.

7   *Cropping.* All these characteristics of the different soil textures determine which crops can be grown and affect their yield potential.

2.3.2   Soil structure

Structure is usually defined in terms of the shape and size of the units or aggregates. Structure, unlike texture, can be altered naturally by weathering (e.g. lumps changed to crumbs by frost action or by alternate wetting and drying), by the penetration of plant roots, by incorporation of organic matter and, very importantly, by cultivations. Soils with a high clay or organic matter content tend to have a more stable soil structure than the sands and silts. Soils containing free calcium carbonate and, to a lesser extent iron oxides, also tend to have a stable soil structure. Poor drainage has a negative effect.

It is important that the farmer avoids damaging the soil structure by causing a compacted layer as this can affect crop root growth and the water-holding capacity, reduce nutrient uptake and make the crop more prone to pests and diseases. All these effects can lower final crop yields.

The structure of the soil can be easily damaged by harmful operations, e.g. heavy traffic in wet conditions or ploughing when the soil is too plastic. Over-working poorly structured silty soils can lead to surface capping. Surface capping can lead to crop failure if it occurs after drilling a small seeded crop and also increases the risk of water erosion.

Tilth is a term used to describe the condition of the soil in a seedbed. For example, the soil may be in a finely-divided state or it may be rough and lumpy. Whether a tilth is suitable or not partly depends on the crop to be grown. In general, a small seeded crop requires a finer tilth than large seeds.

2.3.3 Organic matter

Organic matter in the soil is composed of decomposing residues and residual organic matter or *humus*. The organic matter content of a soil varies with texture and cropping. It usually increases with clay content, e.g. ordinary heavy soils have about 3–4% organic matter compared with 1–1.5% in very light soils. Mineral soils where arable crops are continuously grown usually have an organic matter content below 5% compared with 10–20% under permanent grassland. There are varying amounts of organic matter in the 'peat' soils. True peat contains more than 50% organic matter. On draining, liming and cropping these soils have suffered rapid breakdown or *wastage*. For example, the peaty soils only have 20–35% organic matter.

Organic matter may remain for a short time in the undecayed state. However, the organic matter is soon attacked by the soil organisms, such as bacteria, fungi, earthworms and insects. Sugars, starch and crude protein are broken down first and the complex molecules of cellulose, lignin and waxes last. The remaining residual material—humus—has no fixed chemical composition. It commonly contains 45% lignin, 30% protein and the rest a mixture of fats, waxes and complex carbohydrates. The amount of humus formed is greatest from plants which have a lot of strengthening (lignified) tissue, i.e. straw. Organic matter is broken down most rapidly in warm, moist soils which are well limed and well aerated. Breakdown is slowest in waterlogged, acid conditions.

The dark-coloured humus produced during the breakdown process is very beneficial in restoring and stabilising soil structure. Like clay, humus is a colloid, i.e. it is a gluey substance which behaves like a sponge; it absorbs water and swells up when wetted and shrinks on drying. The humus colloids are not so gummy and plastic as the clay colloids but they can improve light (sandy) soils by binding groups of particles together. This reduces the size of the pores (spaces between the particles) and increases the water-holding capacity. Humus can also improve clay soils by making them less plastic and by assisting in the formation of a crumb structure (lime must also be present). Earthworms help in this soil improvement by mixing the clay and humus material with lime.

Plant nutrients, particularly nitrogen and phosphorus, are released (mineralised) each year for plant uptake when organic matter breaks down. The humus colloids can also hold bases such as potassium and ammonium ions in an available form. Organic matter in the soil may be maintained or increased by growing grass, working-in straw and similar crop residues, farmyard manure, composts and green manures. In areas where erosion by wind and water is common, mineral soils are less likely to suffer damage if they are well supplied with humus. Increasing the organic matter content of a soil will increase its ability to hold water and the amount of water available for crop growth (available water capacity), as well as the cation exchange capacity.

2.3.4   Water in the soil

A productive soil will usually have a very good supply of available water and be well drained. The soil is a mass of irregular-shaped particles forming a network of spaces or channels called the pore space, which may be filled with air or water or both. If the pore space is completely filled with water, the soil is waterlogged. It is then unsuitable for most plant growth because the roots need oxygen for respiration. Ideally, there should be about equal volumes of air and water.

The size of the pores is also very important in determining the amount of water held in a soil. *Macropores* are the largest pores and are important for drainage and aeration. *Mesopores* are medium sized and are the pores containing water available for crop growth. *Micropores* are the smallest pores containing water that is not available for root uptake. A sandy soil has a large proportion of macropores so normally more of the pores contain air rather than water; it is the opposite for the clay soils where there are a larger amount of small pores.

The pore space may be altered by a change in the:

- Structure, e.g. by cultivations.
- Amount of organic matter present. The larger the organic matter content the greater the water-holding capacity.
- Compaction of the soil. Compaction reduces the number and size of the pores. It can also affect rooting depth and hence amount of available water.

Another important factor is the surface area of the particles. Water is held as a thin layer or film around the soil particles. The smaller the particles, the stronger are the attractive forces holding the water. Also, the smaller the particles, the greater is the surface area per unit volume. A comparison for pure materials is set out in Table 2.1. The surface area of the particles in a cubic metre of clay may be over 1000 hectares.

When water falls on a dry soil it does not become evenly distributed through the soil. The topmost layer is saturated first and, as more water is added, the depth of the saturated layer increases. In this layer most of the pore space is filled with water. However, a well-drained soil cannot hold all this water for very long; after a day or two some of it will soak into the lower layers or run away in drains. The soil will eventually reach a stage when the amount of any additional water added by rain or irrigation is matched by the loss of water from the profile through natural or artificial drainage. This is known as *field capacity*. The amount of water which can be held in this way varies according to the texture and structure of the soil. The weight of water held by a clay soil may be half the weight of the soil particles, whereas a sandy soil may hold less than one-tenth of the weight of the particles. The water-holding capacity of a soil is usually expressed in millimetres per centimetre, e.g. a clay soil may have a field capacity of 4 mm/cm in depth.

The ways in which water is retained in the soil can be summarised as follows:

- As a film around the soil particles.
- In the organic matter.
- Filling some of the smaller spaces or pores.

Plant roots can easily take up most of this water, but as the soil dries out the remaining water is more firmly held and eventually a stage is reached when the plant can extract no more water. This is the *permanent wilting point* because plants remain wilted and soon die. (At permanent wilting point there is still some water present in the soil.) This permanent wilting should not be confused with temporary wilting which sometimes occurs on very hot days. This is because the rate of transpiration is greater than the rate of water absorption through the roots. In these cases the plants recover as it gets cooler on the same day. The water that can be taken up by the plant roots is called the *available water*. It is the difference between the amounts held at field capacity and at permanent wilting point. In clay soils only about 50–60% of water at field capacity is available; in sandy soils up to 90% or more may be available. It is the silt loams and peaty soils that have the highest available water capacity.

The difference between the actual amount of water in the soil and the amount held at field capacity is known as the soil moisture deficit (SMD). It is the main factor in determining the need for irrigation (see pages 190–194).

Water in the soil tends to hold the particles together and thus lumps of soil may stick together. When a clay soil is at, or above, half field capacity, it is possible to form it into a ball which will not fall apart when handled. As a clay soil becomes wetter it becomes more like plasticine. When there is enough soil moisture to roll the soil into a 3 mm diameter 'worm' without breaking it is at the *lower plastic limit*. Cultivating the soil at this moisture content will damage the soil structure. Some of the water in soils with very small pores and channels moves through the soil by capillary forces, i.e. surface tension between the water and the walls of the fine tubes or capillaries. This is a very slow movement and may not be fast enough to supply plant roots in a soil which is drying out.

Water is lost from the soil by evaporation from the surface and by transpiration through plants. It moves very slowly from the body of the soil to the surface; after the top 20–50 mm have dried out, the loss of water by evaporation is very small. Cultivations increase evaporation losses. Plants take up most of the available water in a soil during the growing season and air moves in to take its place. Air moves easily where the soil has large pore spaces, but the movement into the very small pore channels in clay soils is slow until the soil shrinks and cracks – vertically and horizontally – as water is removed by plants. The water which enters the soil soon becomes a dilute solution of the soluble soil chemicals. It dissolves some of the carbon dioxide in the soil and so becomes a weak acid.

2.3.5   Soil aeration

Plant roots and many of the soil animals and micro-organisms require oxygen for respiration. The air found in the soil is really atmospheric air which has been changed by these activities (and also by various chemical reactions); as a result it contains less oxygen and more carbon dioxide. After a time this reduction in oxygen and increase in carbon dioxide can become harmful to the plant and other organisms.

Aeration is the replacement of this stagnant soil air with fresh air. The movement of water into and out of the soil mainly brings about the process, e.g. rainwater soaks into the soil filling many of the pore spaces and driving out the air. Then, as the surplus water soaks down to the drains or is taken up by plants, fresh air is drawn into the soil to refill the pore spaces. Additionally, oxygen moves into the soil and carbon dioxide moves out by a diffusion process similar to that which happens through the stomata in plant leaves (pages 4–5).

The aeration process is also assisted by:

- Changes in temperature.
- Changes in barometric pressure.
- Good drainage.
- Cultivations.
- Open soil structure.

Size and number of macropores (pores >0.5 mm) affect the aeration of the soil. Sandy soils are usually well aerated because of their open structure. Clay soils are usually poorly aerated, especially when the very small pores in such soils become filled with water. Good aeration is especially important for germinating seeds and seedling plants.

2.3.6   Soil micro-organisms

There are thousands of millions of organisms in every gramme of fertile soil. They are very important in the breakdown of organic residues and formation of humus. Many different types are found, but the main groups are listed below.

1   Bacteria are the most numerous group and are the smallest type of single-celled organisms. Most of them feed on and break down organic matter. They need nitrogen to build cell proteins, but if they cannot get this nitrogen from the organic matter they may use other sources such as the nitrogen applied as fertilisers. When this happens (e.g. where straw is ploughed in) the following crop may suffer from a temporary shortage of nitrogen. Some types of bacteria can convert (*fix*) the nitrogen from the air into nitrogen compounds which can be used by plants (legumes and the nitrogen cycle, page 23). Numbers of bacteria vary in the soil; they increase near a source of organic matter. In the area next to plant roots (*rhizosphere)* there is an increase in the number of bacteria compared with the rest of the soil. The

rhizosphere bacteria may be important in crop nutrient uptake. Bacteria are found in all soils although fewer are present in acid conditions.

2   Fungi are simple types of plants which feed on and break down organic matter. They are the most important organisms responsible for breaking down complex compounds such as lignified (woody) tissue and waxes. They have neither chlorophyll nor proper flowers. The species and number of fungi in the soil are constantly changing, depending on soil conditions and the type of organic material present. Fungi can live in more acid and drier conditions than bacteria; they are the main organisms for decomposing organic material in acid soils. Some fungi form symbiotic associations with plant roots (*mycorrhiza*); these associations can improve nutrient and water uptake and can protect root systems from some pathogens. Mycorrhiza are important in woodland and organic farming systems where levels of available phosphorus and nitrogen are low. As well as the beneficial fungi there are some disease-producing fungi present, e.g. those causing take-all in cereals or club root in brassicas.

3   Actinomycetes are organisms which are intermediate between bacteria and fungi and have a similar effect on the soil. They need oxygen for growth and are more common in the drier, warmer soils. They are not as numerous as bacteria and fungi. Some types can cause plant disease, e.g. common scab in potatoes (worst in light, dry, alkaline soils).

4   Algae in the soil surface are very small simple organisms which contain chlorophyll and so can build up their bodies by using carbon dioxide from the air and nitrogen from the soil. Algae growing in swampy (waterlogged) soils can use dissolved carbon dioxide from the water and release oxygen. This process is an important source of oxygen for crops such as rice. Algae are important in colonising bare soils in the early stages of weathering. Blue-green algae can fix atmospheric nitrogen.

5   Protozoa. These are very small, single-celled animals. Most of them feed on bacteria and similar small organisms. A few types contain chlorophyll and can produce carbohydrates like plants.

2.3.7 Earthworms

Earthworms have a very beneficial effect on the fertility of soils, particularly those under grass. Several million earthworms per hectare have been recorded under grassland. There are several different kinds found in our soils, but most of their activities are very similar. They live in holes in the soil and feed on organic matter – either living plants or, more often, dead and decaying matter. They carry down into the soil fallen leaves and similar materials. Earthworms do not thrive in acid soils because they need plenty of calcium (lime) to digest the organic matter they eat. Their casts, which are usually left on the surface, consist of a useful mixture of organic matter, mineral matter and lime. The greatest numbers are found in loam soils, under grass, where there is usually a good supply of air, moisture, organic matter and lime. Soil disturbance can have a negative effect on earthworm populations. Numbers are often higher under minimal cultivation systems compared with ploughing. Earthworms are the main food of the mole which does so much damage by burrowing and throwing up heaps of soil.

2.3.8   Other soil animals

In addition to earthworms, there are many species of small animals present in most soils. They feed on living and decaying plant material and micro-organisms. Some of the common ones are slugs, snails, millipedes, centipedes, ants, spiders, nematodes, mites, springtails, beetles and larvae of various insects such as cutworms, leatherjackets and wireworms. The farmer is mainly concerned with those which damage his crops or livestock and those which are predators of crop pests such as the carabid beetles.

2.4   Soil fertility and productivity

Soil fertility is a rather loose term used to indicate the potential capacity of a soil to grow a crop (or sequence of crops). The productivity of a soil is the combined result of fertility and management. The fertility of a soil at any one time is partly due to its natural make-up (inherent or natural fertility) and partly due to its condition (variable fertility) at that time. Natural fertility can have an influence on the rental and sale value of land. It is the result of factors which are normally beyond the control of the farmer, such as:

1   The texture and chemical composition of the mineral matter.
2   The topography (natural slope of the land) which can affect drainage, temperature and workability of the soil.
3   Climate and local weather, particularly the effects of temperature, and rainfall (quantity and distribution).

Soil condition is largely dependent on the recent management of the soil. It can be built up by good husbandry but, if this standard is not maintained, the soil will soon return to its natural fertility level. The application of fertilisers can raise soil fertility by increasing the quantities of plant food in the growth and decay cycle.

Soil management can control the following production factors:

1   The amount of organic matter in the soil (page 23).
2   The amount of water in the soil by drainage and irrigation.
3   The loss of soil by erosion (removal by wind and water).
4   The pH of the soil (Liming, pages 47–51).
5   The amount of plant nutrients in the soil.
6   The soil structure (Cultivations, page 195).

Good management of the above factors should maintain or increase soil fertility and at the same time be commercially profitable.

2.5 Farm soils

The following is a general farming classification based on some of the main soil types – clays, sands, calcareous soils, silts, peats and peaty soils.

2.5.1 Clay soils

These soils have a high proportion of clay and silty material – usually over 60%, at least half of which is pure clay. These so-called heavy soils include the clay, silty clay and sandy clay textures. It is the clay content which is mainly responsible for their specific characteristics. Particles of clay have very important properties:

1 They can group together into small clusters (flocculate) or become scattered (deflocculated).
2 They usually have a high cation exchange capacity, and can combine with various chemical substances (base exchange) such as calcium, sodium, potassium and ammonia; in this way plant nutrients can be held in the soil.

Grouping or flocculation of the particles is very important in making clay soils easy to work. Clay particles combined with calcium (lime) will flocculate easily, whereas those combined with sodium will not. The adhesive properties of clay are very beneficial to the soil structure when the groups of particles are small (like crumbs). Deflocculation can occur when clay soils are flooded with sea water or worked when wet. If the latter occurs they become puddled and if the weather then becomes dry the clay dries into hard lumps or clods. Frost action, and alternating periods of wetting and drying, will help to restore large lumps or clods to the crumb condition. Some air is drawn into cracks caused by shrinkage. These cracks remain when the clod is wetted again and so lines of weakness are formed which eventually break the clod. In prolonged dry weather, these cracks may become very wide and deep which later may be very beneficial for drainage.

2.5.1.1 Characteristics

1 Clay soils feel very sticky and roll like plasticine when wet.
2 They can hold more total water than most other soil types and, although only about half of this is available to plants, crops seldom suffer from drought.
3 They swell when wetted and shrink when dried, so a certain amount of restructuring can take place in these soils depending on weather conditions.

4   They lie wet in winter and so stock should be taken off the land to avoid poaching (the compaction of soils by animals' hooves).
5   They are very late in warming up in the spring because water heats up more slowly than mineral matter.
6   They are normally fairly rich in potash, but are deficient in phosphates.
7   Clay soils usually need large infrequent dressings of lime. Overliming will not cause any troubles such as trace element deficiency.

2.5.1.2   Management

Clays are often called heavy soils because for ploughing and subsequent cultivations, compared with light (sandy) soils, two to four times the amount of tractor power may be required. All cultivations must be very carefully timed (usually restricted to a shorter period than on other soil types) so that the soil structure is not damaged. Minimum cultivations are often used to establish crops on these soils and many farms only plough rotationally (page 195). Autumn ploughing, to allow for a frost tilth, is essential if good seedbeds are to be produced in the spring.

Good drainage is very necessary for arable cropping. Some clay fields are still in 'ridge and furrow', which was set up by ploughing, making the 'openings' and 'finishes' in the same respective places until a distinct ridge and furrow pattern was formed. The direction of the furrows is the same as the fall on the field so that water can easily run off into ditches. If the ridges and furrows are levelled out, a mole drainage system using piped main drains should be substituted (page 188). This change is well worthwhile where arable crops are grown.

In many clay-land areas, especially where rainfall is high, the fields are often small and irregular in shape because the boundaries were originally ditches which followed the fall of the land. The hedges and deciduous trees, which were planted later, grow very well on these fertile, wet soils. Organic matter, such as straw-rich farmyard manure, ploughed-in straw or grassland residues, makes these soils easier to work.

2.5.1.3   Cropping

Because of the many difficulties to be overcome only a limited range of arable crops are grown though yield potential is high. In high rainfall areas the ground is usually left in permanent grass and only grazed during the growing season. Winter wheat is the most popular arable crop; winter beans are also grown. These crops are planted in the autumn when more liberties can usually be taken with seedbed preparation than would be permissible in spring. For the same reason, August-sown winter oilseed rape is also popular. Sugar beet and potatoes are occasionally grown in some districts. However, there can be difficulties in seedbed preparation, weed control and harvesting these crops, especially in a wet season. The best sequence of cropping with these root crops is following a period under grass; then the soil structure is more stable and the soil easier to work.

2.5.2   Sandy soils

These soils are mainly composed of sand (greater than 70%) with very little clay (less than 15%) or silt. The soil texture is either a sand or loamy sand. Because these soils contain very little clay or organic matter they tend to be very weakly structured. They are very easy to work. No natural restructuring takes place on these soils so any compaction has to be removed using cultivations. As sand is inert the cation exchange capacity of these soils is very low. Because of the large particle size of sand these soils contain a large number of macropores which tend to contain air rather than water.

2.5.2.1   Characteristics

1   They can be worked at any time, even in wet weather, without harmful effects.
2   They are normally free-draining, but some drains may be required where there is clay or other impervious layer underneath, or a high water table.
3   These soils are droughty as they have a very low available water capacity. For some crops irrigation is required.
4   They warm up early in spring.
5   These soils are prone to nutrient loss and are very responsive to application of nutrients.
6   They are unstable and can be easily eroded by water (on slopes) and by wind.
7   They have little natural structure of their own and often need subsoiling at regular intervals to loosen compacted layers (pans).
8   Some crop pests and diseases are affected by the soil type, e.g. slugs are not usually a problem, whereas 'take-all' disease can be very serious in second cereal crops on sandy soils.

2.5.2.2   Management

In many ways these are the opposite of clays and are often called light soils because, when working them, comparatively little power is required to draw cultivation implements. These soils are not suitable for minimum cultivation systems and topsoil loosening is required, usually by ploughing. After ploughing few cultivations are required to produce a seedbed, often a one-pass system is used. Adequate amounts of fertiliser must be applied to every crop. Liming is necessary but must be used carefully – 'little and often'. Organic matter, especially as humus, is very beneficial because it helps to hold water and plant nutrients in the soil. On properly limed fields it breaks down very rapidly. This is because the soil micro-organisms are very active in these open-textured soils which have a good air supply.

Sandy soils are very prone to both wind and water erosion. Wind erosion in the spring can destroy a young crop. Various methods are available to reduce the risk of erosion including planting nurse crops, applying organic manures and cultivations (see pages 70 and 195).

2.5.2.3 Cropping

A wide range of crops can be grown, but yields are very dependent on a good supply of water and adequate plant food. Cereal yields are usually low. Market gardening is often carried out where a good sandy area is situated near a large population.

On the lighter sands in low rainfall areas and where irrigation is not possible, less drought-susceptible crops are grown, e.g. rye, carrots, sugar beet, forage maize and lucerne; lupins are grown in a few areas where the soil is very poor and acid. To reduce the risk of drought stress, crops are established early to try and encourage deep rooting, better results are produced from winter- rather than spring-sown cereals. Winter barley is often preferred rather than wheat as it ripens earlier and may be less affected by a summer drought. It is important to ensure there is no soil compaction which would also restrict rooting.

On the better sandy soils, and particularly where the water supply (from rain or irrigation) is reasonably good, the main arable farm crops grown are cereals, peas, sugar beet, potatoes and carrots. Because of the inherent low fertility of this type of land, the farms and fields are usually larger than on more fertile soils. Where there are hedges they are often left tall to act as windbreaks. The trees are usually drought-resistant coniferous types. Livestock can be outwintered on sandy soils with very little risk of damage by poaching, even in wet weather.

2.5.3 Loams

Loams are intermediate in texture between the clays and sandy soils and, in general, have most of the advantages and few of the disadvantages of these two extreme types. These soils have developed over a wide range of parent material; they are called medium-textured soils. The amount of clay present varies considerably, and can be up to 35%. Texture classes include the clay loams (resembling clays in many respects), silty clay loams, and sandy loams (resembling the better sandy soils). A large proportion of the sand is coarse and/or medium sized which makes the soil feel gritty. Climate (i.e. wetness) and topography are the main limitations to crop production.

2.5.3.1 Characteristics

1   Average water-holding capacity and so are fairly resistant to drought.
2   They warm up reasonably early in the spring.
3   They are moderately easy to work.
4   Depending how they were formed, some of the loams can contain stones which can affect sowing and harvesting of some crops.
5   A potentially fertile soil.

2.5.3.2 Management

Loams are moderately easy to work but should not be worked when wet, especially clay loams. Minimum cultivation systems can be successful on these soils.

2.5.3.3   Cropping

Loams are generally regarded as the best all-round soils because they are naturally fertile and can be used for growing any crop provided the depth of soil is sufficient. Crop yields do not vary much from year to year. These soils can be used for most types of arable or grassland farming but, in general, mixed farming is carried on. Cereals, potatoes and sugar beet are the main cash crops, and leys provide grazing and winter bulk foods for dairy cows, beef cattle or sheep.

2.5.4   Calcareous soils

These are soils derived from chalk and limestone rocks and contain various amounts of calcium carbonate, between 5% and 50%. The depth of soil and subsoil may vary from 8cm to over a metre. In general, the deep soils are more fertile than the shallow ones. The ease of working and stickiness of these soils depend on the amount of clay and chalk or limestone present; they usually have a loamy texture. Sharp-edged flints of various sizes, found in soils overlying some of the chalk formations, are very wearing on cultivation implements and rubber tyres, as well as being destructive when picked up by harvesting machinery. In some places the flints are found mixed with clay, e.g. clay-with-flint soils.

The soils overlying chalk are generally more productive than are those over limestone because plant roots can penetrate the soft chalk and explore for water. The limestones are harder and mainly impenetrable. Limestone rock pieces, loosened by cultivations, are a more severe problem to the farmer than the pieces of chalk that work their way to the surface on the downland arable farms.

Dry valleys are characteristic of the limestone and chalk downland. The few rivers rise from underground streams and the deliberate flooding of watermeadows in the river valleys used to be a common practice. Watercress beds flourish along some chalk streams. The farms and fields on this type of land are usually large, especially on the thinner soils. There are very few hedges and the trees are mainly beech and conifers. Walls of local stone form the field boundaries in some limestone areas.

2.5.4.1   Characteristics

1   The soils are free draining except in a few small areas where there is a deep clay subsoil.
2   The soils are often shallow and can be prone to drought (limestone soils).
3   The soils are usually found at fairly high altitudes, above 120m.
4   Stones can affect sowing and harvesting of some crops.
5   Naturally contain low levels of some major nutrients. The alkalinity of the soil can cause some trace elements to be unavailable. Only the deepest and/or sandy outcrops ever need liming.

2.5.4.2 Management

All, except soils with a high clay content, are easy to cultivate. Minimum cultivation systems have been very successful on these well structured soils. As the soils are free draining there is a large window when it is possible to cultivate. Drilling of spring cereals can often take place very early in the spring. Organic matter can be beneficial, but in the alkaline conditions it breaks down fairly rapidly.

2.5.4.3 Cropping

Cereals are good crops for these soils and malting barley has been very successful. Continuous barley and/or wheat production has been common practice, but this has changed partly due to the difficulty of controlling grass weeds and better returns from other crops. Roots such as sugar beet, fodder beet and potatoes (some for seed) are grown on the deeper soils on some farms. Other crops such as oilseed rape, peas, beans, linseed, and leys for grazing, conservation or seed production provide a break from cereals. On the thin limestone soils it is important to establish the winter cereals early so that they are less affected by any summer drought.

As calcareous soils can be found at fairly high altitudes and the fields are often exposed, great care should be taken when growing crops which shed their seed easily, e.g. oilseed rape. Harvesting must be carried out carefully to minimise yield loss.

Areas of black puffy soil (18–25% organic matter) are found on some chalkland farms and they require special treatment; cultivations can be difficult and often trace element deficiencies, e.g. for copper, need correcting.

2.5.5 Silts

The silty soils can contain up to 80% silt. Silty soils include sandy silt loam and silt loam textures. The sand fraction is mainly of very fine sand particles. They have a very silky, buttery feel. These soils are formed from glacial, river, marine and wind-blown deposits and usually have deep stone-free subsoils.

2.5.5.1 Characteristics

1   Most silts need some sort of drainage. In certain coastal areas there has to be a pumped system in order for the water table to be lowered and arable crops grown.
2   Potentially very fertile mainly due to their great depth and very high available water capacity.
3   Good working properties, provided that organic matter is maintained above 3%.
4   Water erosion can be a problem in some areas.

2.5.5.2   Management
Although very fertile, silts can be difficult to manage. The two main problems are capping and compaction.

Capping occurs when heavy rain falls on a very fine seedbed. Silt and clay particles go into suspension in a surface slurry and, as this dries out, it forms an impenetrable layer on the surface of the soil. It is particularly damaging if seeds have been sown but not yet germinated and emerged. Crops with small seedlings, such as sugar beet or brassicas, will sometimes need to be redrilled if capping occurs. Leaving a rougher seedbed and increasing surface organic matter can decrease the risk of capping.

The lighter silts, as with the sands, can be damaged by compaction if worked in unsuitable conditions. Correction by deep cultivations may be necessary.

2.5.5.3   Cropping
On the heavier silts, mainly grass, cereals or fruit are grown; yield potential is high. However, on the drained light silts there are no limitations to crop growth. A wide range of crops can be grown including wheat, potatoes, sugar beet, vining peas, bulbs and field vegetables. River and marine silts in high rainfall areas, that are liable to flood, are best left in permanent grass.

2.5.6   Peat and peaty soils
Peaty soils contain about 20–35% of organic matter, whereas there is over 50% in true peat. These organic soils are usually very black or dark brown in colour and feel silty.

The blanket bog peats have formed in waterlogged upland areas overlying impermeable parent rock where there is a high rainfall. Plants such as mosses, cotton grass, heather, *Molinia* grass and rushes grow. The dead material from these plants is only partly broken down by the types of bacteria which can survive under these acidic waterlogged conditions. This 'humus' material builds up slowly, about 30 cm every century. These peats are very acidic and are very low in nutrients. A certain amount of improvement can be done by drainage, fertilising and liming but because of their situation they are only suitable for permanent grassland or forestry.

There are two important types of lowland peat, the sedge or fen peats and the moss peats. The fen peats developed in marshy land depressions below sea level, where the water coming into the area was rich in nutrients. Successive layers of decaying plant material built up. The main vegetation consisted of reeds, sedges and woody plants. Old bog oaks are sometimes brought up by cultivations. When drained these peats break down to produce very fine particles. These soils are slightly acidic and are low in phosphorus, potassium and some trace elements. The moss peats developed on boggy land with a high rainfall. The main vegetation was moss which grew and decayed very slowly to produce a raised area (raised moss). When drained these peats have a low pH and are very low in nutrients. Moss peat is more fibrous than fen peat.

2.5.6.1 Characteristics

1   These soils are rich in nitrogen, released by the breakdown of the organic matter, but are very poor in phosphates and potash and also trace elements such as manganese and copper.
2   Very fertile soils if drained and fertilised. As well as ditches for drainage those areas below sea level need a pumped system if arable crops are to be grown.
3   The soils are naturally acidic and initially need heavy liming. Overliming can lead to trace element deficiencies.
4   On drainage the peat initially shrinks. Afterwards the organic matter breaks down very rapidly and the soil level can fall by as much as 2.5 cm per year; this is called '*wastage*'. Eventually the mineral subsoil will come close to the topsoil and have a major effect on the characteristics and potential cropping of the soil.
5   Wind erosion or 'blowing' in spring is a serious problem due to the light weight of peat when dry. Several plantings of crop seedlings, together with the top 5–8 cm of soil and fertilisers, may be blown into the ditches. Deep ploughing and/or cultivation to mix the underlying clay/mineral subsoil with the organic topsoil can prevent this. Husbandry techniques such as straw planting, nurse crops or strip cultivation are also used to prevent wind erosion (page 195).
6   Weed control can be very difficult. Weeds grow very vigorously with the high nitrogen levels. Residual or soil-applied herbicides are not normally effective as the chemicals become adsorbed onto the particles of organic matter. Foliar-acting herbicides or mechanical methods are commonly used for weed control on these soils.
7   Peat soils have a very high available water capacity for crop growth. Crops rarely suffer from drought on these soils.
8   Under grassland these soils are very susceptible to poaching.

2.5.6.2 Management

Before reclaiming this land for cropping, some of the peat can be cut away for fuel or sold as peat moss for horticultural purposes. Good drainage must then be carried out by cutting deep ditches through the area. Deep ploughing also helps to drain the soil. Most of the lowland peat is below sea level and so the water in the ditches has to be pumped over the sea walls or into the main drainage channels. The depth of the subsequent water table will affect the potential cropping of the area. Once the peat has wasted to a depth of less than 0.9 m, under drainage may be required if the subsoil is clay.

Heavy applications up to 25 t/ha of ground limestone may be required to reduce the soil acidity. Sometimes a very acid reddish brown layer ('*drummy*' layer) builds up between the organic topsoil and the underlying clay subsoil. Peat soils are easy to cultivate and do not suffer from soil compaction. Under-consolidation of the seedbed is common. The peaty soils with a lower organic matter content suffer less from under-consolidation. Cultivation systems for the organic mineral soils are similar to those for the mineral soils.

2.5.6.3   Cropping
Drained fen or light peat soils are among the most fertile arable soils. Crops such as potatoes, sugar beet, celery, onions, carrots, lettuce and market garden crops are commonly grown. Cereals produce low yields. On light undrained peats, or where the water table remains high, the main crop is grass. On some of these areas willows are grown. Peaty loams are suitable for growing root crops and cereals. Where the depth of peat is very shallow (due to wastage) cropping is similar to that for the underlying mineral soil type.

2.6   Soil improvement

Other than drainage (Chapter 8) one of the most important improvements that can be done to the soil to ensure optimum crop growth is liming of acidic soils. (In chalk and limestone areas there is plenty of free calcium carbonate and liming is not required.)

2.6.1   Liming
Most farm crops will not grow satisfactorily if the soil is very acidic. In acid conditions aluminium, iron and manganese become more readily available. Excessive uptake of aluminium in acid conditions can severally affect crop growth. Some crops are more affected than others. This can be remedied by applying one of the commonly used liming materials.

2.6.1.1   The chemistry of liming
pH and crop growth
To give crops the best opportunity to grow well, the minimum soil pH should be as follows.

Barley, sugar beet, beans, peas and lucerne	6.5 pH
Maize, oilseed rape, oats, wheat, cabbage and carrots	6.0 pH
Potatoes, rye and apples	5.5 pH
Ryegrass and fescues	5.0 pH

Soil reaction
All substances in the presence of water are either acid, alkaline or neutral. The term 'reaction' describes the degree or condition of acidity, alkalinity or neutrality. Acidity and alkalinity are expressed by a pH scale on which pH 7 is neutral, numbers below 7 indicate acidity and those above 7 alkalinity. (pH

of a soil solution is the negative logarithm of the hydrogen ion concentration.)
Most cultivated soils have a pH range between 4.5 and 8.5 and may be grouped
as in Table 2.2. Soil pH can be assessed in the field by using a soil indicator
or more accurately using an electrode in the laboratory.

Table 2.2   pH value of soils

pH	Description
7+	Calcareous
7	Neutral
6.0–6.9	Slightly acid
5.2–5.9	Moderately acid
5.2–	Very acid

Indications of soil acidity (i.e. the need for liming)

1   Crops failing in patches, particularly the acid-sensitive ones such as barley
and sugar beet. The plants look yellow and stunted with a stumpy root
system.
2   On grassland, there are poor types of grasses present such as the bents.
Often a mat of undecayed vegetation builds up because the acidity reduces
the activities of earthworms and bacteria that break down such material.
3   On arable land, weeds such as sheep's sorrel, corn marigold and spurrey
are common where a soil is or has a history of being acidic. Also, in acid
conditions residual herbicide activity may be reduced.
4   Club root disease in the brassicas is aggravated in acid soils. Liming is
used as a method of control.

2.6.1.2   Lime requirement

This is based on the *optimum* pH for the crop to be grown, the soil texture,
organic matter content and current pH. Grassland has a lower requirement than
arable crops. Heavy (clay) soils and soils rich in organic matter require more
lime to raise the pH than other types of soil. For example, to raise the pH from
5.5 to 6.5 on a loamy sand may require about 7 t/ha of ground limestone, but
on a clay soil 10 t/ha of ground limestone may be necessary. The sandy soil,
however, will need to be limed more frequently than the clay soil. Lime should
be applied before growing susceptible crops such as beet or barley but after
growing such tolerant crops as potatoes. (Lime requirements are shown in Table
2.3.)
The main benefits of applying lime are:

1   It neutralises the acidity of the soil by removing the hydrogen ions on the
soil charged sites and replacing them with calcium or magnesium.
2   It supplies calcium (and sometimes magnesium) for plant nutrition.
3   It improves soil structure and makes the structure more stable. In well-
limed soils, plants usually produce more roots and grow better. Bacteria

Table 2.3 Lime requirement of different soils

Cropping	Soil type	pH 6.0	pH 5.5	pH 5.0
Arable cropping	Light	4 t/ha	7 t/ha	10 t/ha
	Medium	5 t/ha	8 t/ha	12 t/ha
	Heavy	6 t/ha	10 t/ha	14 t/ha
	Organic	4 t/ha	9 t/ha	14 t/ha
	Peat	0 t/ha	8 t/ha	16 t/ha
Grassland	Light	0 t/ha	3 t/ha	5 t/ha
	Medium	0 t/ha	4 t/ha	6 t/ha
	Heavy	0 t/ha	4 t/ha	7 t/ha
	Organic	0 t/ha	3 t/ha	7 t/ha
	Peat	0 t/ha	0 t/ha	6 t/ha

are more active in breaking down organic matter and this also usually results in an improved soil structure and the soil can be cultivated more easily.

4   It affects the availability of plant nutrients. Nitrogen, phosphate and potash are freely available on properly limed soils. Too much lime is likely to make some minor nutrients or trace elements unavailable to plants, e.g. manganese, boron, copper and zinc. This is least likely to happen in clay soils and most likely to happen on organic soils.

2.6.1.3 Lime losses
Lime is removed from the soil in many ways.

Drainage
Lime is fairly easily removed in drainage water. About 125–2000 kg/ha of calcium carbonate may be lost annually. The rate of loss is greatest in industrial, smoke-polluted areas, areas of high rainfall, well-drained soils and soils rich in lime.

Fertilisers and manures
Fertilisers containing ammonium, nitrate, sulphate or chloride ions have the greatest acidifying effect. Every 1 kg of sulphate of ammonia removes about 2 kg of calcium carbonate from the soil. Other nitrogen fertilisers have a smaller acidifying affect. (Some organic manures can slightly reduce soil acidity.)

Crops
The approximate amounts of calcium carbonate removed by crops are:

Cereal grain plus straw	50 kg/7 t of grain.
Potatoes	20 kg/40 t of tubers.
Sugar beet	58 kg/40 t of roots.
	230 kg/35 t of tops.
Swedes	100 kg/60 t of roots.
Kale (carted off)	450 kg/60 t of crop.
Lucerne	58 kg/t of hay.

Stock also remove lime, e.g. a 500 kg animal sold off the farm removes about 16 kg of calcium carbonate in its bones, a 40 kg lamb about 1.3 kg of calcium carbonate, and 5000 l milk about 18 kg of calcium carbonate.

2.6.1.4 Materials commonly used for liming soils

Ground limestone or chalk (also called carbonate of lime or calcium carbonate $CaCO_3$). This is obtained by quarrying the limestone or chalk rock and grinding it to a fine powder. It is the commonest liming material used at present. The finer the particle size the quicker the reduction of acidity. Burnt lime (also called quicklime, lump lime or calcium oxide, CaO). This is produced by burning lumps of limestone or chalk rock with coke or other fuel in a kiln. Carbon dioxide is given off and the lumps of burnt lime which are left are sold as lumps, or are ground up for mechanical spreading. This concentrated form of lime is especially useful for application to remote areas where transport costs are high. Burnt lime may scorch growing crops because it readily takes water from the leaves. When lumps of burnt lime are wetted they break down to a fine powder called hydrated lime.

Hydrated lime (also called slaked lime $Ca(OH)_2$). This is a good liming material but it is usually too expensive for the farmer to use.

Waste limes

These are liming materials that can sometimes be obtained from industrial processes where lime is used as a purifying material. These limes are cheap but usually contain a lot of water. Some of the sources are: sugar beet factories, waste from manufacture of sulphate of ammonia, soap works, bleaching, tanneries. Care is needed when using these materials because some may contain harmful substances. Sugar beet waste lime is also a valuable source of plant nutrients, but because of the root disease rhizomania (Table 7.1) its use is now very much restricted.

A comparison of the various liming materials shows that one tonne of burnt lime (CaO) is equivalent to:

- 1.37 t hydrated lime $Ca(OH)_2$.
- 1.83 t ground limestone $CaCO_3$ or at least 2.5 t waste lime (usually $CaCO_3$).

The supplier of lime must give a statement of the neutralising value (NV) of the liming material which is the same as the calcium oxide equivalent.

Magnesian or dolomite limestone

This limestone consists of magnesium carbonate ($MgCO_3$) and calcium carbonate ($CaCO_3$). It is commonly used as a liming material in areas where it is found. Magnesium carbonate has a better neutralising value than calcium carbonate of approximately 20 %. Additionally, the magnesium may prevent magnesium deficiency diseases in crops (e.g. interveinal yellowing of leaves in potatoes, sugar beet and oats).

Calcareous sand or shell sand
This liming material is collected from some beaches where there is a high shell content. The neutralising value is lower than for ground limestone. It is a useful material when other liming products have to come a long distance.

2.6.1.5   Cost
The cost of liming is largely dependent on the transport costs from the lime works to the farm. By dividing the cost per tonne of the liming material by the figure for the neutralising value, the unit cost is obtained. In this way it is possible to compare the costs of the various liming materials. Most farmers now use ground limestone or chalk and arrange for it to be spread mechanically by the suppliers. Where large amounts are required (over 7 t/ha) it may be preferable to apply it in two dressings, e.g. half before and half after ploughing.

2.7   Further reading

Batey T, *Soil Husbandry*, Soil and Land Use Consultants Ltd, 1988.
Bridges E, *World Soils*, Cambridge University Press, 1997.
Curtis L F, Courtney F M and Trudgill S, *Soils of the British Isles*, Longmans, 1976.
Davis B D, Eagle D J and Finney J B, *Resource Management: Soil*, Farming Press, 2001.
Gerard G, *Fundamentals of Soils*, Routledge, 2000.
Kilham K, *Soil Ecology*, Cambridge University Press, 1994.
MAFF Report, *Modern Farming and the Soil*, HMSO, 1970.
Rowell D L, *Soil Science: Methods and Applications*, Longmans, 1994.
Simpson K, *Soil*, Longmans, 1987.
Soil Survey for England and Wales, *Soils and their use in East England, Midlands and North Western England, S. E. England, S. W. England and Wales*, SSEW, 1985.
Wild A, *Soils and the Environment: an Introduction*, Cambridge University Press, 1995.

3

Fertilisers and manures

3.1   Nutrients required by crops

If good crops are to be grown in a field, there should be at least as many nutrients returned to the soil as have been removed and these nutrients should be readily available to the plant as and when needed. This can be achieved by the sensible use of chemical fertilisers (inorganic manures), where possible in conjunction with organic manures such as farmyard manure and slurry or, as in organic systems, the use of natural inorganic fertiliser materials in conjunction with organic manures. It is important that any nutrient supply system minimises the losses of nutrients to the environment as well as avoiding either a deficiency or an excess of individual, required elements. Table 3.1 indicates average figures for nutrients removed by various crops.

Nutrients required by crops can be divided up into major elements and micro (trace) elements:

Major elements	Taken up by plant as
Nitrogen	nitrate
Phosphate	hydrogen phosphate
Potassium	K^+
Sulphur	sulphate
Magnesium	Mg^{2+}
Calcium	Ca^{2+}
Sodium	Na^+

Micro (trace) elements include boron, copper, iron, manganese, molybdenum, zinc and cobalt, some of which are dealt with later in this chapter.

Table 3.1 Nutrients removed by crops (Kg/ha)

Crop (good average yield)	N	P_2O_5	K_2O	Mg	S
Wheat					
grain 7 t/ha	130	55	40	9	35
straw 5 t/ha	17	4	31		
Total	147	59	71	9	35
Barley					
grain 6 t/ha	100	47	34	7	22
straw 4 t/ha	25	3	25		
Total	125	50	59	7	22
Oats					
grain 6 t/ha	100	47	34		
straw 5 t/ha	15	5	59		
Total	115	52	93		
Beans					
grain 4 t/ha	176	44[a]	48	8	40
Potatoes					
tubers 50 t/ha	150	50	290	15	20
Sugar beet					
roots 45 t/ha	80	43	86		
fresh tops 35 t/ha[b]	120	39	202	27	33
Total	200	82	288	27	33
Kale					
fresh crop 50 t/ha[b]	224	67	202		

[a] The response to phosphatic fertilisers is greater than these figures suggest.
[b] If sugar beet tops or kale are eaten by stock on the field where grown, some of the nutrients will be returned to the soil.

Nitrogen is supplied by fertilisers, organic manures, nodule *Rhizobium* bacteria on legumes (e.g. clovers, peas, beans, lucerne), and bacteria in the soil which decompose organic matter and produce nitrates in a process called mineralisation. Free-living bacteria (*Azotobacter*) also fix nitrogen in the soil. Some nitrogen is also produced during thunderstorms by lightning strikes and some is contained in rainfall. Nitrogen can be removed from the soil by bacteria which steal the oxygen from nitrates to leave gaseous nitrogen. This is a process called denitrification. Nitrogen can also be lost by leaching (the process of nitrates in solution moving down away from the root zone and eventually into the drains or bedrock).

Phosphate is supplied by fertilisers and organic matter from manures or plant debris. Soluble fertilisers provide hydrogen phosphate for immediate use by the plant while insoluble fertilisers need to be worked on by bacteria and the action of weak acids in the soil before they can be used. Organic matter phosphate goes through a transition phase of microbial phosphate before becoming hydrogen phosphate in soil solution.

Potassium comes from organic manures and plant debris as well as fertilisers. Within the soil there are three types of potassium: water-soluble potassium which is available to plants, exchangeable potassium (an intermediate stage) and a potassium reserve held in the clay lattices within the soil which, by the action of weathering, becomes available over a period of time.

Sulphur is provided by fertilisers, organic matter and as sulphur dioxide and sulphurous acid in rainfall (acid rain). Organic matter sulphur is converted to sulphur and then sulphate by microbial action. Under anaerobic conditions sulphur can be lost from the soil in the form of hydrogen sulphide (a gas smelling of bad eggs).

Table 3.2 shows the major plant foods for crop growth.

Table 3.2   The need for and effect of some plant nutrients

Plant nutrient	Crops which are most likely to suffer from deficiency	Field conditions where deficiency is likely to occur	Deficiency symptoms	Effects on crop growth	Effects of excess	Time and method of application
Nitrogen (N)	All farm crops except legumes (e.g. beans, peas, clover); especially important for leafy crops (e.g. grass, cereals, kale and cabbages)	On all soils except peats, and especially where organic matter is low, and after continuous cereals	Thin, weak, spindly growth; lack of tillers and side shoots; small yellow/pale green leaves, sometimes bright colours	Speeds up growth of seedlings and roots; hastens leaf growth and maturity; encourages clover in grassland; improves quality	Lodging in cereals; delayed ripening; soft growth susceptible to frost and disease; may lower sugar and starch content	N fertilizers in seedbed or top-dressed
Phosphorus (P)	Root crops (e.g. sugar beet, swedes, carrots, potatoes), clovers, lucerne and kale	Clay soils; acid soils especially in high rainfall areas, chalk and limestone soils and peats; poor grassland	Similar to nitrogen except that leaves are a dull, bluish-green colour with purple or bronze tints	Increases leaf size, rate of growth and yield; makes leaves dark green	Might cause crops to ripen too early and so reduce yield if not balanced with nitrogen and potash fertilisers	Phosphorus fertilisers applied in seedbed for arable crops; 'placement' in bands near or with the seed is more efficient; broadcast on grassland
Potassium (K)	Potatoes, carrots, beans, barley,	Light sandy soils, chalk soils, peat, badly	Growth is squat and growing points 'die-	Crops are healthy and resist disease and	May delay ripening too much; may cause	K fertilizers broadcast or 'placed' in seedbed for

Table 3.2   *(Cont'd)*

Plant nutrient	Crops which are most likely to suffer from deficiency	Field conditions where deficiency is likely to occur	Deficiency symptoms	Effects on crops growth	Effects of excess	Time and method of application
	clovers, lucerne, sugar beet and fodder beet	drained soils, grassland repeatedly cut for hay, silage or 'zero' grazing	back', e.g. edges and tip of leaves die and appear scorched	frost better; prolongs growth; improves quality; balances N and P fertilizers	magnesium deficiency in fruit and glasshouse crops and 'grass-staggers' in grazing animals	arable crops; broadcast on grass-land in autumn or mid-summer
Magnesium (Mg)	Cereals, potatoes, sugar beet, peas, beans, kale	Light sandy soils, chalk soils, often of a temporary nature due to poor soil structure, excessive potash, etc.	Chlorotic patterns on leaves (short of chlorophyll)	Associated with chlorophyll, and potassium metabolism	Unlikely, requirements about same as phosphorus and one-tenth that of nitrogen and potash	Use FYM or slurry; lime with magnesium limestone; Epsom salts; fertiliser containing kieserite
Sulphur (S)	Most crops but especially brassicae, e.g. kale, oilseed rape; grass cut for conser-vation; bread wheat	Light sandy soils– especially if low in organic matter and intensively cropped; modern fertilisers used; no smoke pollution	Yellowing leaves and less vigorous growth, like nitrogen deficiency	Nitrogen made more efficient	Unlikely to occur	Use FYM or sulphur foliar spray or to soil as gypsum

A decision has to be made as to how much and what type of fertilisers should be used for each crop. For phosphorus, potassium and magnesium the amount applied should be based on an analysis of a soil sample.

Soil analysis will show:

1   Soil texture.
2   pH (usual range 4–8). This is a useful guide to the lime requirement needed to bring acid soils up to an optimum level for the particular crop.
3   Available nutrients. This is the level of phosphate, potash and magnesium and it is indicated by index ratings 0–9. A deficiency level is indicated by 0 and an excessively high level by 9 (never reached under field conditions). An index of 2 and 3 is satisfactory for farm crops. As indices drop below

2 more fertiliser is recommended so that the index will go up over a period of time, and as they get above 2 or 3 the recommendations decrease until, at index 4 and over, no phosphorus, potassium or magnesium is required.

The field sampling for analysis must be done in a methodical manner if a reasonably accurate result is to be obtained. Generally, sampling should be carried out every four years, although pH may need to be checked more frequently, especially on light soils where calcium is easily leached. At least 25 soil cores, 15 cm deep (7.5 cm on long-term grassland), should be taken to make up a sample weighing at least 0.5 kg. Unless the area is very uniform, samples should be taken for every four hectares. A 'W' pattern should be walked across the area with sub samples being taken frequently. Headlands, gateways and minor areas of obvious soil differences should be avoided. The analysis is carried out on the fine part of the soil and this must be borne in mind when interpreting the results for very stony soils. In this case the indices are usually too high.

The recommendations for fertiliser use were reviewed in 2000 and published by the Ministry of Agriculture, Fisheries and Food (MAFF) in their Seventh Edition Booklet *Fertiliser Recommendations for Agricultural and Horticultural Crops* (RB 209). This is essential reading for anyone involved in crop production. The new recommendations take more account of the economic use of fertilisers, the value of organic manures and the minimisation of environmental risk associated with the use of all types of fertilisers and manures.

Particularly important are the changes made to the nitrogen recommendations. Nitrogen is the most important element in terms of crop yield and quality, and deficiencies will lead to a serious reduction in profitability. However, excess nitrogen will be wasteful and could lead to the pollution of the environment. It is important therefore that the nitrogen requirements of individual crops should be calculated as accurately as possible. Nitrogen response curves using value of crop yield plotted against cost of applied nitrogen can be used to find the economic optimum rate of nitrogen to apply. This economic optimum application rate takes into account the amount of nitrogen supplied from the soil and can be provided from inorganic and/or organic sources.

The new MAFF nitrogen recommendations use Soil Nitrogen Supply (SNS) as their basis rather than the old soil nitrogen index system. MAFF defines SNS as 'the amount of nitrogen in the soil that becomes available for uptake by the crop from establishment to the end of the growing season, taking account of nitrogen losses'. It is calculated using the equation:

$$\text{SNS} = \text{Soil Mineral Nitrogen (SMN)} + \text{estimate of total crop} \\ \text{nitrogen content} + \text{mineralisable nitrogen.} \qquad [3.1]$$

Although the SMN (and thus the SNS) can be measured using laboratory analysis of soil samples, in most situations the SNS index will be determined by a 'Field Assessment Method' based on previous cropping, previous fertiliser and organic manure use, soil type and winter rainfall.

The SNS index system is shown in Table 3.3.

Table 3.3   MAFF soil nitrogen supply (SNS) index system

Rainfall	Variables used to determine SNS Soil type
Low Rainfall Areas: 500–600 mm annual rainfall, 50–150 mm excess winter rainfall Moderate Rainfall Areas: 600–700 mm annual rainfall, 150–250 mm excess winter rainfall High Rainfall Areas: over 700 mm annual rainfall, or over 250 mm excess winter rainfall	Light sands or shallow soils over sandstone Medium soils or shallow soils (not over sandstone) Deep clay soils Deep fertile silty soils Organic soils Peat soils

The SNS of a field following an arable, forage or vegetable crop will depend on what soil type and in what rainfall area the crop was grown. For example, cereals can leave a field anywhere between SNS Index 0 and 6 depending on where they were grown. A winter wheat crop grown on a light sand soil in a low rainfall area will mean an index 0 field; the same crop grown on a deep fertile silt in a moderate rainfall area will mean an index 2 and grown on a peat soil in all rainfall areas will leave the field at index 5 or 6. The nitrogen requirements for a following crop can then be read from a table using the determined SNS value. For example, oilseed rape at SNS index 0 requires 220 kg/ha of spring N; at SNS index 3 it only requires 120 kg/ha. The SNS system and the recommendation tables can be found in MAFF Book RB 209.

SNS indices for sequences of crops following grassland are more complicated. Below is an example of indices for a medium soil or shallow soil (not over sandstone) in any rainfall area.

Crop	1st crop	2nd crop	3rd crop
All leys with 2 or more cuts annually receiving little or no manure. 1–2 year leys, low N* 1–2 year leys, 1 or more cuts 3–5 year leys, low N, 1 or more cuts	1	1	1
1–2 year leys, high N*, grazed 3–5 year leys, low N, grazed 3–5 year leys, high N, 1 cut then grazed	2	2	1
3–5 year leys, high N, grazed	3	3	2

*Low N = less than 250 kg/ha fertiliser and manure N averaged over 2 years.
High N = more than 250 kg/ha fertiliser and manure N averaged over 2 years or high clover content or lucerne.
The SNS indices above do not take into account any organic manure application nor do they include permanent pasture. RB 209 explains how these can be taken into account.

3.2   Trace elements

The need for trace elements is likely to be greatest:

1   On very poor soils.
2   Where soil conditions, such as a high pH or a very high organic matter, make them unavailable.
3   Where intensive farming (with high yields) is practised.
4   Where organic manures such as farmyard manure and slurry are not used.

The importance of trace elements in plant nutrition is most appreciated when plants are grown in culture solutions circulated past their root systems (hydroponics). This is a form of horticultural crop production that is rapidly increasing, especially where the solutions can be replenished automatically with nutrients.

Supplying trace elements to plants is not always easy and care is required to prevent overdosing which may damage or kill the crop. To facilitate their application and availability to the plants, trace elements such as copper, iron, manganese and zinc can now be used in a chelated form. These chelates are 'protected' water-soluble complexes of the trace elements with organic substances such as EDDHA (ethylene diamine dihydroxyfenic acetic acid) or EDTA (ethylene diamine tetra-acetic acid). These organic substances form a protective 'crab's claw' around the element, effectively sequestering them and protecting them temporarily. They can be safely applied as foliar sprays for quick and efficient action, or to the soil, sometimes as a supplement with a compound fertiliser, for root uptake without wasteful 'fixation' because they do not ionise. They are compatible when mixed with many spray chemicals. Trace elements may also be applied as *frits*, which are produced by fusing the elements with silica to form glass which is then broken into small particles for distribution on the soil.

It is very important, whenever possible, to obtain expert advice before using trace elements. This is because crops that are not growing satisfactorily may be suffering for reasons other than a trace element deficiency, e.g. major nutrient deficiency, poor drainage, drought, frost, mechanical damage, viral or fungus diseases. Trace element deficiencies can be diagnosed by either leaf or soil analysis, depending on the nutrient.

Manganese is the element most frequently found to be deficient in field crops. It is associated with high pH levels and high organic matter, both of which lock up manganese in the soil. It can be remedied using foliar sprays of manganese sulphate or chelated manganese, usually applied just before periods of rapid crop growth.

Boron deficiency is most often seen in root vegetables and sugar beet where it causes a problem of growing point necrosis. Soils can be tested for boron levels and remedial action should be taken before crops are grown. Sprays of boron or boronated fertilisers can be used where necessary. Copper deficiency is a problem of peaty, very light sandy and thin organic chalky soils. Symptoms are most often seen in wheat and barley where it produces shrivelled leaf and ear tips. Chelated copper can be applied directly to crops or copper sulphate applied to soils.

3.3   Units of plant food

3.3.1   Plant food requirements
The kilogram is now the unit of plant food and recommendations are given in

terms of kilograms per hectare (kg/ha). For example, the recommendation for a spring cereal crop may be to apply 100kg/ha N, 50kg/ha P_2O_5 and 50kg/ha K_2O. To convert this into numbers of bags of fertiliser per hectare, it is necessary to use the percentage analysis figures for the fertiliser in question. This is clearly stated on each bag, e.g. 20:10:10 means that the particular fertiliser contains 20% N, 10% P_2O_5 and 10% K_2O, always given in that order. The percentage declarations are calculated from the atomic weights of the individual elements. For example, 100kg of 20:10:10 fertiliser contains 20kg N, 10kg P_2O_5 and 10kg K_2O; therefore a 50kg bag contains 10kg N, 5kg P_2O_5 and 5kg K_2O, i.e. the number of kilograms of plant food in a 50kg bag of fertiliser is half the percentage figures.

In the example of the plant food requirements for the spring cereal crop, i.e. 100kg N, 50kg P_2O_5, 50kg K_2O, this would be supplied by 10 bags of 20:10:10 fertiliser. In some cases, the figures may not work out exactly and a compromise has to be accepted. For example:

	N kg/ha	P_2O_5 kg/ha	K_2O kg/ha
Spring cereals recommendation	100	50	50
Four bags 10:25:25 combine-drilled, supply	20	50	50
Difference	80		

The extra 80kg of nitrogen would be top-dressed and may be supplied by approximately $4^1/_2$ bags per hectare of ammonium nitrate (34.5%), i.e. 80/17.25 = 4.6. If bulk fertiliser is used, each tonne contains 10 times the percentage of each plant food:
1 tonne (1000kg) = 100 × 10kg; therefore, 1 tonne of 20:10:10 contains 200kg N, 100kg P_2O_5 and 100kg K_2O.

In the case of liquid fertilisers, the amount applied will depend on the type of declaration of the manufacturer. Some declare their fertilisers on a weight per volume basis and some on a weight per weight basis. The weight per volume declaration is relatively easy to work out, i.e. so many kilograms of plant food per 100 litres of water. A weight per weight declaration means that the user must know the specific gravity, of the product (SG) e.g.

100 litres of **water** weighs 100kg

but a 28% N solution has an SG of 1.29

therefore 100 litres of 28% fertiliser weighs 129kg

28% fertiliser means 28kg in 100kg product (wt per wt)

therefore there are 28kg of N in 100/1.29 = 77.5 litres of product

or, put another way, the weight of N in 100 litres is 36kg

3.3.2 Kilogram cost

The cost of plant food per kilogram can be calculated from the cost of fertilisers which contain only one plant food such as nitrogen in ammonium nitrate, phosphate in triple superphosphate, and potash in muriate of potash.

If 1 tonne of ammonium nitrate (34.5% N) costs £107 and contains 345 kg N, then

$$\text{1 kg of N costs } \frac{10\,700}{345} \text{ pence, i.e. 31p}$$

If 1 tonne of triple super (45%) costs £135 and contains 450 kg P_2O_5, then

$$\text{1 kg } P_2O_5 \text{ costs } \frac{13\,500}{450} \text{ pence, i.e. 30p}$$

If 1 tonne of muriate of potash (60%) costs £105 and contains 600 kg K_2O, then

$$\text{1 kg } K_2O \text{ costs } \frac{10\,500}{600} \text{ pence, i.e. 17.5p}$$

1 tonne of (9:24:24) compound contains:

N	P_2O_5	K_2O
90 kg	240 kg	240 kg

The value of this, based on the costs of a kilogram of nitrogen, phosphate and potash, is:

N	90×31	= 2790
P	240×30	= 7200
K	240×17.5	= 4200
Total		14 190 = £141.90

Normally, the well-mixed granulated compounds cost more than the equivalent in 'straights'. Bulk blends of 'straights' are usually cheaper. Fertiliser prices have varied considerably in recent years; actual costs at any time should be substituted in the calculations shown.

It is also possible to compare the values of 'straight' fertilisers of different composition on the basis of cost per kilogram of plant food, e.g. the cost of a kilogram of nitrogen in ammonium nitrate (34.5% N) is:

$$345 \text{ kg N/tonne } \frac{£107 \text{ per tonne}}{345} = 31p$$

This compares with the cost of a kilogram of nitrogen in urea (46% N) which is

$$460 \text{ kg N/tonne } \frac{£110 \text{ per tonne}}{460} = 24p$$

Therefore, apart from lower handling costs, the urea is the lower-priced fertiliser although, compared with ammonium nitrate, it does have some limitations.

3.4   Straight fertilisers

Straight fertilisers supply only one of the major plant foods.

3.4.1   Nitrogen fertilisers

The nitrogen in many straight and compound fertilisers is in the ammonium (NH_4^+ cation) form but, depending on the soil temperature, it is quickly changed by bacteria in the soil to the nitrate (NO_3^- anion) form. Many crop plants, e.g. cereals, take up and respond to the NO_3^- anions quicker than the NH_4^+ cations, but other crops, e.g. grass and potatoes, are equally responsive to NH_4^+ and NO_3^- ions.

The ammonium cation, as a base, is held in the soil complex at the expense of calcium and other loosely-held bases which are lost in the drainage water. This will have an acidifying effect on the soil.

Ammonium nitrate and urea fertilisers are produced by spraying a solution of the fertiliser from a vibrating shower head into a 'prilling tower'. As the droplets fall down the tower against a stream of cold air they become round and solid, producing prills 1–3 mm diameter.

Nitrogen fertilisers in common use are:

1   *Ammonium nitrate* (33.5–34.5% N). This is a very widely used fertiliser for top-dressing. Half the nitrogen (as nitrate, NO_3^-) is very readily available. It is marketed in a special prilled or granular form to resist moisture absorption. It is a fire hazard but is safe if stored in sealed bags and well away from combustible organic matter. Because of the ammonium present, it has an acidifying effect.

2   *Ammonium nitrate lime* (21–26% N). This granular fertiliser is a mixture of ammonium nitrate and lime. It is sold under various trade names. Because of the calcium carbonate present it does not cause acidity when added to the soil.

3   *Urea* (46% N). This is the most concentrated solid nitrogen fertiliser and it is marketed in the prilled form. It is sometimes used for aerial top-dressing. In the soil, urea changes to ammonium carbonate which may temporarily cause a harmful local high pH. Nitrogen, as ammonia, may be lost from the surface of chalk or limestone soils, or light sandy soils when urea is applied as a top-dressing. When it is washed or worked into the soil, it is as effective as any other nitrogen fertiliser and is most efficiently utilised on soils with adequate moisture content, so that the gaseous ammonia can go quickly into solution. In dry conditions in the height of summer it is probably better to use ammonium nitrate. Chemical and bacterial action changes it to the ammonium and nitrate forms. If applied close to seeds, urea may reduce germination.

4   *Sulphate of ammonia* (21% N, 60% SO_3). At one time, as a fertiliser, this was the main source of nitrogen. However, sulphate of ammonia is seldom

used now. It consists of whitish, needle-like crystals and it is produced synthetically from atmospheric nitrogen. Bacteria change the nitrogen in the compound to nitrate. It has a greater acidifying action on the soil than other nitrogen fertilisers. Some nitrogen may be lost as ammonia when it is top-dressed on chalk soils.

5   *Sodium nitrate* (16% N, 26% Na). This fertiliser is obtained from natural deposits in Chile and is usually marketed as moisture-resistant granules. The nitrogen is readily available and the sodium is of value to some market garden crops. It is expensive and is not widely used.

6   *Calcium nitrate* (15.5% N). This is a double salt of calcium nitrate and ammonium nitrate in prilled form. It is mainly used on the Continent.

7   *Anhydrous ammonia* (82% N). This is ammonia gas liquefied under high pressure, stored in special tanks and injected 12–20cm into the soil from pressurised tanks through tubes fitted at the back of strong tines. Strict safety precautions must be observed; it is a contractor rather than a farmer operation. The ammonia, as ammonium hydroxide, is rapidly absorbed by the clay and organic matter in the soil and there is very little loss if the soil is in a friable condition and the slit made by the injection tine closes quickly. It is not advisable to use anhydrous ammonia on very wet or very cloddy or stony soils. It can be injected when crops are growing, for example into winter wheat crops in spring, between rows of Brussels sprouts and into grassland. The cost of application is much higher than for other fertilisers, but the material is cheap, so the applied cost per kilogram compares very favourably with other forms of nitrogen. On grassland it is usually applied twice – in spring and again in midsummer – at up to 200kg/ha each time. In cold countries it can be applied in late autumn for the following season, but the mild periods in winters in this country usually cause heavy losses by nitrification and leaching. At one time it was fairly popular in the United Kingdom. However, because the main marketing source ceased, this is no longer the case, although there is no reason why it should not be used again.

8   *Aqueous ammonia* (12% N). This is ammonia dissolved in water under slight pressure. It must be injected into the soil (10–12cm), but the risk of losses is very much less than with anhydrous ammonia. Compared with the latter, cheaper equipment can be used, but it is still usually a contractor operation.

9   *Aqueous nitrogen solutions* (26–32% N). These are usually solutions of mixtures of ammonium nitrate and urea, and are commonly used on farm crops (liquid fertilisers).

Various attempts have been made to produce slow-acting nitrogen fertilisers. Reasonable results have been obtained with such products as resin-coated granules of ammonium nitrate (26% N), sulphur-coated urea prills (36% N) (soil bacteria slowly break down the yellow sulphur in the soil), urea condensates and urea formaldehydes (30–40% N). At present these types of fertiliser are considered too expensive for farm cropping but are used in amenity and production horticulture.

Organic fertilisers such as *Hoof and Horn* (13% N), ground-up hooves and horns of cattle, *Shoddy* (up to 15% N), waste from wool mills, and *Dried Blood* (10–13% N), a soluble quick-acting fertiliser are usually too expensive for farm crops and are mainly used by horticulturists.

It should be noted that a significant proportion of the nitrogen now supplied to farm crops comes from compound fertilisers in which it is usually present mainly as monammonium phosphate (MAP) or diammonium phosphate (DAP), as described in the section on phosphate fertilisers.

3.4.2 Phosphate fertilisers

Phosphate fertilisers can be classed as:

1 Those containing water-soluble phosphorus.
2 Those with no water-soluble phosphorus. The insoluble phosphorus is soluble in the weak soil acids.

By custom and by law, the quality or grade of phosphate fertilisers is expressed as a percentage of phosphorus pentoxide (P_2O_5), a gas equivalent only used in the UK. The rest of Europe uses a declaration based on elemental phosphate. The main phosphate fertilisers used in agriculture are:

1 *Single superphosphate.* This contains 18–20% water-soluble P_2O_5 produced by treating ground rock phosphate with sulphuric acid. It also contains 27% SO_3 plus a small amount of unchanged rock phosphate in addition to gypsum ($CaSO_4$), which may remain as a white residue in the soil. It is suitable for all crops and all soil conditions, but is not widely used now.
2 *Triple superphosphate.* This contains approximately 46% water-soluble P_2O_5 in a granulated form. It is produced by treating rock phosphate with sulphuric acid, followed by phosphoric acid. It also contains a small amount of elemental sulphur. One bag of triple superphosphate is approximately equivalent to $2^1/_2$ bags single superphosphate.
3 *Ammonium phosphates.* These are produced by treating rock phosphate with phosphoric acid and ammonia. They range from 46 to 55% P_2O_5 and also contain useful amounts of nitrogen.
4 *Nitrophosphates.* Contain 20% P_2O_5 and the same amount of N. They are made by treating rock phosphate with nitric acid.
5 *Ground mineral phosphate* (ground rock phosphate). This contains 25–40% insoluble P_2O_5. It is the natural rock ground to a fine powder, i.e. 90% should pass through a '100-mesh' very fine sieve (16 holes/mm^2). It should only be used on acid soils in high rainfall areas and then preferably for grassland. A softer rock phosphate is obtained from North Africa. It is sometimes known as 'Gafsa'. It can be ground to a very high degree of fineness, 90% passing through a '300-mesh' sieve (48 holes/mm^2). This

means that it will dissolve more quickly in the soil, although it should still only be used under the same conditions and for the same crops as ordinary ground mineral phosphate. To be classified as a soft rock phosphate under EC Regulations, at least 55 % of its total P_2O_5 must dissolve in a 2 % solution of formic acid. Ground mineral phosphate fertilisers are now granulated.

6   *Basic slags*. These contain 5–22 % insoluble P_2O_5. They are by-products from the manufacture of steel. However, because of improved methods of steel manufacture, very little of this valuable fertiliser is now available to farmers in the United Kingdom. Small quantities of low-grade slag are sometimes imported.

3.4.3   Potassium fertilisers

These are also known as potash fertilisers which is short for potassium ash. The quality or grade of potassium fertilisers is expressed as a percentage of potassium oxide (K_2O) equivalent.

The main potassium fertilisers used in agriculture are:

1   *Muriate of potash* (potassium chloride). As now sold, it usually contains 60 % K_2O. It is the most common source of potash for farm use and is also the main potash ingredient for compound fertilisers containing potassium. As a straight fertiliser it is normally granulated, but some is marketed in a powdered form. KCl is found in vast quantities all over the world and is mined from rock deposits left by dried-up oceans. It is nearly always found in conjunction with NaCl and the two are separated by a flotation process.

2   *Sulphate of potash* (potassium sulphate). This is made from the muriate and so is more expensive per kilogram K_2O. It contains 48–50 % K_2O and, ideally, should be used for quality production of crops such as potatoes, tomatoes and other market garden crops. However, the cost limits its use. It also contains 27 % SO_3.

3   *Kainit, sylvanite and potash salts*. These are usually a mixture of potassium and sodium salts and, depending on the source, magnesium salts. They contain 12–30 % K_2O and 8–20 % sodium (Na). They are most valuable for sugar beet and similar crops for which the sodium is an essential plant food.

3.4.4   Magnesium fertilisers

1   *Kieserite* (26 % MgO). This fertiliser is quick acting and is particularly useful on severely magnesium-deficient soils where a magnesium-responsive crop such as sugar beet is to be grown. It also contains 50 % SO_3.

2   *Calcined magnesite* (80 % MgO). This is the most concentrated magnesium fertiliser, but it is only slowly available in the soil. To help to maintain the

magnesium status of a light sandy soil which is naturally low in magnesium, 300 kg/ha should be applied, say, every 4 years in a predominantly arable cropping situation.

3   *Epsom salts* (17 % MgO). This is a soluble form of magnesium sulphate and is used as a foliar spray where deficiency symptoms may have appeared on a high-value crop. It also contains sulphur.

3.4.5   Sodium fertilisers

Sodium is not an essential plant food for the majority of crops. However, for some, notably sugar beet and similar crops, it is highly beneficial and should replace at least half the potash requirements. The adverse effects it has on weak-structured soils such as the Lincolnshire silts should be noted but, on other soils, this should not be a problem. Agricultural salt (sodium chloride, 37 % Na) is the main sodium fertiliser used. It is now available in a granular form.

3.4.6   Sulphur fertilisers

Sulphur is an important plant nutrient (involved in the build-up of amino acids and proteins in the plant), a deficiency of which can limit the response of the plant to nitrogen. Sulphur deficiency has become more pronounced in some, but not all, crops in the last decade. The natural build-up of sulphur in the plant is now much less because purer fertiliser is being used and the plant itself is growing in a less polluted atmosphere. It is, however, generally only necessary in areas away from industry and then perhaps for certain crops such as second and third cut grass for silage and oilseed rape.

The main sulphur fertilisers (which should be applied in the spring) are gypsum (calcium sulphate), potassium sulphate and sulphur contained in compound fertilisers.

3.4.7   Phosphorus and potassium plant nutrients

Increasingly, plant nutrients are now expressed in terms of the elements P (phosphorus) and K (potassium) instead of the commonly used oxide terms P_2O_5 and K_2O respectively. Throughout this book the oxide terms are used, but these can be converted to the element terms by using the following factors:

$$P_2O_5 \times 0.43 = P, \text{ e.g. } 100 \text{ kg } P_2O_5 = 43 \text{ kg P},$$

$$K_2O \times 0.83 = K, \text{ e.g. } 100 \text{ kg } K_2O = 83 \text{ kg K.}$$

3.5   Compound fertilisers

3.5.1   Compound fertilisers, complex fertilisers and blending

The main constituents for compound fertilisers used in the United Kingdom are urea, mono and diammonium phosphate and potassium chloride. These compound fertilisers, or compounds, supply two or three of the major plant foods (nitrogen, phosphorus and potassium). Other plant foods, e.g. trace elements, as well as pesticides, can also be added, although this is not commonly done now. The exception to this is sulphur which is increasingly being offered as part of compound fertilisers to overcome the deficiencies in certain parts of the country.

Because of the use of more concentrated basic ingredients, compound fertilisers have become much more concentrated in the last 50 years. For example, in 1948 the total N, P, K content averaged 24% and in 2000 the total N, P, K content averaged 50% with some concentrations at 60%, e.g. 10% N, 25% P_2O_5, 25% K_2O. Approximately 75% of all fertilisers now used are complex compounds or blended compounds. They can exist as either solid or fluid materials.

Complex fertilisers are normally made by drying a wet slurry, containing the appropriate raw materials, on a fluid bed system to produce granules, each containing the declared nutrients in the correct ratio, size 2–5 mm diameter.

Blending of the straight fertilisers to make a mixture is becoming more popular again. It should not be compared with the farm mixture which was quite common 40 years ago. In modern blends it is important that the individual single ingredients are, as far as possible, matched in physical characteristics (granular size and density). This is to avoid segregation out (separation) of the ingredients. It is a dry mix.

High quality blended fertilisers can be made to specific plant food ratios. They are cheaper than the complex fertilisers, but generally the handling and, most important, spreading qualities are not as good. However, new blending technology, involving blending towers and computer control systems, means that high quality blends are now being produced without the need for fillers and with accurate matching of ingredients.

3.5.2   Plant food ratios

Fertilisers containing different amounts of plant food may have the same plant food ratios. For example:

	Fertiliser	Ratio	Equivalent rates of application
(a)	12 : 12 : 18	$1 : 1 : 1\frac{1}{2}$	5 parts of (a)
(b)	15 : 15 : 23	$1 : 1 : 1\frac{1}{2}$	= 4 parts of (b)
(c)	15 : 10 : 10	$1\frac{1}{2} : 1 : 1$	7 parts of (c)
(d)	21 : 14 : 14	$1\frac{1}{2} : 1 : 1$	= 5 parts of (d)
(e)	12 : 18 : 12	$1 : 1\frac{1}{2} : 1$	5 parts of (e)
(f)	10 : 15 : 10	$1 : 1\frac{1}{2} : 1$	= 6 parts of (f)

Some examples of compound fertilisers and possible uses are shown in Table 3.4.

Table 3.4   Some compound fertilisers

		N	P_2O_5	K_2O	Applied rate of material
Compound (N : P : K	Crop	kg/ha	kg/ha	kg/ha	kg/ha
12 : 12 : 18	Potatoes	150	150	225	1250
20 : 10 : 10	Spring cereal	100	50	50	500
0 : 24 : 24	Autumn cereal	0	60	60	208

3.5.3   Ways in which fertilisers are supplied
3.5.3.1   Solids

1   *50 kg bags.* Although a small amount of fertiliser is supplied as individual bags, most of it is delivered on 30 bag (1.5 tonne) pallets to facilitate handling. Special fork-lift equipment is required, as manhandling into spreaders is slow and can be dangerous.

2   *Big bags*, e.g. 500 and 1000 kg bags with a top-lift (hook) facility are easy to load into spreaders, but they can be difficult to stack. 'Dumpy' bags 750 kg on pallets stack well but they may be not quite so easy for loading spreaders. It is estimated that 90% of all fertiliser now used on farm is delivered in 500 kg bags. Care must be taken to avoid damage to the bags which can result in the loss of fertiliser, and in lumpy or sticky material reducing the efficiency of the spreader.

3   *Loose bulk* (1 tonne occupies about 1 m^3). This system is not widely used now. The fertiliser can be stored in dry, concrete bays and covered with polythene sheets. It can be moved into spreaders by tractor plus hydraulic loaders or augers.

3.5.3.2   Fluids
In the United Kingdom as a whole the fluid fertiliser share of the market is about 9%. However, in the typical arable areas where the system has, until recently, been concentrated, liquid nitrogen has a share in the range of 10–15%, and fluid compounds account for between 5 and 10% of the total tonnage of compound fertiliser used.

3.5.3.3   Solutions
These fertilisers are non-pressurised solutions of the same raw materials that are used for solid fertilisers. They should be distinguished from pressurised solutions, such as aqueous ammonia and anhydrous ammonia. At the time of writing, there is very little price differential in the kilogram cost of plant food in the solid or fluid form. Fluid compounds are based on ammonium polyphosphate or ammonium phosphate, urea and potassium chloride, whilst ammonium nitrate and urea are the main constituents for liquid nitrogen fertilisers.

The concentration of liquid nitrogen is, for all practical purposes, the same as solid nitrogen fertiliser as long as both urea and ammonium nitrate are used together in the same solution. Fluid phosphorus and potassium compounds are only about two-thirds the concentration of solids. Potash is the main constraint on the concentration of the compounds as it is the least soluble of the important plant foods. However, fluids have bulk densities in the range 1.2–1.3 kg/l compared with 0.9–1.0 kg/l for solid fertilisers (i.e. 100 l = approximately 125 kg). The higher bulk density plus quicker handling and application of fluids will compensate for the lower concentration. It is an obvious consideration when fertiliser work rates are assessed on the weight of nutrients applied in a given time and not the weight of the product. However, the cost of storage may be higher and special care must be taken during handling, storage and application to avoid environmental pollution. The Fertiliser Manufacturers' Association has produced a Code of Practice for Transport, Handling and Storage of Fluids, 1995, as well as one for solids.

Fluid solutions are stored in steel tanks (up to 60 t capacity) on the farm. A cheaper form of storage – glass-reinforced plastic tanks – is available as an alternative to mild steel. Storage capacity is generally based on up to 33% of the annual farm requirements.

There is a range of equipment available for applying fluid fertiliser. The broadcasters, which range from a 600 l tractor mounted to 2000 l capacity trailed applicators, have spray booms of 12–24 m wide. Although the broadcaster is specially designed for fertiliser, by changing the jets it can be used for other agricultural chemicals, and so the fluid system fits in well with tramlining because only two pieces of equipment have to be matched. Equipment is also available for the placement of fertiliser for the sugar beet, potato and brassica crops, and a combine drill attachment can be used for sowing the fluid and cereal together. For top-dressing, and to minimise scorch, a special jet can easily be fitted which produces larger droplets for a better foliar run-off. Alternatively, a dribble-bar, which dribbles the fertiliser on to the soil, can be used.

Volume of product to apply
Example, if 100 kg nitrogen/ha is recommended, liquid nitrogen, N 37 kg per 100 l is used, i.e. $100/37 \times 100$ l/ha = 270 l/ha.

Units of sale
Whilst solid fertilisers are applied by weight (kg/ha) and sold by weight as £ per 1000 kg, fluid fertilisers are applied by volume (l/ha) and sold by volume as £ per 1000/l (1 m^3).

3.5.3.4 Suspensions
These are fluids containing solids, and they can be produced with a concentration almost up to that of the granular solid. This is achieved either by crushing the solid raw material and adding to it up to 3% of a semi-solid attapulgite clay to help minimise the settling of the soluble salts, or by reacting phosphoric acid and ammonia followed by the addition of potassium chloride and extra

nitrogen. For long-term storage, crystallisation may be a problem and occasional agitation is necessary to prevent the crystals from growing whilst in store. Special applicators are also needed and this is why suspensions are, at present, usually applied by a contractor. Suspensions are developing in North America (where, in fact, fluids are more widely used) and to some extent on the Continent.

3.5.3.5 'Distressed' fertiliser

This is fertiliser which has been damaged, usually in transit from overseas. It is bought by distributors and sold at a heavy discount to farmers who normally make it into a liquid fertiliser, simply by dissolving it in water.

There is no difference in plant growth following the application of fertiliser in solution, suspended or solid form. The benefit of fluid fertilisers seems to be that they can be delivered more quickly and accurately to the plants.

3.6   Application of fertilisers

The main methods used are:

1   *Broadcast distributors* using various mechanisms such as:
     (a) *Pneumatic types*. The principle involved is that the fertiliser is metered into an airflow which conveys it through flexible tubes to individual outlets placed over deflector plates to distribute the fertiliser evenly. Hopper capacity is from 1 to 6 tonnes with a spreading width of 12–24 m (selected to match tramline widths). Application rates vary from 5 kg/ha (for broadcasting seeds) to 2500 kg/ha.
     (b) *Spinning disc types*. These consist of a hopper placed above a rotating disc. The fertiliser is fed on to the disc from where it is distributed. Wide areas can be covered, but the accuracy of distribution can vary quite considerably (see below).
     (c) *Oscillating spout type*. A large nozzle situated at the back of the hopper moves rapidly from side to side throwing the fertiliser to the required width.
     The spread patterns of all broadcast fertiliser spreaders should be checked regularly. This can be done by driving the spreader through a line of collection boxes and then measuring the amount of material in each box. The quality and uniformity of the fertiliser material also has an effect on spread pattern. For nitrogen fertilisers, a standardised Spread Pattern (SP) Rating has been adopted by the Fertiliser Manufacturers' Association (FMA). The Silsoe Research Unit in Bedfordshire measures the 'throw and flow' characteristics of nitrogen materials. Ratings are on a scale of 1–5 with SP5 being the best.
2   *Combine drills*. Fertiliser and seed (e.g. cereals) from separate hoppers are fed down the same or an adjoining spout. A star-wheel feed mechanism

is normally used for the fertiliser and this usually produces a 'dollop' effect along the rows. In soils low in phosphate and potash, this method of placement of the fertiliser is much more efficient than broadcasting. It is known as combine drilling and is sometimes referred to as 'contact placement'. Because of possible scorch, combine drilling should only be used for cereal crops.

3   *Placement drills.* These machines can place the fertiliser in bands 5–7 cm to the side and 3–5 cm below the row of seeds. It is more efficient than broadcasting for crops such as peas and sugar beet. Other types of placement drills attached to the planter are used for applying fertilisers to the potato crop and some brassicas.

4   *Broadcast from aircraft.* This is useful for top-dressing cereals, especially in a wet spring. It can also be used for applying fertilisers in inaccessible areas such as hill grazings. Highly concentrated fertilisers should be used, e.g. urea.

5   *Liquids injected under pressure* into the soil, e.g. anhydrous and aqueous ammonia.

6   *Liquids (non-pressurised)* broadcast or placed.

3.7   Organic manures

3.7.1   Farmyard manure (FYM)
This consists of dung and urine, and the litter used for bedding stock. It is not a standardised product, and its value depends on:

1   *The kind of animal that makes it.* If animals are fed strictly according to maintenance and production requirements, the quality of dung produced by various classes of stock will be similar. But in practice it is generally found that, as cows and young stock utilise much of the nitrogen and phosphate in their food, their dung is poorer than that produced by fattening stock.

2   *The kind of food fed to the animal that makes the dung.* The more proteins and minerals in the diet, the richer will be the dung.

3   *The amount of straw used.* The less straw used, the more concentrated will be the manure and the more rapidly will it break down to a 'short' friable condition.

Straw is the best type of litter available, although bracken, peat moss, sawdust and wood shavings can be used. About 1.5 tonnes of straw per animal are needed in a covered yard for six months, and 2–3 tonnes in a semi-covered or open yard.

4   *The manner of storage.* There can be considerable losses from FYM because of bad storage, although it is appreciated that expensive, elaborate storage is no longer viable these days.

Dung from cowsheds, cubicles and milking parlours should, if possible, be put into a heap which is protected from the elements to prevent the washing out and dilution of a large percentage of the plant food which it contains. Dung made in yards should preferably remain there until it is spread on the land, and then, to prevent further loss, it is advisable to plough it in immediately.

Farmyard manure is important chiefly because of the valuable physical effects on the soil of the organic matter it contains. It improves the friability, the structural stability and the water-holding capacity of the soil. It is also a valuable source of plant foods, particularly nitrogen, phosphate and potash, as well as other elements in smaller amounts. An average dressing of 25 t/ha of well-made FYM will provide about 40 kg N, 50 kg P_2O_5 and 100 kg K_2O in the first year. At least one-third of the nitrogen could be lost before it is ploughed in, although this depends on the time of application. All the plant food in FYM is less readily available than that in chemical fertilisers.

Application. The application of FYM will be dealt with under the various crops.

3.7.2   Liquid manure and slurry

The widespread use of cow cubicles and self-feed silage clamps and the need to dispose of dirty water has meant that many dairy farmers produce slurry rather than the traditional FYM from strawed yards. Similarly, intensive pig units with their mechanised waste management systems find it easier to deal with liquids rather than solids.

Slurry must not be allowed to pollute watercourses. At 10000–20000 mg/l it has a high Biological Oxygen Demand (BOD). This is the measure (in milligrams per litre) of the amount of oxygen needed by the micro-organisms to break down organic material (pollution of water is caused when the micro-organisms multiply, and so extract oxygen, to deal with the organic material). Problems can also arise from the nuisance of smells and possible health hazards over a wide area when the slurry is applied. If slurry is stored in anaerobic conditions, dangerous, obnoxious gases (mainly hydrogen sulphide) are produced and are released when it is being spread. The relevant environmental Authorities have a duty to prevent the pollution of streams and rivers with farmyard effluents. Legislation and the Code of Good Agricultural Practice should prevent any problems in this respect, but it does mean that the farmer has to find the best possible way to utilise the slurry produced on the farm.

When applying slurry, relevant recommendations from the Code of Good Agricultural Practice should be followed:

1   A single slurry application should not exceed 50000 litres/ha. At least a three-week interval should be allowed between applications. When applying slurry through irrigation lines, the precipitation should not exceed 5 mm/hour. Rain guns should be avoided.

   Too much slurry can:

   (i) restrict aeration leading to partial oxidation products such as methane and ethylene which are toxic to plants;

(ii) weaken soil structure by the sealing of the soil surface.

2   An untreated strip of at least 10m width should be left next to all watercourses, and slurry should not be applied within at least 50m of a spring, well or bore-hole that supplies water for human consumption or is to be used for farm dairies.

3   Slurry should not be applied:

(i) to fields which are likely to flood in the month after application;

(ii) to fields which are frozen hard; surface run-off must be avoided;

(iii) when the soil is at field capacity;

(iv) when the soil is badly cracked down to the field drains, or to fields which have been piped, moled or subsoiled over a drainage system in the preceding 12 months.

Table 3.5 indicates the average amounts and composition of slurry produced by livestock.

1   For slurry diluted 1 : 1 with water, the figures in Table 3.5 should be divided by 2.

2   To estimate the dilutions of slurry, a comparison should be made of the volume of slurry in the store with the expected volume of undiluted slurry.

Table 3.5 Nature of slurry produced by livestock during housing period

Livestock	Undiluted excreta	Nutrients produced		
		N kg/m^3	P$_2$O$_5$ kg/m^3	K$_2$O kg/m^3
Dairy cow	9.6	48	19	48
Fattening pig	1.5	10.5	7.5	6.0
Poultry (1000 hens)	41	660	545	360

The figures in Table 3.5 can only be used as a basis and they apply to slurry collected in an undiluted form, e.g. under slatted floors or passageways where washing and rainwater are excluded. The slurry on most farms is obviously diluted and it is not easy to be certain of its composition, even when samples are analysed. Rota-spreaders and similar machines can handle fairly solid slurry, and modern vacuum tankers and pumps are very efficient in dealing with slurry with less than 10% dry matter.

The injection of slurry into the soil using strong tines fitted behind the slurry tanker can reduce, but not eliminate, wastage and offensive odours. It can also reduce the volatisation of nitrogen into the air as ammonia gas. However, it is expensive and it can spoil the surface of a grass field where there are stones present.

It is highly desirable that the valuable nutrients in organic manures should be utilised as fully as possible. This is best achieved if the slurry can be applied at a time when growing crops can utilise it, normally in the spring. This means that storage is usually necessary. This is expensive, whether the slurry is stored in a compound (lagoon) or storage tank. The annual cost of storage and handling of slurry is usually more than the plant food value of the slurry.

Nitrogen losses resulting from applying slurry in the autumn/early winter (ground conditions permitting) to the following year's maize field, for example, can be reduced by the use of nitrogen inhibitors such as dicyandiamide (e.g. Didin : Enrich). These inhibit the activity of the nitrifying bacteria which delays the change of ammonia to nitrate. Denitrification and leaching are therefore reduced. The cost of inhibitors is approximately £45/ha with slurry applied at 50000 litres/ha. It is claimed that the increased yield with the maize crop following the use of an inhibitor can be valued at £70/ha.

In addition to forage maize, slurry is best applied to:

1 *Grass*. At least a six-week, preferably longer, interval should be allowed between a slurry application and taking the crop for silage or hay. This will avoid any disease problems and it should minimise any possible contamination of the conserved crop. Slurry can be applied to grazing swards but an interval of at least four weeks should elapse before animals are allowed access. Animals should then be carefully monitored for symptoms of hypomagnesaemia (staggers), as the high levels of potash in slurry could lead to reduced herbage magnesium content.

2 *Kale*. As this crop (like maize) is not normally sown until late April or May, there is usually a good opportunity to apply slurry from the winter accumulation. There should be no need for any phosphate and potash fertiliser following a slurry application, although extra nitrogen will normally be necessary. This fertiliser recommendation also applies to forage maize.

Although it is not so usual, slurry can also be used on cereals and other forage crops. New band spreaders have been developed which make it possible to spread across the whole width of tramlines. These also reduce the ammonia emissions by 30–40% compared with traditional slurry tankers.

Slurry can be separated into solid and liquid fractions by screening or centrifugal action. The solid part (12–30% dry matter (DM), can be handled like FYM, and the liquid portion is much less likely to cause tainting of pastures; the disease risk is also reduced. Slurry separators are, however, expensive and they are not easy to justify. Table 3.6 shows the constitution of cattle slurry.

Table 3.6 The constitution of cattle (dairy) slurry

1000 kg slurry DM, unseparated (kg)			Mechanically separated slurry			Weeping wall separated		
N	P_2O_5	K_2O	N	P_2O_5	K_2O	N	P_2O_5	K_2O
3.0	1.2	3.5	4.0	1.2	3.5	3.0	0.5	3.0

See also MAFF Booklet RB209.

3.7.3 Pig slurry

The composition of pig slurry depends on the feeding system, i.e. dry meal feed has a 10% DM; liquid feed has a 6–10% DM and whey 2–1% DM (see Table 3.7). Supplementation to the diet means that pig slurry can contain large amounts of copper and zinc. These elements will build up slowly in the soil, but eventually there could be crop toxicity problems. It should be noted that sheep are particularly susceptible to copper poisoning.

Table 3.7 Comparison of pig slurry

	% dry matter	Total nitrogen (kg/m^3) N	Available nutrients (year of application) (kg/m^3) P$_2$O$_5$	K$_2$O
Pig slurry	2	3.0	0.5	1.8
	4	4.0	1.0	2.3
	6	5.0	1.5	2.7

Note: The availability of nitrogen will depend on the timing and method of application, the winter rainfall and the soil type (see also MAFF Booklet RB209)

3.7.4 Poultry manure
Poultry manure refers to:

1   Fresh poultry manure.
2   Broiler manure.
3   Deep litter manure.
4   Dried manure.

Table 3.8 shows the nutrient value of poultry manure.

Table 3.8 Comparison of poultry manures

	% dry matter	Available nutrients (year of application)* (kg/tonne) N	P$_2$O$_5$	K$_2$O
Battery layers	30	8	7.8	8.1
Broilers	60	13.5	15	16

*assumes soil incorporation within 24 hours.

- *Fresh poultry manure* from battery cages or from slatted or wire floors is free from litter. It is semi-solid, but rather sticky and, to reduce any public nuisance from smell, it should be spread as soon as possible. To reduce nitrogen losses, the preferred time is in spring and summer. Application rates are similar to those for farmyard manure. Copper and zinc toxicity could be a problem if very high rates of fresh poultry manure are made.
- *Broiler manure* is the droppings mixed with litter. This produces a bulky manure, relatively dry and friable and easily handled. It should be treated like farmyard manure.

3.7.5   Sewage sludges (biosolids)

As a means of disposal, most Water Companies offer sewage sludge to farmers and growers at a fairly nominal cost. There is a Department of the Environment (DoE) Code of Practice for the use of sewage sludge in agriculture and its recommendations should be followed. There is also the 'Safe Sludge Matrix' which has been produced by agreement between Water UK and the British Retail Consortium and which goes beyond the cropping and grazing restrictions contained within the Code of Practice.

Apart from the objectionable smell, raw, untreated sludge contains pathogenic organisms and heavy metals which, apart from being a possible health hazard, can contaminate both the soil and the growing crop. It could be used on combinable or animal feed crops until the end of 2001 but had to be injected or incorporated quickly and its analysis had to be known before application. However, most sewage sludge is now treated and, as such, it is quite valuable, e.g. 20 tonnes dried digested sludge can contain 75 kg N, 90 kg P_2O_5 and 30 kg K_2O available in the year of application. New Class 'A' biosolids are now being produced by digestion, thermal drying and pelleting. These can be used safely on all crops including fresh salad vegetables and fruit.

3.7.6   Seaweed

Seaweed is sometimes used instead of farmyard manure for crops such as early potatoes in coastal areas, e.g. Ayrshire, Cornwall and the Channel Islands. However, it is expensive to handle which is why it is not used to any great extent now. It can be a useful source: 10 tonnes contain about 50 kg of N, 10 kg P_2O_5 and 140 kg K_2O; it also contains about 150 kg salt. The organic matter in seaweed breaks down rapidly because it is mainly cellulose. It should therefore be collected, spread and ploughed in immediately. In this way the loss of potassium, particularly, will be reduced.

3.7.7   Cereal straw

Cereal straw is lignified material and can be a useful source of soil organic matter for maintaining or improving soil fertility. It is particularly beneficial for light, sandy and silty soils in which organic matter breaks down rapidly. However, the amount of organic matter in a soil is closely related to the texture of the soil, and repeated incorporations of straw on various soil types over many years have had little or no effect on the percentage of organic matter in the soils.

The average yield of straw (4 t/ha) supplies about 15 kg N, 5 kg P_2O_5 and 35 kg K_2O, and so the soil should benefit when the straw is ploughed in or otherwise worked into the soil. However, with the sudden influx of straw, the soil bacterial population will start to multiply to cope with the problem of breaking down the straw. In so doing, much of the nitrogen in the soil gets used up for the bodily needs (the protein) of the bacteria. The carbon : nitrogen ratio increases to 50 or 60: 1 from the normal 10: 1. The protein is released as the straw decomposes and eventually the ratio returns to 10: 1. But in the meantime, unless extra nitrogen is added (20–25 kg/ha), the following crop can suffer from shortage of nitrogen. This extra nitrogen is only necessary under naturally low nitrogen conditions.

Incorporation of straw is easier to achieve if it is chopped and spread from the combine. Ploughing is the best method, preferably using fairly wide furrows (30 cm) and 15–20 cm deep. There is seldom any great advantage in premixing; increased costs can result. Tines and/or discs can only do the job well with chopped straw, and the result is not as satisfactory as good ploughing. Trash and clods left near the surface encourage slugs which can cause damage to following crops by eating the seed and young plants. Repeated shallow incorporations can build up high organic matter levels in the surface layers of some soils. These can be beneficial to the development of young crop plants, but can also cause problems with the adsorption of soil-acting herbicides. If the straw is incorporated at a wet time, anaerobic conditions may develop and the decomposing straw will release toxic substances, such as acetic acid, which can check or kill seedling crop plants.

3.7.8 Green manuring
This is the practice of growing and ploughing in green crops to increase the organic matter content of the soil. It is normally only carried out on light sandy soils. White mustard is a very commonly grown crop for this purpose. Sown at 9–17 kg/ha it can produce a crop ready for ploughing within 6–8 weeks. Fodder radish is becoming more popular for green manuring and, like mustard, it can also provide useful cover for pheasants. However, it has been shown that a short ley has very little benefit in the way of building up organic matter in the soil. There must, therefore, be even less effect from quick-growing crops, which break down equally quickly in the soil.

3.7.9 Green cover cropping
This describes the practice of growing a green crop with the primary objective of absorbing soil nitrogen over the winter months to prevent its leaching. This can benefit the environment as well as saving on expensive fertiliser nitrogen.

The cover crop is sown in late summer to hold any nitrogen in the autumn and winter, to be released in the spring for the following spring-sown crop. A number of crops are at present under trial and it would appear that those taking up most nitrogen in November (and this month can be significant) do not necessarily hold it until March. For example, because of their rapid growth, white mustard and forage rape are very effective in trapping autumn nitrogen, but natural senescence and frost kill mean that the nitrogen can be lost before any following crop can take it up. Rye and ryegrass hold nitrogen well in the autumn and winter and retain it until the spring. They can also be used for early cropping in the spring.

3.7.10 Waste organic materials

Various waste products are used for market garden crops, partly as a source of organic matter and partly as a means of releasing nitrogen slowly to the crop. They are usually too expensive for ordinary farm crops.

- *Shoddy* (waste wool and cotton) contains 50–150 kg of nitrogen per tonne. Waste wool is preferable, and the recommended application rate is 2.5–5 t/ha.
- *Dried blood*, ground *hoof and horn* and *meat and bone meal* are also used; the plant food content is variable.

3.8 Residual values of fertilisers and manures

The nutrients in most manures and fertilisers are not used up completely in the year of application. The amount likely to remain for use in the following years is taken into account when compensating outgoing farm tenants.

Up to 70% of nitrogen in soluble nitrogen fertilisers (e.g. ammonium nitrate and some other compounds) is used in the first year. For phosphate in soluble form, e.g. triple superphosphate and compounds containing phosphate, allow two-thirds after one crop, one-third after two and one-sixth after three crops. For phosphate in insoluble form, e.g. bone meal and ground mineral phosphate, allow one-third after one crop, one-sixth after two and one-twelfth after three. For potash, e.g. muriate or sulphate of potash and compounds containing potash, allow a half after one crop and a quarter after two crops. For lime one-eighth of the cost is subtracted each year after application.

3.9 Fertilisers and the environment

When applying fertilisers, it is economically and environmentally important to use the optimum amounts and not just those required for a maximum yield. Accuracy of application, timing and handling are also important in order to reduce the losses, either by leaching or run-off. Only small amounts of phosphorus will be lost by run-off, but with potassium, losses can be greater both by leaching and run-off. Nitrogen as nitrate is the nutrient which is most likely to be lost by leaching and this is affected by soil type, cropping, cultivation timing and rainfall.

There are now directives from the European Community (EC) concerning nitrates in drinking water; a maximum of 50 parts per million has been set. In 1990 Nitrate Sensitive Areas (NSAs) were designated to study the effects of husbandry on nitrate leaching. Initially ten areas were chosen – from Somerset across to Lincolnshire – where the water exceeded or was at risk of containing more than 50 parts per million nitrates. Although the scheme was voluntary, most farmers in the designated areas joined, receiving payments for complying with the regulations. There are 212 NSA contracts in England alone but it has now ended and no more contracts have been issued.

The NSA scheme involved small changes in husbandry; only the economic optimum, or less, of nitrogen for some crops was allowed, but the timing and amount applied at any one time were restricted. Application of organic manures was also affected and a careful record had to be kept. To hold the nitrogen in the soil, winter fallows were avoided by planting cover crops. There were restrictions on timing of cultivations on grassland. Hedgerows and/or woodland could not be removed unless replaced by an equivalent area.

Superseding the NSAs and coming into force in 1998 was the designation of Nitrate Vulnerable Zones (NVZs) and the NVZ Action Programme Regulations. There were originally 68 designated NVZs covering 600 000 ha of land in England and Wales and the scheme is run by the Environment Agency. The regulations require the careful management of fertiliser and manure use and require farmers to keep careful and accurate field records. They also impose restrictions on nitrogen applications (organic and manufactured) by specifying closed periods (when nitrogen cannot be applied) and 12 month limits on total nitrogen use. These apply to both grassland and arable areas. There are also spreading and storage controls. The European Commission does not consider that England has fully implemented the 1991 Nitrates Directive. To comply either all or four-fifths of England will have to follow the NVZ's regulations.

3.10 Further reading

Archer, J, *Crop Nutrition and Fertiliser Use*, 2nd ed., Farming Press, 1988.

Codes of Good Agricultural Practice for Air, Soil and Water, MAFF 1999.

Simpson, K, *Fertilisers and Manures*, Longmans, 1986.

Fertiliser Recommendations for Agricultural and Horticultural Crops, RB 209, 7th ed. 2000, The Stationery Office.

Managing Livestock Manures Booklet 1 – Making better use of livestock manures on arable land. ADAS Gleadthorpe.

Managing Livestock Manures Booklet 2 – Making better use of livestock manures on grassland. ADAS Gleadthorpe.

Managing Livestock Manures Booklet 3 – Spreading system for slurries and solid manures.

The Air Code, DEFRA Publications 1998.

The Soil Code, DEFRA Publications 1998.

The Water Code, DEFRA Publications 1998.

4

Climate and weather

4.1 Introduction

Climate has an important influence on the types of crops which can be grown in the United Kingdom. It may be defined as a seasonal average of weather conditions. Weather is the state of the atmosphere at any one time. It is the combined effect of such conditions as heat or cold, wetness or dryness, wind or calm, clearness or cloudiness, pressure and the electric state of the air. It has a major influence on the yield and quality of the crops grown.

The climate of the United Kingdom is mainly influenced by:

1 Its distance from the equator (50–60 °N latitude).
2 The warm Gulf Stream which flows along the western coasts.
3 The prevailing south-westerly winds.
4 The numerous 'lows' or 'depressions' which cross from west to east and bring most of the rainfall.
5 The distribution of highland and lowland; most of the hilly and mountainous areas are on the west side or run from north to south through the middle of the country.
6 Its nearness to the continent of Europe, from where hot winds in summer and very cold winds in winter can affect the weather in the southern and eastern areas.

Local variations are caused by altitude, aspect and slope.

Altitude: Height above sea level can affect climate in many ways. The temperature drops about 0.5 °C for every 90 m rise above sea level. Every 15 m rise in height can shorten the growing season by two days (one in spring and one in autumn) and it may check the rate of growth during the year. High land is more likely to be buffeted by strong winds and will receive more rain from the moisture-laden prevailing winds, which are cooled as they rise upwards.

Aspect: The direction in which land faces can affect the amount of sunshine absorbed by the soil and thus the soil temperature. In this country the temperature of north-facing slopes may be 1 °C lower than on similar slopes facing south.

Slope: When air cools it becomes heavier and will move down a slope and force warmer air upwards. This is why frost often occurs on the lowest ground on clear still nights whereas the upper slopes may remain free of frost. 'Frost pockets' occur where cold air collects in hollows or alongside obstructing banks, walls, hedges, etc. (Fig. 4.1). Frost-susceptible crops such as early potatoes, maize and fruit should not be grown in such places.

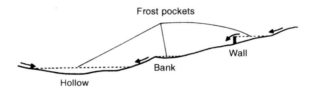

Fig. 4.1   Frost pockets are formed as cold air flows down a slope.

Crop production can be regarded as essentially the conversion of solar radiation, water and soil nutrients into useful end-products. The influences of various aspects of weather and climate on potential crop production can be numerous and varied. For instance, sunshine and rainfall could be considered as primary controls of production because of their immediate influence on rate of crop metabolism, nutrient uptake, turgidity, biochemical processes within the plant and crop structure, as well as their effect on soil and air temperature. In addition there will be an effect from topography, geographical position and proximity to bodies of water. Other climatic elements such as frosts, wind and humidity will also have an effect on production because of their influence on such things as crop disease and direct physical damage.

4.2   Solar radiation and rainfall

Solar radiation in itself is important because of its role in photosynthesis. In the absence of any other restrictions, crop production would be determined by the interception of sunlight. The intensity of solar radiation changes with season, latitude, aspect, slope and the amount of cloud cover and pollution. More light falls onto the south of England than in Scotland and the difference between the UK and, say, the south of France is even more marked. Yields of sugar beet in the Paris basin are often 25–30% higher than in East Anglia, whilst the yield of apples in Provence can be 100% higher than in the West Midlands. Other factors may be involved but a large part of the difference is due to the intensity and the duration of sunlight. The efficiency of use of this energy by crops is not good, with a maximum potential of about 8%, but, in commercial field crops, it is nearer 1%. The *interception* of light will vary according to ground cover; early in the season much of the sunlight will fall onto bare ground because of the small size of the plants and late in the season, with senescence and subsequent leaf loss, the interception of energy will again fall. During the crop's 'grand growth stage' interception of light will be near 100%.

Water supply is vital for crop growth. Without enough water the yield and quality of crops are compromised in several ways. Cell division and cell growth are the first processes to be affected. Eventually photosynthesis will slow down due to the closing of the stomata. Leaves will wilt and their ability to intercept sunlight will be adversely affected and it will be more difficult for the plant to take up nutrients in solution from the soil. The soil water status underneath a crop will depend on the difference between rainfall and evapotranspiration, and there is far more variability in rainfall than there is in evaporation. Growers must therefore be able to compensate for this variability by storing water when in excess and applying it by irrigation when in deficit.

In the UK rain comes mainly from the moist south-westerly winds and from the many 'depressions' which cross from west to east. Western areas receive much more rain than eastern areas; this is partly due to the west to east movement of the rain-bearing air and also because most of the high land is along the western side of the country. As the moisture-laden air rises over this high ground it is cooled, its ability to hold moisture decreases and this moisture is deposited as rain.

The average annual rainfall on lowland areas in the west is about 900 mm (35 in) and in the east is about 600 mm (24 in). Larger differences can be seen by comparing eastern lowland with western highland–average annual rainfall for parts of Cambridgeshire is 500 mm (20 in) whilst in the highest, western facing districts of the Lake District it is 5000 mm (200 in)!

4.3   Air and soil temperature

The optimum temperature for the metabolic processes within a temperate plant, especially photosynthesis, is about 25 °C. Therefore, in most seasons and especially during autumn and early spring, plants will not be metabolising at their optimum rate in the UK. Small increases or decreases in temperature from year to year will not have a significant effect on production, but it has been shown that, in very hot summers e.g. those of 1975 and 1976, the photosynthetic rate and the duration of the Green Area Index (GAI) of crop plants was significantly decreased. These summers were characterised by temperatures above 25 °C. Temperature has a large effect on the growing season, with growth of most crops beginning when temperatures rise above 5–6 °C and ceasing in the autumn when temperatures drop below this level. This can give growing seasons of as little as 180 days in the higher areas of Scotland and more than 350 days in the south west of England. These differences are very important when considering the potential for grass growth on livestock farms. Very high temperatures running up to harvest can also increase the rate of senescence in cash crops and hence reductions of harvestable yield.

Soil temperatures are more relevant to plant growth during the early part of the growing season. Germination is at its fastest at soil temperatures between 15 and 25 °C but for our autumn- and spring-sown crops these conditions are unlikely to be found. Winter wheat will germinate at temperatures as low as 3 °C as long as the soil is not waterlogged and it is not uncommon to find soil temperatures of 10–15 °C at 5 cm depth when most drilling occurs in September and early October. The gradual increase of mean UK temperatures due to global warming is discussed later in this chapter. Temperature changes are mainly due to:

1   The seasonal changes in length of day and intensity of sunlight.
2   The source of the wind, e.g. whether it is a mild south-westerly, or whether it is cold polar air from the north or from the Continent in winter, and hot, dry wind from the south in summer. Wind direction depends mainly on the position of areas of high and low pressure in the immediate vicinity of the British Isles, although local winds, e.g. offshore breezes and katabatic winds from high ground, can occur independently of the cyclonic system.
3   Local variations in altitude and aspect.
4   Cloud cover during the night. Night temperatures are usually higher when there is cloud cover which prevents too much heat escaping into the upper atmosphere.

The soil temperature may also be affected by colour – dark soils absorb more heat than light coloured soils. Also, damp soils can absorb more heat than dry soils. The presence of stones can also increase mean soil temperatures by acting as heat reserves in a similar way to storage radiators in buildings. The average January air temperature in lowland areas along the western side of the country is about 6 °C and about 4 °C along the eastern side. The average July temperature in lowland areas in the southern counties is 17 °C and 13 °C in the north of Scotland. Temperature can also have an effect on the agronomy of the crop. For example it can affect the rate of breakdown of residual herbicides applied to the soil, it can affect the efficiency of use of plant growth regulators and it can affect the volatilisation of ammonia from applied urea.

4.4   Other aspects of climate and weather

Advection is the horizontal transfer of heat. The effects of advection are most noticeable in the farming areas close to the western coasts of the British Isles.

The Gulf Stream and the prevailing south-westerly winds sweeping up over the Atlantic mean that air temperatures over the sea to the south-west of the UK are often 10–15 °C higher than those on land. The effects of this are noticeable over most of the UK but are at their most intense over west-facing coastal areas. These are often known as early cropping areas, where potatoes and vegetable crops can be sown during winter and harvested in late spring, e.g. Scilly Isles, Pembroke, Cornwall. Often more than one crop per year can be achieved as there is virtually year-round plant growth and little or no chance of frosts.

The likely incident of frosts in an area will have a large effect on the types of crops grown. Most areas of England and Wales will be subject to periods of night frosts during the winter months. Autumn-sown crops are bred to exhibit a certain level of frost hardiness and will survive most winters unscathed. In areas of Scotland where temperatures are lower and may never exceed 0 °C, even during the day, the crops are mostly spring-sown and this is certainly the case in the huge wheat growing areas of the US and Canada. High value crops such as soft fruit are most at risk from night frosts occurring in the spring when the flower buds are emerging. This can be partially overcome by the use of shelter, artificial protection with mulches or the use of irrigation. Other weather phenomena which can affect crop production are gales, floods and hailstorms.

The epidemiology and life-cycles of many diseases are heavily influenced by weather. Potato blight is a classic example with the disease requiring very specific humidity and temperature conditions to spread within the crop. These conditions can be monitored and predicted, and an early warning system can be used by farmers to help to control the disease effectively with fungicides. Many of the most serious fungal diseases on mainstream crops, such as cereals and oilseed rape, are favoured by wet, windy and humid conditions, whereas in a dry spring and summer, disease levels will usually be much lower. The introduction of a philosophy of integrated crop management means that much more use of weather data to try and predict disease incidence and spread will need to be used in the future. This can also apply to crop pests.

Plants growing in a given climate have a potential production which is modified (often limited) by the soil in which they are grown and by individual weather conditions experienced during the growing season. A very important effect of climate on soil is that of the soil condition. The soil processes which can be affected by climate include:

1   Littering and humification – the accumulation of raw organic matter and its subsequent conversion to humus. The breakdown of organic matter is quicker in higher temperatures because of the increased biological activity of the soil.
2   Decalcification – the removal of calcium carbonate from soil profiles by, among other things, leaching. This will be worse in light soils combined with high rainfall.
3   Gleying – the reduction of iron under anaerobic conditions with mottling of ferric concretions. Found in heavy soils with poor drainage characteristics in areas where rainfall amounts mean long periods of waterlogging of the soil.

4   Salinisation and alkalinisation – the build-up of salt in the soil and the accumulation of sodium (Na^+) ions on the exchange sites. Usually associated with the use of saline water for irrigation but can occur naturally where high temperatures give high evaporation from the soil surface. Salts are moved to the surface by capillary action and there is not enough rain to wash them away from the rooting zone. The overuse of water for irrigation in some areas has also led to the ingress of sea water into aquifers used for irrigation.

Perhaps the most important interaction between climate and soil is that of soil type and rainfall. At the one extreme this interaction can allow a crop to fulfil its full potential production and at the other extreme it can severely limit this potential, to the point where near crop failure can occur. Two identical soil types in terms of structure and texture can have very different characteristics depending on where they are located. Take, for example, a sandy loam situated in South Devon and a similar sandy loam in Suffolk. The total rainfall in Devon could be as high as 1000 mm whereas in Suffolk this would be nearer 500 mm. The soil in Devon could well be under permanent grassland because of the difficulty in finding available work days for cultivations in autumn and spring. If it was cultivated, leaching of nitrogen and calcium would be a major concern but water supply to crops would be adequate through the summer and justification for an irrigation system would be difficult. The Suffolk soil would probably be under arable or horticultural production. There would be plenty of available work days but irrigation would be vital for high value crops so that yield potential and quality could be maintained.

4.5   Climate change

The climate history of the earth is one of dramatic change cycles over millions of years. Many periods of ice ages alternating with periods of warmer climates have helped to shape the landscapes that are seen today and have determined the course of rivers, the shape of the oceans and the types of soil deposition, especially in temperate areas of the world. There are changes in the weather from year to year and there are decades where the global temperatures are higher or lower than average. There are extremes of weather and climate, often bringing with them disaster and fatality – hurricanes, floods, tidal waves, snowstorms, heat waves and droughts. But these are the phenomena caused by the natural 'pulse' of the planet's weather and climate systems; the inevitable result of the chaos existing in the atmospheric circulation and the ocean's currents.

One of the most important events influencing the world's climates is El Niño. Every three to five years a large area of warmer water appears in the Pacific Ocean off the coast of South America. This area lasts for one to three years. The physical effects of El Niño are twofold. Firstly the evaporation from the warmer ocean increases, changing the dynamics of the atmospheric circulation across the globe, and secondly a slow moving wave (called a Rossby wave) moves out from the epicentre of the event. The resulting tidal surges cause problems for low-lying areas and prevent ground water from draining effectively into the oceans.

Far more sinister are the changes caused by man's activities on the planet. These are mostly industrial activities but also involve lifestyle influences. The three most important changes, and those which are exercising the minds of scientists and politicians all over the world, are global warming, ozone depletion and the increase in acid rain.

In order for the earth to maintain a fairly even average temperature over its surface, the incoming solar radiation must be balanced by an equal outgoing thermal radiation from its surface and atmosphere. The effect of naturally-occurring carbon dioxide, dust particles and water vapour in the atmosphere is to trap some of the thermal radiation while releasing enough heat to balance the sun's input and to keep the average surface temperature at about 15 °C. This state of relative stability has lasted for millions of years although it has been interrupted by the ice ages. This phenomenon is known as the *natural greenhouse effect*. Because of human activities such as burning of fossil fuels and deforestation we are now experiencing an *enhanced greenhouse effect* due to the increase in carbon dioxide emissions, and in those of other gases, into the atmosphere. This has the effect of reducing the thermal emissions back into space and, to restore the heat balance, the earth's surface temperature must rise. It has been speculated that a doubling of the CO_2 content of the atmosphere will mean an average temperature rise of about 5 °C, i.e. the average surface temperature will be 20 °C instead of 15 °C. The level of CO_2 has risen by approximately 30 % since the early 1800s but there are now serious global attempts to reduce greenhouse gas emissions.

The effects of global warming are a mixture of good and bad. Because of the increased CO_2 levels certain crops (mainly the C3 types) will respond with increased yields. Certain parts of the planet will change their cropping patterns to include crops that otherwise could not be cultivated while other parts will become more habitable. To offset this, however, it is likely that existing arid and semi-arid zones will find it increasingly difficult to produce food because of lack of useable water. These areas are often already overpopulated with insufficient food and this can only get worse under the anticipated climate change.

It is likely that many crops can be adapted by genetic manipulation to thrive in the new conditions. However, there is more concern about long-term crops such as forests and plantation crops, whose life cycles are over a much longer time scale and so may adapt less well. There is also concern about thermal expansion of the oceanic water mass, the increased rate of polar ice cap melting and the consequent rise in sea levels. Many areas of comparatively fertile agricultural land at or below sea level could be flooded unless serious attempts are made to improve sea defences.

Ozone is an important gas of the upper atmosphere. It helps to prevent dangerous ultraviolet radiation from reaching the earth's surface and is also one of the natural greenhouse gases. Large amounts of ozone have been destroyed by the use of chlorofluorocarbons (CFCs) in refrigerators, insulation and aerosols. This allows more UV radiation to reach the earth and, because CFCs are much more effective greenhouse gases than ozone, adds to the problems of global warming. Thankfully the problem has been recognised and concerted efforts by industry have meant that alternatives to CFCs have been developed and are now widely used.

Following the industrial revolution and the burning of high-sulphur fossil fuels, high levels of sulphur dioxide were being released into the atmosphere. Reactions between water and sulphur dioxide produced weak acids in solution which fell to the earth as acid rain. The effects of acid rain were often felt downwind of the areas causing the problem and large areas of vegetation were destroyed. The control of sulphur dioxide emissions in the US and Europe has meant a dramatic reduction in acid rain incidence, but some parts of the world continue to burn high-sulphur fossil fuels and it is likely that these countries will not change their practices for some time.

4.6   Further reading

Houghton, J, *Global Warming – The Complete Briefing*, 2nd ed., Cambridge University Press, 1997.

Barry R G and Chorley R J, *Atmosphere: Weather and Climate*, 6th ed., Routledge, 1992.

5

Weeds

5.1 The impact of weeds

Weeds are plants which are growing where they are not wanted. Even crop plants can be serious weeds in other crops, e.g. volunteer potatoes or weed beet. Weeds can significantly affect crop yields by shading and smothering the crop as well as by competing for plant nutrients and water. The competitiveness of weeds varies both between species and at different times of the year. Some crops are also more affected by weed competition than others. Potatoes produce a dense crop canopy which can be very effective at competing with weeds compared with sugar beet which is precision drilled at a low population, and has many weeks when there is little crop cover. As well as affecting crop yield, weeds also affect crops in other ways, e.g.

1   They can spoil the quality of a crop and so lower its value, e.g. wild oats in seed wheat; black nightshade berries in vining peas.
2   They can act as host plants for various pests and diseases of crop plants, e.g. charlock is a host for flea beetles and club root which attack brassica crops; fat hen and knotgrass are hosts for virus yellows and root nematodes of sugar beet; common couch is a host for take-all and eyespot of cereals.
3   Weeds such as bindweed, cleavers, couch and thistles can affect combine harvesting. All the green material other than grain going through the harvester reduces the work rate. These weeds can also cause a crop to lodge which again will affect harvesting.
4   Weed contamination at harvest can increase the cost of drying and cleaning the grain.
5   Thistles, buttercups, docks and ragwort in grassland can reduce the grazing

area and feeding value of pastures. Some grassland weeds may taint milk when eaten by cows, e.g. buttercups and wild onion.

6   Weeds such as ragwort, horsetails, nightshade, foxgloves and hemlock are poisonous and if eaten by stock are likely to cause poor growth or even death. Fortunately, most stock animals normally do not eat poisonous weeds in the field although they will if such plants are conserved in hay or silage.

When planning a weed control programme it is important to have knowledge of the weed life cycles. Some of the most important weed species such as cleavers, chickweed, wild oats and blackgrass are annuals. Only a small number of biennials are important weeds and are more of a problem in permanent crops such as spear thistle and ragwort in grassland. There are a number of important perennial weeds. Seed production is not necessarily the main method of propagation for perennials. Underground stems (rhizomes) or roots are very important for the spread of weeds such as common couch and creeping thistle.

5.1.1   Success of plants as weeds
There are a number of reasons why some plant species have been more successful as weeds than other plants.

Seed production
Some weed species are able to produce thousands of seeds per plant. Examples of prolific seed producers include poppies and mayweed. The seed reservoir (weed seed bank) in some soils can be as high as $40\,000/m^2$. Not all the seed produced in one year will germinate the next, the percentage emergence may only be around 2–6% of the weed seedbank. Many species have some sort of seed dormancy mechanism that has to be broken before they will germinate. Once dormancy has been broken environmental conditions must also be correct for germination; this accounts for some of the variation in weed populations between years. Losses of seed and seed viability are taking place all the time. Depth of burial in the soil, number and type of cultivations and weed species affect the rate of decline. The seed viability of some broad-leaved species such as fumitory declines very slowly compared with the rapid decline of some grasses such as barren brome.

Seed spread
As well as being able to multiply in one field, weeds can spread from field to field. Before the implementation of some of the current very strict seed standards one important method of spread of some weeds such as wild oats, corn cockle and cornflower was in the seed. The weed sets its seed at the same time as the cereal crop is harvested. The weed seed is then resown with the cereal seed. Commercial seed is now virtually weed free but weed seed can still be spread around fields by farm machinery. Other methods of spread include wind, water

and birds, depending on the characteristics of the seed. Some seed, such as docks, can remain viable in slurry for many weeks and so can be spread around the different fields where the slurry is applied.

Time of germination
Only a few weed species such as chickweed and annual meadow grass germinate all the year round. Most have specific germination periods. Continuous cropping of the same crop, at the same time of year, will tend to encourage those weeds that germinate at the same time. Examples include autumn-germinating blackgrass in winter cereals, early summer-germinating black nightshade in forage maize and cranes-bill (late summer germination) in early-sown winter oilseed rape.

Same family as crop
Some important weed species are in the same family as the crop plant, e.g. charlock in oilseed rape. Weed control using herbicides is then often difficult as there is not enough selectivity in the herbicide activity to kill the weed without injuring the crop.

Weed competitiveness
Weed species differ in how they compete with a crop. Some weeds such as speedwell compete early in the growing season and then die back before harvest. Others, such as cleavers, compete later and carry on growing until harvest. Cleavers is a very vigorous plant and much more competitive than speedwell. Examples of weed competitiveness in a cereal crop:

- Very competitive weeds
  - cleavers;
  - wild oats.
- Moderately competitive weeds
  - mayweeds;
  - common chickweed;
  - common poppy;
  - blackgrass;
  - rough-stalked meadowgrass.

- Least competitive weeds
  - field speedwell;
  - ivy-leaved speedwell;
  - field pansy;
  - annual meadowgrass.

5.1.2 Assessing weed problems in the field
The seriousness of a weed problem can be judged by:

1   The possible weeds which are likely to appear, for example in a root crop. A record of the weeds present in the field in recent years can be a very good guide when selecting a suitable soil-acting residual herbicide. Weather and soil conditions affect weed populations emerging each year.

2   The correct identification of weed seedlings growing up in a crop; young grass seedlings are particularly difficult to identify correctly.
3   The numbers of weeds present, e.g. per square metre or hectare.
4   Proper field inspection. Weed populations tend to be very patchy over the field. It is very time-consuming to map weeds manually. Currently, there are no automated systems available for weed mapping. One option for some of the major weeds, because weeds spread by only a few metres per year, is to assess the weeds in early summer in the previous crop to produce a map. This map can then be used for patch spraying.

The effects on yield will depend on the type of weed and its aggressiveness and on the density and vigour of the crop. Other effects, which must be considered, include ease of control in the different crops being grown.

5.2   Weed types and identification

For optimum control of weeds, both culturally and chemically, it is very important to be able to identify the weeds correctly. There are two main groups of weeds, the monocotyledons or grass weeds and the dicotyledons or broad-leaved weeds.

5.2.1   Grass weeds

There are several features which can aid identification (see Table 1 in Appendix 9). Annuals are spread by seed, whereas perennials are spread either by seed or by vegetative parts. Some of the stem/root characteristics help with identification such as presence of stolons or rhizomes (Fig. 1.17).

- *Leaves*. Some grasses have folded leaves in the leaf sheath which help in recognition. Leaf shape and size are important, as is the presence of *auricles*. Size and shape of the *ligule* usually aid identification of species in the same family (pages 397–8).
- Presence or absence of *leaf/stem hairs*. A few grasses have very hairy leaves, such as barren brome and Yorkshire fog. Wild oats only have hairs on the leaf margin.
- *Stem base coloration*. A few important grasses have coloured stem bases; ryegrass is a deep red, blackgrass has a purplish blotch and awned canary grass is pink.
- *Ear characteristics* are distinctive for most of the grasses as detailed below. Normally control is carried out before ear production. Ear number is important when assessing effectiveness of control that year; weed mapping for patch spraying and checking for development of resistant plants is necessary.

The following are some important characteristics of the major grass weeds.

Common couch (Fig. 5.1) has been a problem for a very long time. It can spread rapidly from pieces of rhizomes and can cause serious yield losses as well as perpetuating diseases and making harvesting difficult. Pre-harvest glyphosate is recommended in many crops including cereals, peas, beans, linseed and oilseed rape, but not in crops for seed. If there are several clones of couch in a field, then viable seed may be produced.

Onion couch (false or tall oat-grass) (Fig. 5.2) is a more serious problem to deal with than couch. It is commonly found in cereal-growing areas on the thin limestone soils. It has bulbous rhizomes and senesces too early for the pre-harvest glyphosate technique in some crops, e.g. wheat. It may have to be dealt with by spraying regrowth in the stubble. It can produce much viable seed. Continuous winter cereals and minimum cultivations encourage it. Some of the wild oat herbicides can help reduce bulbils and seeding.

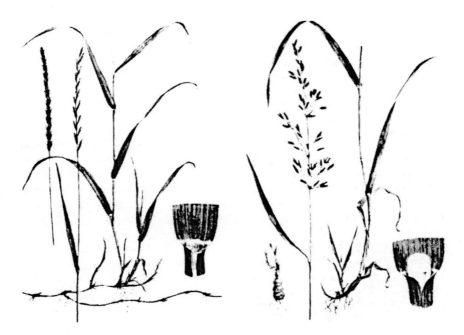

Fig. 5.1   Common couch.   **Fig. 5.2**   Onion couch.

Watergrass (*creeping bent*) (Fig. 5.3) is not so difficult to deal with. It is a perennial that is spread by seed and surface rooting stolons which can be controlled by good ploughing or spraying when it is green in the stubble.

Black bent (Fig. 5.4) is common in some areas. It has rhizomes like ordinary couch and can be dealt with in the same way, but it does produce much more viable seed than couch.

Fig. 5.3   Creeping bent.   Fig. 5.4   Black bent.

Meadow-grasses (Fig. 5.5) are some of the most commonly occurring grass weeds and can cause problems when present in large numbers. They can be effectively controlled by some of the grass weed herbicides, often at reduced rates. Rough stalked meadowgrass is much more competitive than is annual meadowgrass. Rough stalked meadowgrass is a perennial and spreads by both seed and stolons. It is a common weed on mixed arable and grass farms.

Wild oats (Fig. 5.6) are found on a large percentage of cereal-growing farms. There are two main types of wild oat in the UK. The spring wild oat is the most common, germinating mainly in the spring with a small amount of seed germinating in the autumn. The winter (autumn) germinating type is found mainly in southern and eastern areas; it mainly germinates in the autumn. Once established in a field, no matter what control measures are taken, wild oats are likely to persist for a very long time because of dormant seed in the soil. This dormancy problem is made worse by stubble cultivations after harvest which bury the seed. Most of the shed seeds, if left on the surface, are destroyed or disappear. Deeply buried seed, which may have fallen down cracks on clay soils, for example, can remain viable for decades. Seed germinates from up to 25 cm depth of soil. The seed can arrive in a field in many ways, e.g. in the crop seed, dropped by birds, in farmyard manure made with infested straw, and from the combine harvester.

    The first wild oat to appear in a field should be removed by roguing. Hand

Fig. 5.5   Annual meadowgrass.   **Fig. 5.6**   Common wild oat.

roguing can be justified when numbers are fewer than 500 plants/ha. Later, if numbers are allowed to increase, it may be necessary to use herbicides. It is difficult to decide when this can be justified. It can be based on yield response or likely grain contamination or to prevent a build-up of numbers in the future. Left uncontrolled, wild oats can increase threefold each year. There are now fields where wild oats are resistant to many of the commonly applied products.

Cereal yield losses may be very high, 25% is possible with moderate infestations. Premiums can be lost for seed, malting, bread-making and Intervention purchases. Straw carrying wild oat seeds may be difficult to sell. The presence of wild oats on a farm may incur a high dilapidations claim at the end of a tenancy.

Blackgrass (Fig. 5.7) is a very important annual grass weed on many cereal farms (100 plants/m^2 can reduce yield by 1 t/ha). It used to be associated with heavy, wet soils where winter cereals were grown, but is now widespread on most types of soil where autumn-sown cereals predominate. Blackgrass produces a very great number of viable seeds (individual plants can produce up to 150 heads and many thousands of seeds). Most seed germinates in the first three years but some can remain dormant for up to nine years. The seeds germinate mainly in early autumn, but in heavily infested fields spring germination can also be important. A high percentage control must be achieved each year in continuous cereals to contain the problem; this usually means that cultural practices are necessary to supplement herbicides. Vigorous crop competition is very important. Ploughing, to bury the seed deeply, is preferable to shallow cultivations. Blackgrass only germinates in the top 5 cm soil. Spreading the seeds to clean fields should be avoided by using weed-free seed and thorough cleaning of the combine harvester. Delaying drilling or changing to spring cereals can be helpful, but the resultant lower yields may not be acceptable. A number of herbicides are available (see Table 2 in Appendix 9) and the cost can be justified in cereals where there are more than two blackgrass plants/m^2. Care has to be taken with choice of herbicide as there is an increasing number of fields where blackgrass is resistant to the main products.

Barren brome or sterile brome (Fig. 5.8). This became a serious problem on many cereal farms after the very dry years in the mid-1970s. It spread from hedgerows and waste corners and was encouraged by increased early sowing of winter cereals, established using minimal cultivations. The seeds are anything but barren and germinate very readily in the autumn. The seed has little dormancy. Ploughing can be very effective in controlling this weed if the seed is buried more than 13 cm. Stubble cultivation, in a damp autumn, can also aid control. Where only headlands are infested they are sometimes left for drilling in spring after the germinated brome seedlings have been destroyed. There are only a limited number of effective products for use in cereals, although a number of herbicides used in other crops are effective.

The incidence of other brome species, such as rye brome and meadow brome, is increasing and such weeds can be more difficult to control than is barren brome.

Fig. 5.7   Blackgrass.   Fig. 5.8   Barren brome.

Awned canary grass is a relatively new problem in the UK although it is more common in other parts of the world. It is not widespread but it can cause serious yield losses and harvesting problems where it does occur. The heads resemble large Timothy heads. At the seedling stage the plant looks very like blackgrass. The seedling has a pink stem base and when broken a red sap is often released. The seed has a limited amount of dormancy. Deep ploughing and spring cropping can help reduce weed incidence. Care must be taken with herbicide choice as only a few of the commonly used grass weed herbicides control awned canary grass (Table 2 in Appendix 9).

Loose silky bent is a local problem on sandy or light loam soils in eastern and southern England and many parts of Europe. As with many other annual grass weeds deep ploughing and spring cropping can be very effective methods of control. If autumn drilling then it is preferable to grow a tall strawed/competitive crop.

5.2.2   Broad-leaved weeds

Control, both chemical and mechanical, of annual broad-leaved weeds is usually most effective at the seedling stage. In order that the correct product is chosen it is important to identify the weed correctly. Many broad-leaved weeds have very characteristic cotyledons (seed leaves) or true leaves which aid identification. Size, shape, colour, if stalked, presence of hairs or prominent leaf wax all help recognition. Figure 5.9a–5.9d shows the characteristics of some common broad-leaved weeds.

Cleavers (Fig. 5.9a) is the most competitive weed in winter cereals and winter oilseed rape, more competitive than wild oats. It can be confused at the seedling stage with ivy-leaved speedwell (Fig. 5.9a). Note that cleavers has a notched cotyledon and the cotyledons are not stalked. Cleavers germinates at fairly low temperatures mainly in the autumn and winter.

Charlock (Fig. 5.9a) is a brassica showing the typical large kidney-shaped cotyledons. It is the shape and presence or absence of hairs on the true leaves that identifies the different brassicas. Charlock used to be a major weed problem in cereals but with the introduction of more effective chemical control methods it is now less important.

Speedwells (Fig. 5.9b) have increased in recent years. Ivy-leaved speedwell is mainly autumn germinating whilst common field speedwell germinates all the year round. All the speedwells have spade-shaped cotyledons except ivy-leaved speedwell. It is important to be able to distinguish between these weeds, as chemical control is different for each of them. Speedwells are low growing and are not very competitive.

Fat hen (Fig. 5.9b) is in the same family as sugar beet and so can sometimes be difficult to control in beet. It germinates in the spring and is often associated with the application of dung. It has a very distinct waxy/mealy appearance on the cotyledons and first true leaves. The stalk and underside of the cotyledons are red.

Polygonums (Fig. 5.9b) including knotgrass, redshank and black bindweed are a very important group of spring germinating weeds. Knotgrass is the first to germinate; it has narrow cotyledons in the shape of a 'V'. All the polygonums seedlings have a red stem.

Mayweeds (Fig. 5.9c) germinate in most months, although spring is the peak period. Unlike many broad-leaved weeds, mayweeds remain green all season

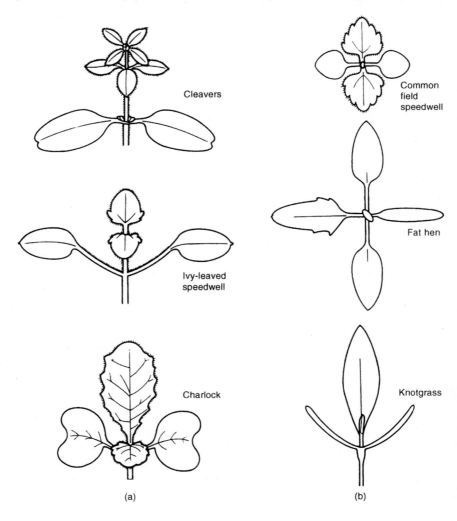

Cleavers

Common field speedwell

Ivy-leaved speedwell

Fat hen

Charlock

Knotgrass

(a)   (b)

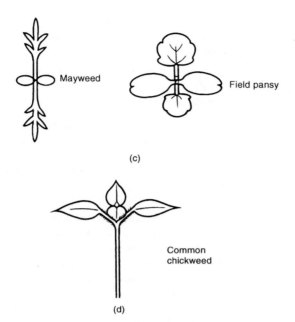

Fig. 5.9   Identification of common broad-leaved weeds (a) large cotyledons (b) medium-sized cotyledons (c) small cotyledons (d) similar-shaped cotyledons and true leaves.

and so compete with the crop for a long time. They are prolific seeders and the seed can remain dormant for many years. Mayweeds have small cotyledons and very distinctive shaped first true leaves.

Field pansy (Fig. 5.9c) has become more abundant in recent years partly due to some commonly applied herbicides having little control. It is, however, a very poor competitor. The seedlings have distinct small, notched cotyledons and hairless first true leaves.

Chickweed (Fig. 5.9d) is the most common broad-leaved weed in arable crops. It favours fertile soils, germinating all the year round with a peak in the autumn and the spring. It readily overwinters. This weed is very successful partly because it can produce more than one generation of plants a year and thus a lot of seed. At the seedling stage it is a bright green colour and has true leaves with the same shape as the cotyledons

5.3   Control of weeds: general

The control of weeds with herbicides is standard practice on most farms as detailed in Tables 2–4 in Appendix 9. Nevertheless, it is worth remembering that other good husbandry methods can still play an important part especially where there is a herbicide resistance problem. Non-chemical methods of weed control are the only ones available to organic farmers.

5.3.1   Methods of weed control

1 Crop hygiene. It is particularly important to ensure that only clean seed is sown when using home saved seed. Hand roguing can be a very effective method of preventing weeds becoming a problem in the first place. Avoidance of machinery contamination can also help to prevent weed increase, e.g. wild oats.

2 Cultivations. Ploughing can be very effective at containing or reducing some annual grass weed problems such as blackgrass and barren brome. Weeds such as wild oats that can germinate from plough depth are unaffected by ploughing; sometimes populations are worse after ploughing compared with the use of minimum cultivations. In some years, if there is enough soil moisture, stubble cultivations and the 'stale' seedbed technique can be a useful aid to weed control. The soil is cultivated ready to sow the next crop, weeds are allowed to grow and are then killed before planting the crop. Inter-row cultivations are still used in some crops, particularly if there is a difficult weed to control such as weed beet in sugar beet. There is some interest in the use of in-crop weeders as a method of reducing herbicide inputs.

3 Cutting, e.g. bracken, rushes, ragwort and thistles. This weakens the plants and prevents seeding. The results are often disappointing if not repeated. However, control of spear thistle can be very effective from cutting as long as the timing is correct, although creeping thistle is more difficult to control.

4 Drainage. This is a very important method of controlling those weeds which thrive in waterlogged soils. Lowering the water table by good drainage will help to control weeds such as rushes, sedges and creeping buttercup.

5 Rotations. Growing leys and various arable crops that are planted at different times of the year usually leads to a different weed flora. There is also an opportunity to use different types or groups of herbicides. This method is still useful if there are difficult weeds such as barren brome, volunteer potatoes and weed beet to control.

6 Maintenance of good fertility. Arable crops and good grass require a high level of fertility, i.e. the soil must be adequately supplied with lime, nitrogen, phosphates, potash and organic matter. Under these conditions crops can compete strongly with most weeds.

7 Crop seed rates. High seed rates and good crop establishment all help to reduce the impact of weeds.

8 Burning. The use of propane burners can be an effective method of weed control; it is mainly used in organic systems

9 Mulches. There are a number of mulching materials that can be used to prevent weed growth in some small-scale plantings, e.g. amenity plantings, fruit and vegetable crops. Black polythene is the most commonly used material. As well as preventing weed growth it also helps to conserve moisture.

10 Chemical control. There are now over a hundred different herbicides and a thousand products on the market. The following is a summary of the main chemicals and methods that are used.

5.3.2 Herbicides

Most of the chemicals used have a *selective* effect, i.e. they are substances which stunt or kill weeds and have little or no harmful effects on the crop in which the weeds are growing. A severe check of weed growth is usually sufficient to prevent seeding and to allow the crop to grow away strongly. There are a limited number of *non-selective* or *total* herbicides such as glyphosate. They kill or check all vegetation and are usually used in non-cropped areas before the crop emerges, or as crop dessicants.

Herbicides are usually sold under a wide range of proprietary names which can be very confusing, but the common name of the active material must always be stated on the container. In the text of this book the common name of the chemical is used when referring to herbicides and the trade name may also be given where there is only one proprietary product. The selectivity of a herbicide depends on a number of factors including the chemical itself and its formulation, the amount of the active ingredient applied as well as the quantity of carrier (water, oil or solid).

The chemicals now commonly used as herbicides can be classified according to their chemical group and biochemical mechanism of action. More simply they can be classified by their basic mode of action as follows:

1 *Contact herbicides.* These will kill most plant tissue by a contact action with little or no movement through the plant; shoots of perennials may be killed but regrowth from the underground parts usually occurs. Effective control relies on good coverage of green plant material by the herbicide. Some examples of contact herbicides are phenmedipham, sulphuric acid, bentazone, ioxynil, bromoxynil, diquat and paraquat. Contact herbicides have very little or no residual action in the soil.

2 *Soil-acting or residual herbicides.* These chemicals act through the roots or other underground parts of the plant after being applied to the soil surface or worked into the soil (the more volatile types). Some examples are atrazine, simazine, prometryn, chloridazon, trifluralin, propyzamide, pendimethalin, triallate, isoproturon and chlorotoluron. These herbicides vary in their persistency in the soil from weeks to months. Some herbicides are also absorbed by foliage, e.g. linuron.

3 *Translocated herbicides* are those which can move through the plant before acting on one or more of the growth processes. Some of these herbicides

can be very effective at controlling perennial weeds, e.g. the control of couch by glyphosate. Good spray cover is not so important with this type of herbicide.

Herbicides work by affecting one or several different biochemical pathways or physiological processes. The speed of action reflects the mechanism of action. Some of the earliest developed herbicides are similar to substances (hormones) produced naturally by plants. Susceptible plants initially show distorted growth and then will take a few days to die back. They are mainly used for controlling broad-leaved weeds in cereals and grassland. The more important ones are MCPA, 2,4-D and mecoprop.

A large number of herbicides affect photosynthesis. Symptoms are usually seen as a rapid yellowing or *chlorosis* of the leaves followed by leaf death or *necrosis*. Examples include paraquat (a non-selective, contact bipyridilium) and isoproturon (a residual urea).

Other herbicides limit or stop cell division and elongation of the growing point. These chemicals can be very slow acting as seen with propyzamide (a residual amide) in oilseed rape.

Several of the newer chemicals, such as the suphonyl ureas, are very specific and only affect the production or synthesis of single compounds such as amino acids. Again these herbicides tend to be slower acting than those that affect photosynthesis. Examples include metsulfuron-methyl, triflusulfuron-methyl and rimsulfuron.

5.3.3   Herbicide choice
Herbicide choice will be affected by a number of factors including:

1. *Weeds present and their growth stage.* The efficacy of herbicides is often affected by the growth stage of the weed. Control of annual broad-leaved weeds is usually most effective when the weeds are small. Control of annual grass weeds is usually more effective before the 4-leaf stage. The range of species controlled is also greatest when the weeds are small. Herbicide recommended rates are often lower when the weeds are small. A weed growth stage key has now been produced which is used on chemical product guides.

    Description of weed growth stages:

Pre-emergence.	Plants up to 50 mm across/high.
Early cotyledons.	Plants up to 100 mm across/high.
Expanded cotyledons.	Plants up to 150 mm across/high.
One expanded true leaf.	Plants up to 250 mm across/high.
Two expanded true leaves.	Flower buds visible.
Four expanded true leaves.	Plant flowering.
Six expanded true leaves.	Plant senescing.
Plants up to 25 mm across/high.	

2. *Crop/variety and crop growth stage.* Some herbicides are only recommended for use on a limited range of crops, e.g. phenmedipham on beet and strawberry crops only, whereas triallate is recommended for wild oat control on a wide range of crops. Other herbicides are only recommended for use on some varieties of crops as there can be problems with crop damage, examples include metribuzin in potatoes and chlorotoluron in cereals.

   Crop growth stage is important mainly because of crop damage. The hormone herbicides such as mecoprop are not recommended after the early stem extension stage of cereals. Other examples include the timing of application of ethofumesate to a beet crop, depending on the rate of chemical being used and mixtures with other chemicals. Crop growth stage is also important with some of the contact acting herbicides. If the crop is too advanced, and is shading the weeds, then it is very difficult to get spray coverage on to the weeds.

3. *Soil type and condition.* In order for the residual chemicals to work effectively the soil tilth should not be too cloddy. Activity is affected by the amount of soil moisture present as this will affect movement of the chemical to the germinating weeds. Persistency of the chemical in the soil will be affected by the rate of chemical applied, and speed of breakdown. The main method of breakdown is by microbial activity, which is affected by soil moisture and temperature.

   The rate of application of some residual chemicals is affected by the soil texture. Some are not recommended on sandy soils as there can be too much leaching down to where the crop is growing, which can lead to crop damage. Chlorpropham, for example, cannot be used on sands or very light soils.

   Herbicide choice is restricted on soils with a high organic matter or those that have a high adsorption coefficient. Residual activity is reduced on soils with a high organic matter (usually greater than 10%). Trash on the soil surface can also have this effect; the chemical becomes attached to the charged sites on the organic matter (*adsorbed*) and is then unavailable for weed uptake. The activity of residual herbicides can be affected by soil pH. Flupyrsulfuron-methyl activity can be reduced in alkaline soils.

4. *Weather conditions.* Weather conditions, including rainfall and temperature, can affect activity and/or efficacy of a treatment, as well as crop damage. Control of many weeds using foliar applied chemicals is most effective when the weather conditions are optimal for weed growth. Some herbicides such as paraquat are very rainfast whereas glyphosate requires at least six hours dry weather after application to give the best results. Other foliar acting herbicides, if applied to a crop under stress, e.g. metamitron on beet, can increase the risk of crop damage.

5. *Tank mix compatibility.* Some chemicals only have limited compatibility with other pesticides; this can be due to a problem with the formulations or with the activity of the chemical being affected.

6. *Cost.* Herbicides, when they are first marketed, are expensive partly to

cover the very high development costs. Once they are off patent costs usually fall, as seen with isoproturon and glyphosate. With low crop prices, cost can significantly affect product choice and rate used.

7. *Following crops.* There are a few restrictions on following crops and intervals between applying the chemicals and following crops; always check the label. Some chemicals also have recommendations to plough before planting the next crop.

8. *Water buffer zone requirements.* Many pesticides have a buffer zone requirement when spraying next to water courses. There are not as many herbicides as other pesticides that are affected in this way, but it can influence choice of chemicals. A Local Environmental Risk Assessment for Pesticides (LERAP) should be carried out; this takes into account type of water course, rate of chemical used and type of nozzles on the sprayer. Undertaking a LERAP can reduce the buffer zone requirement.

9. *Resistance.* Herbicide resistance is an increasing problem particularly on mainly cereal farms. Herbicide resistance is the inherited ability of a weed to survive rates of chemicals that normally control the weed. Blackgrass, wild oats and ryegrass are grass weeds that have developed a problem of resistance to some of the most commonly applied grass weed herbicides. Once herbicide resistance has been diagnosed then chemical choice will be severely affected.

It is vital to check the manufacturer's recommendations before buying and applying herbicides. *Note:* Under the Control of Pesticides Regulations 1986 it is illegal to use any pesticide except those that are officially approved.

5.4   Weed control in cereals

Over the last 50 years there has been a change in types of cereals grown and their husbandy. This has changed the relative importance of the weeds present. When the main cereals were spring sown, spring germinating broad-leaved weeds were the main competitors. Currently, with winter cereals being the major cereal crop, autumn-germinating weeds, such as cleavers and blackgrass, are most important. Establishment using minimum cultivations and continuous winter cereals has encouraged the grass rather than the broad-leaved weeds. There have been many new herbicides available to cereal growers over the years which might also have affected the relative importance of some weeds. The main grass weed problems are now blackgrass, wild oats, annual meadowgrass and couch. The most common broad-leaved weeds in cereals are chickweed, speedwells, mayweed, cleavers, dead nettle, pansy, charlock and the polygonums. Cleavers are regarded as the most important economically, because the herbicides selected for their control significantly add to the cost of weed control.

5.4.1   Economics of weed control in cereals

Researchers have tried to produce *economic thresholds* for control of weeds in cereals. This is the weed population that if not controlled will reduce cereal yields and quality by more than the cost of the chemical. The difficulty with thresholds for weeds is that factors other than yield have to be included, such as effect on the future weed seed bank, harvesting and drying costs. The other problem is the time required to assess the weed populations. The most competitive weeds in cereals are cleavers and wild oats, followed by mayweeds, blackgrass, chickweed and poppy. The least competitive weeds include dead nettle, speedwell and field pansy. Most growers and advisers work on a *safe threshold*, which takes into account seed carryover, ease and cost of weed control in following crops, variations in the competitiveness of the weeds and effectiveness of control.

5.4.2   Grass weed control

Grass weed control can prove more difficult in cereals than broad-leaved weeds. The cost of grass weed herbicides, especially wild oat chemicals, is usually higher than herbicides for broad-leaved weeds. A limited range of herbicides are available, some residual and the others foliar acting. Many of the herbicides, other than the main wild oat herbicides, control some broad-leaved weeds. The most commonly applied herbicide for blackgrass control is isoproturon and for wild oats fenoxaprop-P-ethyl (in wheat), triallate and tralkoxydim. Reduced rates of specific wild oat herbicides may only be required in the spring if herbicides such as isoproturon have been used in the autumn.

There may be a restriction on the use of some of these herbicides. Due to traces of isoproturon being found in water courses there is now an IPU stewardship scheme in place to reduce the risk. Other chemicals can, for instance, be toxic if used on particular cereal crops/varieties, or at specific growth stages in the cereal plant. Some of the herbicides will not mix with other chemicals. Dose rate recommendations are often based on the assumption that all farm sprayers and weather conditions may not be perfect. In ideal situations, or where weed numbers are small, lower rates may be satisfactory, but manufacturers will only consider compensation claims for poor control when the recommended rates have been used. Advisers, with wide experience of treatments and results, and BASIS qualified, are best able to give sound advice.

There are now many cases of resistance of blackgrass, wild oats and ryegrass to some commonly-used grass weed herbicides. The build-up of resistance is faster with products such as fenoxaprop-P-ethyl than standard herbicides such as isoproturon. Where resistance has been confirmed it is very important to use a mixture of control measures rather than just relying on herbicides. Factors such as cultivations, rotations, stubble hygiene, delayed drilling, crop competition and hand roguing can all help to reduce the risk. Herbicide programmes that include mixtures or sequences of chemicals with different modes of action should be used. Good control of resistant blackgrass has been obtained using sequences of triallate and flupyrsulfuron-methyl. Where possible it is better to treat blackgrass, wild oats and ryegrass when the weeds are small. Repeat treatments with the same herbicide should be avoided. It may be worth putting fields with high populations of resistant weeds into set-aside, where glyphosate can be a used.

5.4.3   Broad-leaved weeds

Perennial broad-leaved weeds are more difficult to control than annual weeds. This is especially the case with thistles, field bindweed, wild onion and docks, mainly because the foliage usually appears after the normal time for spraying annuals. Due to the mass of green foliage, field bindweed can be very troublesome, causing lodging and difficulty when combining. However, spraying glyphosate on the nearly mature crop at least one week before harvest can effectively control all perennial broad-leaved and grass weeds. The weeds must be green and actively growing (grain moisture less than 30%) and sprayer booms set to give good weed coverage. Tramlines, high clearance wheels and a sheet under the tractor minimize crop damage. This technique is approved because of the very low mammalian toxicity of glyphosate, but the straw should not be used as a horticultural growth medium or mulch. Stubble treatment is often less effective.

Annual broad-leaved weeds are usually much easier to kill when germinating or as young seedlings. The safest time for spraying a crop is clearly set out in the product literature and must be followed. Some herbicides can be applied at the 3–4-leaf stage, but many of the 'hormone' types, e.g. MCPA and 2,4-D, should be applied between the 5-leaf stage and jointing. If applied too soon (before the spikelets of the young ears have been determined), then some ears are likely to be deformed. If sprayed too late (when or after the cells are dividing to form the pollen and ovules), some of the upper spikelets become sterile and give the ear a 'rat-tailed' appearance and reduced yield accordingly.

Weed growth stages in relation to herbicide efficiency are explained on the product labels. Timing of control of broad-leaved weeds in cereals is not as critical as for annual grasses though again better control is achieved when the weeds are small. If dealing with grass weeds in the autumn in the winter cereal crop, then broad-leaved weeds are normally also controlled. Autumn weed control in a competitive winter barley crop can often mean no requirement for further treatment in the spring except for wild oats or cleavers.

Some broadleaved weeds, such as field pansy, are more easily controlled by a number of the autumn herbicides, e.g. those including diflufenican (DFF). If an autumn treatment is used, it may be possible to reduce chemical rates in the spring. Some of the broad-leaved herbicides can be applied over a long period without crop loss, unlike the 'hormone' weed killers. Metsulfuron-methyl (broad spectrum) and fluroxypyr (mainly for cleavers), depending on the crop, can be applied between crop GS 12 and late stem extension. (Note that metsulfuron-methyl is not recommended for application before 1 February.)

5.5 Weed control in other combinable crops

5.5.1 Oilseed rape

Winter oilseed rape is a very competitive crop. A fairly high weed population, e.g. 50–100 weeds m^2, can have little effect on yield of a vigorous crop. Many farmers grow oilseed rape as a cleaning crop. This is because there is a wide range of herbicides available which will control weeds that are otherwise difficult to control elsewhere in the rotation, e.g. brome. Cleavers and other brassicae are difficult to control and should be dealt with in other crops. Volunteer cereals are one of the main problems, especially if the rape is established using minimum cultivations after winter barley. Several graminicides, e.g. fluazifop-P-butyl (a fop) and cycloxydim (a dim) will control the cereals and other grass weeds. Note that grass weed resistance to these products is increasing and may limit choice of chemical. Currently there are no grass weed resistance problems with herbicides such as trifluralin, metazachlor and propyzamide.

In well-established crops, the time of weed removal is not as important as in backward thin crops. Most broad-leaved weed herbicides are applied in the autumn. A number are very persistent and the manufacturer's recommendations should be followed concerning the cultivations before the next crop. There is only a limited range of products which can be used in spring rape. Trifluralin (incorporated) is commonly applied.

5.5.2 Peas

Both dried and vining peas are uncompetitive crops. Poor weed control can reduce yields as well as affecting harvesting, drying and crop quality. The main method of weed control is with herbicides. Care must be taken in the choice of chemicals, as there are restrictions on variety and soil types. The majority of crops are treated with a pre-emergence residual herbicide. Those commonly used include pendimethalin + cyanazine, simazine + trietazine, terbuthylazine + terbutryn and fomesafen + terbutryn. There are only a limited number of post-emergence broad-leaved weed herbicides approved for use in peas. These include bentazone, MCPA + MCPB, and cyanazine. Post-emergence treatments are used if the pre-emergence treatment fails, or the peas are growing on an organic soil. The most effective timing for control of volunteer oilseed rape is often post-emergence. Care must be taken with post-emergence treatments as they can cause crop damage. The amount of leaf wax should be analysed using the crystal violet test. Application of post-emergence products is limited by crop growth stage.

A number of graminicides are approved for post-emergence use in peas; application should be restricted if resistance to blackgrass or wild oats has been confirmed.

5.5.3   Field beans

Initially, field beans are susceptible to weed competition, although later they are a very competitive crop. Most growers rely on herbicides for weed control, although in-crop harrowing can be used as an aid to keep the weeds under control. Most herbicides used are residual and are applied pre-emergence. Simazine is the standard chemical applied to the winter bean crop as long as the latter is sown deeper than 7.5 cm. In spring beans, several of the pea herbicides are approved pre-emergence. Only bentazone, applied post-emergence, is recommended for broad-leaved weed control in field beans. This chemical only controls a limited range of weeds.

A number of graminicides are approved for grass weed control but again their use may be restricted if grass weed resistance has been confirmed. Alternative grass weed herbicides include triallate and propyzamide.

5.5.4   Linseed

Linseed competes poorly with weeds. Any weed problem not controlled during the growing season can be desiccated at harvest using diquat, glyphosate or glufosinate-ammonium (as in oilseed rape, dried peas and field beans). Glyphosate can also be used pre-harvest for the control of perennial weeds. As linseed is a relatively minor crop, there are at present only a small number of chemicals which are approved for use. These include amidosulfuron, bentazone, bromoxymil + clopyralid and metsulfuron-methyl for broad-leaved weed control. Several graminicides can be used to control non-resistant grass weeds.

5.6   Weed control in root crops

5.6.1   Potatoes

Potatoes are a very competitive crop once they meet across the rows, but early weed emergence, which is not controlled, can reduce yields. Weeds can also affect potato quality and ease of harvesting. Weed control by cultivations may be impossible in wet seasons and there can be considerable loss of valuable moisture and damage to crop roots in dry seasons. Cultivations can also produce clods in some soil conditions. Consequently, in most potato crops, weeds are now controlled by herbicides; often a mixture of contact-acting and residual products.

Pre-planting
Perennial weeds should, if possible, be controlled by glyphosate applied pre-harvest in the previous cereal, oilseed rape, pea, or bean crop (the weeds must be green and actively growing) or by treatment in the autumn.

Pre-emergence of potato shoots
Diquat + paraquat or glufosinate-ammonium will kill most emerged seedling weeds. These chemicals work by contact action. Other residual herbicides can also be used, e.g. linuron, monolinuron or pendimethalin. Metribuzin is more persistent, but there are varietal restrictions. On organic soils it can be incorporated into the ridge to improve the control of weeds.

Post-emergence of potato shoots
There are very few products which can be used post-emergence in potatoes. Metribuzin can be applied to some crops and varieties, as can the foliar-acting herbicides, bentazone and rimsulfuron. Several graminicides are approved for grass weed control e.g. cycloxydim.

Haulm destruction
Chemicals are normally applied to destroy the haulm; this encourages skin set and reduces the spread of blight to the tubers. The following may be used: diquat (but not in dry conditions), sulphuric acid or glufosinate-ammonium.
    Volunteer potatoes from tubers and seed are an increasing problem, particularly in close rotations. Glyphosate, pre-harvest in other arable crops, can aid control.

5.6.2   Sugar beet (fodder beet, mangels)
Couch and other perennial weeds should be controlled in the year prior to planting the crop. Weeds can seriously reduce beet yields if not controlled, particularly in the first eight weeks following the sowing of the crop. Traditionally, weeds were controlled by inter-row cultivations and hand hoeing. Following this, band spraying combined with inter-row cultivations were used.
    Band spraying can be a slow operation so now annual weeds are mainly controlled by overall herbicide treatments applied pre- and/or post-emergence of the crop, using the low-dose technique. It is important that the correct type and amount of herbicide are applied (according to the weed growth stage) and the soil conditions are suitable, i.e. fine and moist. This technique relies on treatment when the weeds are at the cotyledon stage. On average, a sugar beet crop may have four applications of herbicides. Some inter-row cultivations using a steerage hoe are still carried out mainly where there is a problem with weed beet.

Pre-sowing
Pre-sowing treatments, using contact-acting products like paraquat, may be required to kill weeds that have not been controlled during seedbed cultivations.

Pre-emergence
The majority of crops receive a pre-emergence treatment. Choice of chemical will depend on potential weed problems and soil type. Chloridazon +/– ethofumesate or metamitron are commonly used.

Post-emergence

It may be satisfactory or necessary to leave all annual weed control to this time. However, bad weather conditions may delay spraying and many weeds could grow past the susceptible stages. Using the low-dose technique, there is less restriction on timing of the post-emergence herbicides in relation to crop growth stage. Mixtures of herbicides with different modes of action are commonly used, e.g. the contact herbicide, phenmedipham with or without metamitron or other residual types such as ethofumesate or the foliar acting triflusulfuron-methyl. A clopyralid spray can be used to give better control of established thistles and mayweeds, as well as suppressing volunteer potatoes. Graminicides can be used to control wild oats and some grasses and severely check common couch although it is preferable not to use these products if resistant grass weeds are present. Glyphosate can be applied with a rope-wick or roller machine to control weed beet.

A standard herbicide programme in sugar beet is one pre-emergence product followed by at least two low-dose post-emergence treatments. More recently this programme has been adapted in order to reduce costs further. This so-called FAR system uses very low rates of herbicides applied at the first signs of any weeds emerging. (FAR stands for phenmedipham, activator plus residual, e.g. phenmedipham+ethofumesate+metamitron.) Typically, four applications are required at weekly intervals. Care is required in product choice and rate as crop damage can occur on very light soils and some residuals are not effective on soils with a high organic matter. The product recommendations should be followed very carefully.

5.6.3   Brassicae (Brussels sprouts, cabbages, cauliflower, kale, forage rape, swedes and turnips)

Weeds in horticultural brassicas are mainly controlled with herbicides and some cultivations. Chemical choice is very limited in these minor crops; always check the recommendations. Some of the chemicals have off-label recommendations. Although approved, manufacturers do not endorse off-label uses and such treatments are made entirely at the risk of the user. Weeds in transplanted crops are controlled with contact herbicides, e.g. paraquat, or by cultivations before the crop is planted. Little weed control is required in the quick-growing forage brassicas, whereas swedes and kale are very sensitive to weed competition.

Couch and other perennial weeds can be controlled with glyphosate in the previous crop pre-harvest or on the stubble. A limited range of graminicides is recommended for post-emergence grass weed control. Trifluralin, incorporated into the seedbed, plus propachlor or metazachlor pre-emergence, will give good control of a wide range of annual weeds (but not charlock) in many of these crops. Clopyralid is useful for controlling mayweeds and thistles in all crops. Where any of these crops, e.g. kale, are direct drilled into a chemically destroyed grass sward, annual weeds are unlikely to be a problem.

5.7 Weed control in grassland

5.7.1 Permanent grassland and long-term leys

In grassland, it is often difficult to decide which plants should be re_
weeds. Several weeds can produce a reasonable yield but not necessarily of
high quality. Thistles, rushes, bracken, tussock grass and poisonous weeds such
as ragwort, horsetail and hemlock should be destroyed. In arable crops annual
weeds cause most damage but in established grassland biennial and perennial
weeds are responsible for most of the trouble. The presence of the weeds can
bring about a reduction in the yield, nutrient quality and palatability of the
sward. Stock do not like grazing near buttercups, thistles and wild onions. Some
weeds are poisonous, e.g. ragwort and horsetail, and some can taint milk if
eaten, e.g. buttercups and wild onion.

Weeds in grassland are encouraged by such factors as:

1   Bad drainage, e.g. rushes, sedges, horsetails and creeping buttercup.
2   Lime deficiency, e.g. poor grasses (Yorkshire fog and the bent grasses)
    and sorrels.
3   Low fertility. Many weeds can live in conditions which are too poor for
    the better types of grasses and clovers.
4   Poaching (treading in wet weather). The useful species are killed and weeds
    grow on the bare spaces.
5   Over-grazing. This exhausts the productive species and allows poor,
    unpalatable plants such as Yorkshire fog, thistles and ragwort to become
    established.
6   Continuous cutting for hay encourages weeds such as soft brome, yellow
    rattle, knapweed and meadow barley grass. These weeds seed before the
    crop is harvested.

Chemicals are a useful aid to controlling grassland weeds, but should not be
regarded as an alternative to good management. Inexpensive products such as
MCPA, mecoprop, 2,4-D and dicamba check many weeds, although treatment
may need repeating. Newer products such as those with clopyralid, fluroxypyr
or triclopyr can be more effective, although more expensive. These chemicals
are useful for spot treatment. The main chemicals used to control broad-leaved
weeds in swards where clover is important are MCPB, 2,4-D and benazolin.
Asulam is used to control bracken and docks; it also kills some grasses.
Glyphosate, when applied with a rope-wick or roller machine, will selectively
kill any tall weeds which are growing actively, e.g. ragwort.

Table 5 in Appendix 9 is a guide to the control of the more troublesome
weeds in grassland.

5.7.2 New sown leys

When seeding a grass sward without a cover crop, weeds, especially annuals
such as charlock, chickweed and fat hen, can be a problem. Additionally, grass

sown in an arable rotation can have more problems with arable weeds such as blackgrass, cleavers and speedwell. Mowing or early grazing can kill annual weeds with upright stems. However, chickweed should be sprayed with, for example, mecoprop, or a benazolin mixture where clovers are present and the crop is at a safe growth stage.

Where grassland has to be reseeded, and ploughing is not possible or desirable, the old sward can be killed by a glyphosate spray from June to October when it is 25–60 cm in height and the green leaves are growing actively. The treated sward can be grazed or conserved about 10 days after application with the feeding value retained. Alternatively, two to three weeks after applying glyphosate and when the sward is brown and desiccated, the next crop can be direct drilled. However, if there is a thick mat of decaying surface vegetation it should be lightly rotavated; otherwise the new seeds may be killed by toxic substances.

Problems can also arise when seedling weed grasses, such as meadow-grass or blackgrass, establish with the sown seeds. Few herbicides can be used for grass control in new leys, but an exception is ethofumesate. As a residual herbicide ethofumesate can also be applied in one or two doses to established grass between mid-October and February to control many weeds such as chickweed, annual meadow-grasses, blackgrass, sterile and soft brome, wall barley and volunteer cereals. Reseeding is not a cheap operation and so it is important that weed control is not skimped. Weeds can affect crop establishment and the longevity of the sward.

5.8   Spraying with herbicides: precautions

This is a skilled operation and should only be carried out by trained operators. A recognised certificate of competence is required.

The following precautions should be taken when spraying:

1   Make a careful survey of the field to determine the weeds to be controlled. Choose the most suitable and safest chemical and the best time for spraying. The risks associated with the use of any herbicides should be assessed as defined in the HSE regulations (COSHH – *Control of Substances Hazardous to Health*). Appropriate measures should be taken to control any risk.

2   Check carefully the amount of chemical to be applied and the volume of water to be used (220–330 l/ha is a common range). Make sure the chemical is thoroughly mixed with the water before starting. Soluble and wettable powders should be mixed with some water before adding to the tank. Use the agitator if necessary. The rate of application is mainly controlled by the forward speed of the tractor (use a speedometer), the size of nozzle and, to a lesser extent, by the pressure (follow the maker's instructions). Always use clean, preferably not hard, water and always use a filter. An

accurate dipstick is necessary when refilling the tank if it is not emptied each time.

3   Wear the correct protective clothing. Read carefully the instructions issued for that product when handling the concentrate and when spraying.

4   Do not spray on a windy day, especially with the hormone type of herbicide and if the spray is likely to blow into susceptible crops or gardens. It can be helpful to keep the boom as low as possible and to use a plastic spray guard. To avoid spray drift, it is preferable to use a high volume application (larger droplets) than a very low volume mist.

5   Make sure that the boom is level and that the spray cones or fans meet just above the level of the weeds to be controlled.

6   Spray the headlands first. If using a wide boom, it is advisable to use markers.

7   Wash out the sprayer thoroughly. Follow the guidelines for safe disposal.

8   Only use approved tank mixes. Follow the manufacturer's recommendations for mixing.

9   Check buffer zone requirements when spraying next to water courses. (If necessary undertake a Local Environmental Risk Assessments for Pesticides (LERAP) assessment.)

5.9 Further reading

Anderson W P, *Weed Science, Principles and Applications*, West, 1996.

British Crop Protection Council, *Using Pesticides: A Complete Guide to Safe Effective Spraying*, BCPS, 1999.

Caseley J C, Atkin R K and Cussans G W, *Herbicide Resistance in Weeds and Crops*, Butterworth-Heinemann, 1991.

Cooper M R and Johnson A W, *Poisonous Plants and Fungi*, HMSO, 1988.

Cremlyn R J, *Agrochemicals, Preparation and Modes of Action*, Wiley, 1991.

Garthwaite D G and Thomas M R, *Pesticide Usage Survey Report 159. Arable Farm Crops in Great Britain*, MAFF, 1999.

Gwynne D C and Murray R B, *Weed Biology and Control, Batsford Technical*, 1985.

Hance R J and Holly K, *Weed Control Handbook*, Blackwell, 1990.

Hanf M, *The Arable Weeds of Europe*, BASF, 1983.

Hubbard C E, *Grasses*, Penguin, 1984.

LIAISON, Central Science Laboratory, interactive pesticide database, updated daily.

Mathews G A, *Pesticide Application Methods*, Blackwell Science, 2000.

Whitehead R, *The UK Pesticide Guide*, CABI publishing, published annually.

Williams J B and Morrison J R, *The ADAS Colour Atlas of Weed Seedlings*, Wolfe Medical, 1987.

Research Reports, HGCA, AAB, BCPC.

Trade literature websites product guides and grass and broad-leaved weed identification guides.

6

Pests of farm crops

6.1 Insects and nematodes

Pests are responsible for millions of pounds of damage to agricultural crops every year. Some pests are a regular problem and occur most years, although not always at population levels that cause economic damage. Other pests only occur occasionally depending on several factors including rotation, weather and crop growth stage. Some pests are very specific to individual crops whereas others damage a large number of different crops. Some animals are successful as pests partly because they can reproduce in large numbers, very quickly, with many generations a year, have an effective method of spread and adapt to changing environmental conditions. The group of animals that contains some of the most important crop pests is the insects. Nematodes are the next most important group, followed by the molluscs, birds and mammals.

Before discussing the various methods used to control pests, it is important to understand something of their structure, general habits and type of damage that they cause.

6.1.1 Insects

The main groups of insect pests include the plant bugs (e.g. aphids and capsids), butterflies and moths, beetles, including weevils as well as sawflies and true flies. Insects are invertebrates; i.e. they belong to a group of animals which do not possess an internal skeleton. A hard external covering – the exoskeleton – supports their bodies. It is composed chiefly of chitin, and is segmented so that the insect is able to move.

From the diagram of the external structure of an adult insect (Fig. 6.1) it can be seen that the segments are grouped into three main parts:

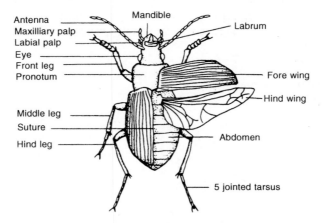

Fig. 6.1   Structure of an insect.

1   The head, on which is found:
   a   the antennae or feelers carrying sense organs for touching and smelling;
   b   the eyes; a number of simple and a pair of compound eyes are present in most species;
   c   the mouth parts (Fig. 6.2); two main types are found in insects:
     i The biting type used for grazing on foliage;
     ii The sucking type – insects in this group suck the sap from the plant and do not eat the foliage.

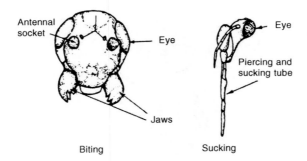

Fig. 6.2   Insect mouth parts.

2   The thorax, which carries:
   a   the legs; there are always three pairs of jointed legs on adult insects;
   b   the wings; these are found on most, but not on all, species.
3   The abdomen; this has no structures attached to it except in certain female species where the egg-laying apparatus may protrude from the end.

6.1.1.1   Life cycle of insects

Knowledge of the life cycles of insects can be of great help in deciding on the best stage at which the insects will be most susceptible to control methods. Most insects begin life as a result of an egg having been laid by the female. What emerges from the egg, according to the species, may or may not look like the adult insect.

There are two main types of life cycle:

1   The *complete* or four-stage life cycle (Fig. 6.3).

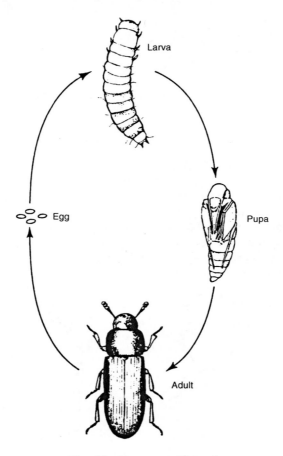

Larva

Egg

Pupa

Adult

Fig. 6.3   Four-stage life cycle.

a   The egg.
b   The larva or the immature growth stage, entirely different in appearance from the adult. This is the active eating and growing stage. The larva usually possesses biting mouthparts, and it is at this stage, in many insects, that they are most destructive to the crops (Fig. 6.2).

c   The pupa – the resting state. The larva pupate and undergo a complete change, or metamorphosis from which emerges –

d   The adult insect – this may feed on the crop, e.g. flea beetle, but in many cases it is the larval stage that causes the major crop damage, not the adults.

There are distinct differences in larval appearance for the different insect groups. Identification characteristics include size, presence or absence of legs on the chest, presence or absence of false or prolegs on the abdomen and the size, development and colour of the head (Fig. 6.4).

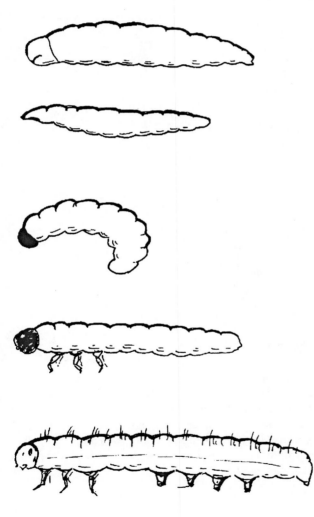

Fly larvae
No distinct head
No legs
Usually pale colour
e.g. Wheat bulb fly

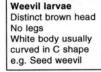

Midge larvae
No distinct head or
mouth parts
No legs
White, yellow or
orange
e.g. Orange blossom
midge

Weevil larvae
Distinct brown head
No legs
White body usually
curved in C shape
e.g. Seed weevil

Beetle larvae
Distinct head
3 pairs of front legs
e.g. Wireworm

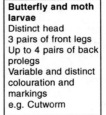

Butterfly and moth larvae
Distinct head
3 pairs of front legs
Up to 4 pairs of back
prolegs
Variable and distinct
colouration and
markings
e.g. Cutworm

Fig. 6.4   Characteristics of insect larvae.

2   The *incomplete* or three-stage life cycle (Fig. 6.5).
    a   The egg.
    b   The nymph – this is very similar in appearance to the adult, although
         it is smaller and may not possess wings. It is the active eating and
         growing stage.
    c   The adult insect – invariably this stage will also feed on and damage
         the crop, e.g. aphids.

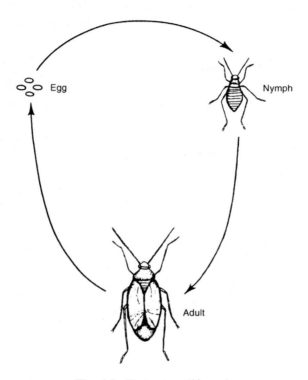

Fig. 6.5   Three-stage life cycle.

Most insects and/or larvae and nymphs depend on the crops on which they
feed for part or all of their existence. The crop is the host plant, whilst the
insect is the parasite to the host. Not all insects are harmful to crops; some are
beneficial in that they prey on or parasitise crop pests. The ladybird and
lacewing are particularly useful because, both as the larva and adult, they feed
on aphids which are responsible for transmitting certain virus diseases in plants
as well as causing physical damage to plants.

6.1.2   Nematodes (eelworms)

Nematodes have non-segmented elongated worm-shaped bodies surrounded by a tough cuticle. They vary from 0.1 mm to 2.0 mm in length with an average size of 1.0 mm. They are usually too small to be seen with the naked eye.

As far as is known, most species of nematodes are free-living and beneficial, but there are a number of important species which are either parasitic within crop plants (endoparasitic) or feed on the surface tissues of crop roots (ectoparasitic). Most nematode pests only attack or cause damage on specific host species. The nematodes themselves have a limited ability to move very far. The most important means of spread is by movement of contaminated soil or water, such as on farm machinery. The mouth parts of these plant-feeding nematodes consist essentially of a cavity – the stoma – in which is positioned the mouth spear or stylet (Fig. 6.6). It is this which pierces the cellular tissue for sucking out the cell contents of the parasitised plant.

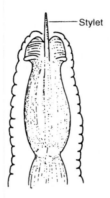

— Stylet

Fig. 6.6   Mouth parts of plant-feeding nematodes.

The life cycle is relatively simple, consisting of an egg hatching into a larval form. There are several juvenile stages (shedding a cuticle each time) before reaching the sexually mature male or female stage. In most cases, part of the life cycle takes place in the soil. With many pest species there is a stage in the life cycle (e.g. a cyst containing eggs) which may remain dormant, sometimes for several years, only becoming active again when conditions are suitable. Some species, however, have more than one generation in the year, and are only inactive during the winter months. Both the juvenile and adult stages are collectively the cause of damage to the host plant. Colonies develop, creating progressive damage. These symptoms, supported by plant and soil analysis, allow the identification of the species involved and the formation of future control strategies. No curative control is yet available to check damage to an already infected crop.

The major nematode pests are the root-attacking cyst nematodes. Other agri-culturally important nematodes are the stem nematodes. These can have a wide host range but on each host they cause varying forms of necrosis and deformity of the shoot systems. These stem nematodes are often spread in contaminated seed or plant debris.

6.2   Other pests of crops

6.2.1   Molluscs

Slugs are the most important mollusc pests of field crops. Most crops can be affected. There are several species of slugs including the field, garden and keeled slugs. Slugs can feed on crops throughout the year although activity is reduced during periods of drought or frost. Crops growing on heavy clay soils are most at risk; previous crop and soil cultivations and consolidation can affect activity. Each slug is able to lay up to 500 eggs; these eggs usually hatch either in early autumn or late spring depending on weather conditions. On hatching, the young slugs resemble the adults except in size. Initially the young feed on organic matter in the soil, later on plant material both underground and on the soil surface.

6.2.2   Birds

Generally, birds are more helpful than harmful, although this will depend on the district and type of farming carried out. To the grassland farmer birds are not such a problem as they are to the arable farmer. Although birds will eat some cereal seed, most of them help the farmer by eating many insect pests and weed seeds; the diet of some also includes mice, young rats and other rodents.

The wood pigeon certainly does far more harm than good. Not only will it eat cereal seed and grain of lodged crops, it also causes considerable damage to young and mature crops of peas and brassicae. The only effective ways of keeping this pest down are by properly organised pigeon shoots and the use of various bird scarers.

6.2.3   Mammals

There are several mammals which can cause damage to crops.

Rabbits and deer can be very serious pests; they eat many growing crops, particularly young cereals. The worst pest of all is the brown rat, which eats and damages growing and stored crops. The mouse is another serious pest; it damages many stored crops. The local rodent officer will give advice on methods of extermination.

The harmless mammals, as far as crops are concerned, are: hedgehogs which eat insects and slugs, foxes which kill rats and rabbits and squirrels which eat pigeons' eggs.

6.3   Types of pest damage

It is often the damage to a crop that is noticed in the field first before a pest is found and identified. The type of damage and the crop are important aids when identifying a pest and deciding on whether control measures are required.

1   Total plant loss: Some pests eat out the seed before it even germinates. For example, slugs can hollow out the grain of cereals, or mice can eat all the seeds in a row of peas. At other times the seed germinates but then subsequently dead plants are found eaten off at ground level or just below the soil surface. Slugs as well as leatherjackets, cutworm, swift moth larvae, vine weevil larvae and the cockchafer grubs can cause this type of damage.
2   Loss of the central growing shoot: Some pests feed on the central growing shoot causing it to die, producing dead hearts. Frit fly in grass and cereals and wheat bulb fly produce these symptoms. The dead stem has to be split open to find the larvae.
3   Holes and/or leaf damage: Leaf damage can sometimes be superficial and not worth controlling. Adult weevils cause notching around the outside of crop leaves whereas slugs cause leaf shredding. Flea beetle adults cause characteristic holing in the leaves. This damage can severely reduce crop vigour in very dry conditions at the seedling stage.
4   Distorted leaves: Some aphids and capsids feeding on crop leaves can cause the crop leaves to become distorted.
5   Loss of flowers and/or buds: Some pests enter flowers and/or flower buds and their feeding damage can cause the buds to abort. Pollen beetle causes this type of damage especially in spring oilseed rape.
6   Stunted growth: Nematodes such as potato cyst nematode and stem nematode in red clover cause severe crop stunting.
7   Transmission of viruses: Several animals act as vectors for some important virus diseases. Examples include free-living nematodes transmitting spraing in potatoes, aphids transmitting barley yellow dwarf virus in cereals and mites that transmit ryegrass mozaic virus.
8   Damage to harvested crop: Several pests can damage the crop just before harvest or even after harvest. Grain aphids can affect wheat grain filling so that the crop is thin and shrivelled. Slugs and wireworm can eat into potatoes so that the quality is severally affected and they become unsaleable. Grain weevils and mites are pests that can cause significant damage to grain in store.

See Table 6.1 at end of this Chapter for further details on crop pests and crop damage.

6.4   Methods of pest control

6.4.1   Non-chemical control methods

Although chemical treatments are currently the most common method of pest control, there are several methods of pest control that can reduce the risk of damage and reduce the need for pesticide treatment. With the increasing number of cases of pesticide resistance, alternative methods have a major part to play.

1   *Rotations*
    Having a more diverse cropping rather than mono-cropping, e.g. with cereals, can have a variable effect on some crop pests. Introducing grass into an arable rotation can increase the risk of frit fly and leatherjackets whereas continuous cereals encourage wheat blossom midges. A short rotation with potatoes, e.g. one year in four or less, encourages potato cyst nematode (PCN).

2   *Time of sowing*
    a a crop may sometimes be sown early enough so that it can develop sufficiently to withstand an insect attack, e.g. wheat bulb fly in the wheat crop.
    b a crop can be sown late enough to avoid the peak emergence of a pest, e.g. aphid transmitted barley yellow dwarf virus (BYDV) in winter cereals or drilling carrots at the end of May after the first major generation of carrot fly.

3   *Cultivations*
    Ploughing exposes pests such as wireworms, leatherjackets and caterpillars, which are then eaten by birds. Well-prepared seedbeds encourage rapid germination and growth which will often enable a crop to grow away from the pest attack. Ploughing can have a negative effect on a number of beneficial organisms.
    Non-ploughing techniques have been found to reduce aphids as BYDV vectors in cereals as well as reducing the risk from frit fly and yellow cereal fly; slugs are usually more of a problem.

4   *Encouragement of growth*
    Good quality seed should be used which will germinate quickly and evenly. It is also important that the crop is not checked to any extent, for instance, by lack of a plant food. A poor growing crop is far more vulnerable to pest attack than a quick growing crop. A top-dressing of nitrogen, just as a crop is being attacked, may sometimes help. Aphids are often more frequently found on thick, well-fertilised crops.

5   *Clean farming*
    Weeds can be alternate hosts to a number of insects both beneficial (e.g. carabid beetles) and pest species (e.g. nematodes). It is important to control these weeds in other crops so that the pest life cycle can be broken.
    Contaminated soil should be returned to the same field after riddling/cleaning crops such as potatoes or sugar beet. If not, the spread of pests such as cyst nematodes can be increased.

Grain stores should be thoroughly cleaned before harvest to reduce the incidence of grain pests.

6 *Resistant varieties*
Some crop varieties show resistance to certain pests, as found in potatoes. There is a wide variation in susceptibilities to slug damage and the cyst nematodes between varieties.

7 *Biological control*
A parasite or predator is used to control only the target species. This is a very important method of control of many pests in glass houses and has reduced the reliance on pesticides. Currently, in the field there are few examples. Examples include:
a predatory mites are used successfully to control the red spider mite in cucumber production under glass;
b *Bacillus thuringiensis* for bacterial control of caterpillars. The crop is sprayed and the bacterium invades the target organism; it is very effective at controlling caterpillars including those of the cabbage white butterfly.

8 *Barriers*
Use of fleece or net crop covers will stop egg laying of some pests such as carrot fly. On a field scale it is an expensive although effective option.

9 *Legislation*
Some countries have introduced legislation in order to avoid the introduction of certain pests. This has been successful in the UK against the introduction of Colorado beetle. Stem nematodes can be spread in the seed of crops such as red clover. Testing of the seed will reduce the risk.

10 *Trap cropping*
Currently trap cropping is being studied. A susceptible crop plant species is sown in strips before the main crop. This crop is used to attract mobile predators before they attack the crop; these pests can then be killed.

6.4.2 Chemical control methods

There are a number of ways pesticides are applied:

1 Sprays and dusts.
2 In granular form.
3 Baits for controlling soil pests such as leatherjackets, slugs and snails.
4 Seed dressing – mainly for the protection of cereals against wireworm, brassica crops against flea beetle and soil-borne pests in sugar beet. Usually the insecticides are combined with a fungicide.

Gases, smokes and fumigants are commonly used in greenhouses chiefly against aphids, and in granaries for the control of beetles and weevils.

Basically, there are two ways in which pesticides kill pests:

1   *By contact*
    The pest is killed when it comes in contact with the chemical, such as
    when:
    a   it is directly hit by the spray or dust;
    b   it picks up the pesticide as it moves over foliage which has been treated;
    c   it absorbs vapour;
    d   it passes through soil which has also been treated.
2   *By ingestion*
    As a stomach poison, the pest eats the foliage treated with the pesticide,
    or the chemical is used in a bait. As a systemic compound it is applied to
    the foliage or to the soil around the base of the plant. It gets into the sap
    stream of the plant and thus the pest is poisoned when it subsequently
    sucks the sap.

Most pesticides kill by more than one method, which makes them very effective.
But many of them are extremely toxic to animals and humans and, by law,
certain precautions must be observed by the persons using them, both when
handling and applying. Many of these chemicals have a statutory requirement
for a 5 m buffer zone when spraying next to water courses.

6.5   Classification of pesticides

6.5.1   Insecticides

1   Naturally occurring insecticides: *Nicotine** is used in horticulture against
    sucking insects. It affects the pest's nervous system.
2   Organochlorides: These insecticides are all stomach and contact poisons.
    Many of the products have high mammalian toxicity, are not very specific
    and have detrimental environmental effects. There are now few approved
    products; most have been banned and withdrawn from the market now that
    safer products are available.
3   The organophosphorus compounds
    As a group, the organophosphorus compounds are also dangerous to use
    and many are subject to the Poisons Act. This group of insecticides mainly
    affects the pest's nervous system. They are generally non-persistent and
    do not accumulate in the environment. They should be handled strictly in
    accordance with the manufacturer's instructions.
        Examples of some of the organophosphorus insecticides in common
    use are:
    a   *non-systemic*
        Pirimiphos-methyl: for control of stored grain pests;
        Chlorpyrifos: controls a wide range of pests in a wide range of crops,
        e.g. cabbage root fly, caterpillars and cutworm in brassicae, leather-
        jacket, blossom midge and frit fly in cereals;

b   *systemic*
  Dimethoate: this is used for the control of aphids on many agricultural crops. This insecticide is not very selective and will also affect many beneficial insects.

4   The carbamate compounds.
  These compounds also affect the pest's nervous system. They tend to work quickly and have a reasonable rate of breakdown. Examples include:
  • *Carbosulfan:* a systemic insecticide for many soils pests, mainly in root crops.
  • *Pirimicarb*: for control of aphids. This insecticide is very specific and does not affect beneficial insects.

5   Synthetic pyrethoids
  These chemicals have a particularly high insecticidal power whilst generally being safe to humans and farm animals. They are efficient contact insecticides with a rapid knockdown action and some are thought to act as antifeedants. These properties are being seen as valuable in the chemical control of plant viruses by controlling the insect vectors, e.g. cypermethrin and λ-cyhalothrin.

  Persistency is a constant cause of concern for food producers, and research is continually directed towards finding safer, less persistent compounds.

6.5.2 Molluscicides

• *Metaldehyde*. This is used as a mini-pellet for the control of slugs and snails. It works by dehydrating the molluscs.
• *Methiocarb*. This is used as a mini-pellet for the control of slugs and snails. It is a stomach-acting carbamate.

6.5.3 Nematicides (including soil sterilants)

Nematicides, such as *fosthiazate*, have been developed for the control of nematode pests of crops. Fosthiazate is a soil-applied contact acting nematicide that controls potato cyst nematodes. Certain chemicals in this group are classified as soil sterilants, e.g. *Dazomet*. It is important to remember that a time interval must be observed between the last application of the pesticide and the harvesting of edible crops as well as the access of animals and poultry to treated areas.With some pesticides this interval is longer than others. This is another reason for very careful reading of the manufacturer's instructions.

---

*An asterisk marks those chemicals included in the Health and Safety (Agriculture) (Poisonous Substances) Regulations. Certain precautions (including the use of protective clothing) must by law be taken when using these chemicals. It is advisable to read the joint MAFF/HSE publication *Pesticides: Code of Practice for the Safe Use of Pesticides on Farms and Holdings*, 1998.

For up-to-date information on pesticides available, and the regulations and advice on the use of these chemicals, reference should be made to the *UK Pesticide Guide*, published annually by CABI/BCPC.

Table 6.1 includes information on the major pests attacking farm crops and their control.

6.5.4   Integrated pest management
With increasing awareness of the side effects of pesticides and the increasing number of cases of pesticide resistance, integrated pest management (IPM) is an important method of pest control. Whilst it may involve the use of pesticides, their use is minimised in an attempt to protect and enhance the activities of beneficial insects (natural enemies and pollinating insects) and extend the life of the pesticides that are available.

Economic thresholds have been calculated for many pests. This is the population of pests which if controlled will give a yield return that will pay for the cost of pesticide and application. Where available these have been included in the tables. Another important tool to help with decision making for application of a pesticide is the use of forecasts of pest populations. A number of organisations run forecasting services for some important crop pests such as cutworm, pea moth and carrot fly. These forecasts can be used to reduce insurance or prophylactic spraying. When pesticides must be applied the most selective and safest should be chosen.

6.6   Further reading

Alford D V, *Pest and Disease Management Handbook,* Blackwell Science, 2000.
Alford D V, *A Textbook of Agricultural Entomology*, Blackwell Science, 1999.
Dent D, *Integrated Pest Management*, Chapman and Hall, 1995.
Gratwick M, *Crop Pests in the UK*, Chapman and Hall, 1992.
Jones F G W and Jones M G, *Pests of Field Crops,* Edward Arnold, 1984.
Mathews G A, *Pesticide Application Methods,* Blackwell Science, 2000.
Whitehead R, *The UK Pesticide Guide,* CABI publishing, published annually.
Proceedings of the Brighton Crop Protection Conferences – Pests and Diseases, BCPC, biennial.

Appendix

Table 6.1 Major pests and their control

Crop attacked	Pest	Description	Life cycle	Symptoms of attack	Control	Notes
Cereals	*Adult:* Species of **Clickbeetle** *Larva:* **Wireworm**	*Adult:* Brown, 6–12 mm long *Larva:* Growing to 25 mm long, yellow colour	Larvae hatch out during summer from eggs laid just below soil surface, mainly in grassland. They take 4–5 years to mature, and after pupation in the soil, the adult appears in early autumn	Yellowing of foliage followed by the disappearance of successive plants in a row. This is caused by wireworm moving down the row. Larvae eat into the plants just below soil surface. They are usually found in soil around the plants	Good growing conditions to help the crop grow away from an attack. Treated seed should be used in susceptible fields, e.g. with tefluthrin or imidacloprid. Damage is reduced when the soil is well consolidated	Do not confuse wireworm attack with other pests such as nematode. Wheat and oats are more susceptible than barley; they should not be grown where the wireworm count is over 2 million/ha
	Adult: **Cranefly** (Mainly *Tipula* sp. and *Nephrotoma* sp.) *Larva:* **Leatherjacket**	*Adult:* Is the 'Daddy longlegs' *Larva:* Leaden in colour, 30 mm long	Eggs laid on grassland or weedy stubble in the autumn from which the larvae soon emerge. They feed on the crop the following spring, pupating in the soil during the summer	Crop dies away in patches, root and stem below ground having been eaten. Larvae found in soil	If possible, plough the field before September to prevent the eggs being laid. Control chemically with chlorpyrifos. Rolling and nitrogen top-dressing help crop to recover	Treatment may be worthwhile if 5 or more leatherjackets are found /m row
	Wheat bulb fly	*Larva:* Whitish-grey,	Eggs laid on bare soil in July, August.	Central shoot of plant turns yellow	If possible, avoid sowing wheat where	Winter wheat and barley are attacked

Table 6.1 *(Cont'd)*

Crop attacked	Pest	Description	Life cycle	Symptoms of attack	Control	Notes
		up to 12 mm long. Larva has a blunt tail end	Eggs start hatching from early January. Larvae feed on the crop until May. Pupation then follows either in the soil or plant	and dies in early spring, 'deadheart' symptoms. Larva found in base of tiller	the field has lain bare from late summer. Protection includes: (1) Seed treatment, e.g. tefluthrin (2) Spray at egg hatch, e.g. chlorpyrifos	especially if drilled deep and late after a susceptible crop. Wheat bulb fly is a problem mainly of central and eastern counties. Use a seed treatment if late drilled and >1.25 million eggs/ha. If early drilled and egg count >5 million/ha use an egg hatch spray
	Frit fly	*Larva:* Whitish, up to 4 mm long	Three generations in the year; the first when in spring eggs are laid on spring oats and larvae feed on crop in May and June. The second generation damages the oat grain, whilst the third generation overwinters on grass, but when the latter is	In early summer, the central shoot of the oat plant turns yellow and dies 'deadhearts', but the outer leaves remain green; blind spikelets and shrivelled grains are caused by second generation; autumn cereals can have a kink above the	Sow spring oats before mid-April, and try and get them past the 4-leaf stage as quickly as possible. Allow a 6 week interval between ploughing grass and sowing the winter cereal. Chlorpyrifos may be applied as a	Spring oats and maize are particularly susceptible. Mainly a problem in rotations with grass. Chemical control likely to be cost effective if 10 % of plants are damaged at the 1 to 2 leaf stage

	Larva		Damage	Treatment	Notes
		ploughed for autumn cereals the larvae move on to the cereals	coleoptile at single-shoot stage, shoot then turns yellow	preventative spray on the previous grass crop. Otherwise treat at the 1 to 2 leaf stage if damage threshold reached	Winter barley rarely attacked; spring-sown cereals virtually immune
Yellow cereal fly	*Larva:* Yellowish; slender, about 6mm long, pointed at both ends	Eggs laid near wheat plant in October and November, hatching early in new year; larvae move down between outer leaves to feed on main tiller. Pupation in early summer, adults appear in June and a month later they migrate to hedgerows before returning to wheat field in late autumn	Circular or short spiral band at base of tiller producing brownish scar and then death of shoots – 'deadhearts'	Early-sown wheat in the eastern counties at greatest risk. Crop loss is usually low. Chemical treatment using a pyrethroid in November can give good control	
Gout fly	*Larva:* Legless, yellowish-white, 8mm long	Two generations in the year, the most important being the first. Larvae hatch and feed in plant	Leaf sheath surrounding the ear is swollen and twisted. Poorly developed grain emerges	Early drilled crops are most at risk. A chlorpyrifos spray at egg hatch can be effective	Barley is chiefly affected in southern counties of England. Crops can compensate for up to 25% damage
Cereal aphids	Various species; mainly	Winged females found feeding on	Damage is either direct feeding or	The grain aphid causes most concern	Aphids carry virus diseases from

Table 6.1 (Cont'd)

Crop attacked	Pest	Description	Life cycle	Symptoms of attack	Control	Notes
		bird cherry, grain and rose grain aphids. Green, yellowish or reddish brown in colour	cereal crops in May and June. Wingless generations produced which continue feeding during summer. Most species move back to winter quarters (woody hosts and some grasses and cereals, depending on aphid species) in autumn, although some may be found on young cereal crops at the end of the year	transmission of virus – BYDV. The grain aphid causes empty and/or small grain; by puncturing the grain in the milk-ripe stage, the grain contents seep out. This also reduces the weight of the grain	in the summer. Spray only when threshold levels are reached. Apply predator safe products such as pirimicarb. Various pyrethroids are approved for use in the autumn	infected to clean plants. See barley yellow dwarf virus disease (Table 7.1). Threshold for grain aphids in the summer is when two thirds of the ears are infected
	Orange wheat blossom midge	*Adult:* 3 mm orange midge *Larva:* up to 2.5 mm bright orange	Adults appear in May and June. Adult if present can be seen at dusk. Eggs are laid in wheat ears. Eggs hatch within a week and the larvae start feeding on the developing grain	Larvae can cause yield loss due to shrivelled grain. Hagberg Falling Number can also be reduced	Treat within a week of finding threshold populations of adults. Use an approved product such as chlorpyrifos	Mainly a pest of wheat though can affect barley and rye. Thresholds for feed wheat is one midge found per three ears. Thresholds for milling wheat is one midge per six ears

Pest	Appearance	Habits	Damage	Control	Notes
Stem and bulb nematode	Too small to be seen without magnification	Live and breed in the plant. If the plant dies, nematodes become dormant in dead tissue or soil, becoming active again when conditions are suitable	Twisting and swelling, and this normally prevents plants from elongating and producing an ear	Resistant varieties. Rotation to starve out the nematodes. Clean seed	Attacks oats and rye
Cereal cyst nematode	Dark-brown lemon-shaped cysts about 1 mm long	Live and breed in the roots. White-looking cysts (female cysts containing large numbers of eggs) are found on roots. Later these cysts (now dark brown) become free in the soil to infect the host plant again	Crop shows patches of stunted yellowish-green plants. Root system very bushy. Cysts visible on roots from June onwards	Avoid growing oats too often in the field. Grow resistant varieties when necessary	Oats mainly affected. Populations of this pest are declining, damage is rarely seen
Slugs and snails	Several species cause damage. The field slug is the most common. Lightish-brown in colour, about 40 mm long	Most active in moist and humid conditions. An attack can be more serious when the seed is direct-drilled if the slit has not been properly covered or if the seedbed is loose, cloddy and trashy. Populations	Wheat grain damaged by being eaten in the ground before it germinates. Young cereals can be completely grazed off by a severe autumn attack	Baits containing metaldehyde, thiodicarb or methiocarb spread evenly over the field prior to drilling. Extra cultivations in preparing the seedbed help to check the pest. Rolling and early	Winter wheat chiefly attacked. Often worse after oilseed rape

Table 6.1 (*Cont'd*)

Crop attacked	Pest	Description	Life cycle	Symptoms of attack	Control	Notes
			and damage are highest on heavy textured soils		nitrogen top-dressing can help a damaged crop to recover	
Stored grain	**Saw-toothed grain beetle**	*Adult:* Dark brown, 3 mm long *Larva:* White and flattened	Eggs are laid on the stored grain; larvae feed on the damaged grain. Pupation takes place in the grain or store	The grain heats up rapidly; it becomes caked and mouldy. This is seen with the appearance of the beetles	Clean store thoroughly before use. Pirimiphos-methyl can be applied as the grain is fed into the store. Check store temperature regularly	
	Grain weevil	*Adult:* Reddish-brown, about 3 mm long with an elongated snout	During autumn the weevils bore into the stored grain to lay their eggs. The larvae feed inside the grain where they also pupate	Hollow grains. Sudden heating of the grain. Weevils found a few feet below the surface of stored grain	See the saw-toothed grain beetle	
	Mites	Adults and larvae are too small to be seen with the naked eye	Eggs are laid on the stored grain. The larvae hatch out and go through several nymph stages. The time taken to	Causes grain to heat up. Can taint flour	Grain should be dried and cooled as quickly as possible. Grain should be stored at 14% moisture content or below.	Storage mites are highly allergenic

Crop	Pest	Description	Life cycle	Damage	Control	Remarks
			complete the life cycle is affected by environmental conditions		Chemical control as for other storage pests	
Maize	Frit fly	*Larva:* Whitish, up to 4 mm long	As for frit fly on oats	Twisting of leaves surrounding the growing point. In severe attacks this is killed, leading to secondary tillers	Can be controlled with a spray of chlorpyrifos	Maize after grass at greatest risk of damage
	Wireworm	See wireworm on cereals			Currently no approved seed treatments in maize	
Beans	Black bean aphid (black fly)	Very small oval body, black to green colour	There are many generations in the year. In summer winged females feed on the crop; wingless generations are then produced which continue to feed. Eventually a winged generation flies to the spindle bush on which eggs are laid for overwintering	On all summer host plants, colonies of black aphids are seen on the stem leaves (and on the flowers). The plant wilts; it can become stunted and with a heavy infestation it may be killed	Apply an insecticide, e.g. pirimicarb, when risk is high and or threshold populations reached	Mainly a pest of spring beans. It also attacks sugar beet and mangels. Populations vary each year. Warning systems are available. The threshold for treatment is 10% of plants infected on headland
	Bean beetle (Bruchid beetle)	*Adult:* 3–5 mm mainly black	Adult lays eggs on the surface of the	Circular holes in seed caused by emerging	Apply an approved insecticide, e.g.	

Table 6.1 (*Cont'd*)

Crop attacked	Pest	Description	Life cycle	Symptoms of attack	Control	Notes
		Larva: 6 mm creamy white with a brown head	developing pod. The larvae hatch and burrow a hole into the pod and enter developing seed. Larvae pupate *in situ*	adults. Value of seed for export, seed and human consumption are affected	deltamethrin when adults are active in the crop and the first pods are developing	
Peas and beans	**Pea and bean weevil**	*Adult:* light brown with lighter stripes, 6 mm long. *Larva:* White with brown head	During early spring eggs laid in the soil near plants. Larvae feed on roots, whilst adults feed on leaves. Pupation takes place in the soil in midsummer	Seedling crops checked. U-shaped notches at the leaf margins caused by adults. Larvae eat root nodules	Some pyrethroid sprays (e.g. deltamethrin) are approved for use when adults first appear	Spring-sown crops are more affected than winter-sown. A monitoring system is now available to give recommendations when spray treatments are warranted
Peas	**Pea moth**	*Adult:* Dull greyish-brown, about 6 mm long. *Larva:* Yellowish-white with darker head; up to 14 mm long	Eggs laid June–mid-August, hatch in a week. Larvae enter pods and feed on peas until fully grown. Larvae leave pod and make way to soil; pupate in spring and adult emerges in early summer	Holes in peas caused by larvae. Heavy infestations can significantly affect yield and quality	One or more sprays of a synthetic pyrethroid insecticide. The timing is important. There is a pea moth monitoring service available. Growers can use pheromone trap to aid spray decision-making	1 Early and very late sown crops suffer less damage. 2 Dried peas for harvesting most vulnerable but only worth treating those for human consumption or seed

Crop	Pest	Description	Life cycle	Damage	Treatment	Threshold
	Pea aphid	Large green aphid up to 4 mm long	Adults and eggs overwinter on leguminous crops. Adults migrate to peas from mid-May. Numbers peak in June/July	Direct feeding causes distortion of plant and pods and direct yield loss. Can also transmit virus	Treatment with an approved pyrethroid is worth-while only if thresholds reached	A predictive model has been developed. Treat when 20% of plants affected
Oilseed rape	Cabbage stem flea beetle	*Adult:* 4 mm, metallic green/black *Larva:* Up to 8 mm, body white with black head	Adult moves into rape crop in September and lays eggs. Larvae invade plant from October to March	Larvae burrow into leaf petioles and cause stunting and leaf loss. In severe cases can cause plant loss	Use an approved pyrethroid	Treat when 5 or more larvae per plant across field or 50% of leaf petioles showing damage. On backward crops treat when 3 or more larvae are found per plant or 30% of plants have damaged petioles
	Pollen beetle (Blossom beetle)	Metallic-greenish black in colour, up to 3 mm long	Adults emerge from hibernation during spring to feed on buds and flower parts. Eggs laid and similar damage caused when larvae emerge	Damaged buds wither and die, and the number of pods set is reduced	Treat with approved pyrethroid if threshold numbers reached. Best results obtained by treatment at green/yellow bud stage. Extreme caution should be taken to ensure that no serious damage occurs to pollinating insects	Spring oilseed rape is most susceptible. Spray thresholds are more than 15/plant on a normal crop or 5/plant on backward crops

Table 6.1 *(Cont'd)*

Crop attacked	Pest	Description	Life cycle	Symptoms of attack	Control	Notes
	Bladder pod midge	*Adult:* 2 mm greyish-brown *Larva:* Whitish-cream up to 2 mm	Females can insert eggs in pods through holes in pods made by seed weevil. Larvae feed on developing seed and walls of pod; several larvae are found per pod. The larvae pupate in soil after 4 weeks	Adults lay eggs in pods; larva cause pods to swell and ripen prematurely. Seed is shed early	As for seed weevil. A wide rotation will help reduce build-up of this pest	Winter oilseed rape most affected
	Cabbage seed weevils	*Adults:* Lead-grey in colour, about 3 mm long *Larva:* Up to 5 mm, creamy white with a brown head	From hibernation near previous year's seed crops adults lay eggs in young pods. Larvae feed on seeds in developing pods; they leave the pods and fall to the ground where they pupate in the soil	Seeds destroyed in pods by larva. Damage also caused by adult which makes holes for the pod midge to enter	Spray with an approved insecticide, e.g. alpha-cypermethrin, when one or more weevils per plant is seen at flowering for autumn-sown crops, and late yellow bud stage for spring crops	As for pollen beetles
Brassicae (cabbage, kale, swedes,	**Flea beetle** (Various species)	*Adults:* dark coloured 1.5–3 mm Jump like a	Adults emerge from hibernation during late spring to feed on crops. Eggs are laid,	Very small round holes are eaten in the cotyledons, first true leaves and stems of	Sow the crop either early or late, i.e. avoid April and May. Good growing	Sugar beet, mangels and oilseed rape can be attacked on occasions

turnips)	flea	but larvae do little damage. Pupation takes place in the soil during the summer	the plants. In severe cases plants are killed. Damage worst in dry conditions	conditions to get the crop quickly past the seed leaf stage. Only a very limited number of insecticides are approved for use, e.g. deltamethrin	A very serious pest of vegetables. A weather-based forecasting system is now available
Cabbage root fly	*Adult:* 6-7 mm grey/black *Larva:* Up to 10 mm creamy white	There are usually 2 or 3 generations a year. First generation eggs are laid during May and second and third generations during July and September	Affected plants are often stunted and may even die especially if invaded when young	Very few chemicals are now approved. Many chemicals have been withdrawn recently. Chlorpyrifos can be used in cabbages but not swedes or turnips. Crop covers can be effective on small areas	
Sugar beet, fodder beet and mangels **Mangold fly**	*Adult:* Greyish brown 5–6mm *Larva:* Yellow-white, up to 8 mm long	White oval-shaped eggs are laid in the underside of leaves in May. Larva bore into the leaf tissue and after about 14days they drop onto the soil where they pupate	Blistering of leaf which can become withered. Retarded growth and in extreme cases death of the plant	Good growing conditions to help the crop pass an attack. Imidacloprid seed treatment is very effective against this pest. Post-em sprays include lambda-cyhalothrin	This pest is now not so common. Spray threshold is when the number of eggs and larvae exceed the square of number of leaves
Aphids (black	The green-fly	During spring	A severe infestation	Imidacloprid seed	For black bean aphid

Table 6.1 *(Cont'd)*

Crop attacked	Pest	Description	Life cycle	Symptoms of attack	Control	Notes
bean and **peach/potato aphid)**		(peach potato aphid) has an oval-shaped body of various shades of green to yellow	winged aphids migrate to the summer host crops. They move from one plant to another, thus transmitting the virus from an unhealthy to a healthy plant. Black bean aphids do not introduce virus but spread what is already present	can cause the death of the plant, but chiefly it will mean a bad attack of virus yellows	dressing is currently a very effective treatment. Pirimicarb can be applied post emergence against non-resistant populations	the threshold for spraying is when 10% of plants are affected by aphid colonies
Beet cyst nematode		See cereal cyst nematode		Crop failing in patches. Plants which do survive are very stunted in growth. Yield losses can be very high	Wide rotations are effective at reducing the risk	If neccessary the soil can be tested for nematodes
Wireworm		See wireworm on cereals		The roots of seedling plants are bitten off	Tefluthrin or imidacloprid seed treatments will give some protection	Wireworm is traditionally a problem in grass rotations. There are an increasing number of cases of wireworm in arable rotations on chalk soils

Crop	Pest/disorder	Cause	Life cycle	Damage	Control	Notes
	Docking disorder – migratory nematodes	Caused by needle and stubby-root nematodes	The free-living nematodes move to germinating seedlings. Nematodes are most active in moist soils	Causes irregularly stunted plants with fangy root growth or even root galls. Damage usually in patches or along rows. A problem of sandy soils	Improve soil structure. Apply a granular insecticide at sowing, e.g. aldicarb	About 15% of national crop at risk
Carrots	**Carrot fly**	*Adult:* About 7 mm long; shiny black body with reddish/brown head and yellowish wings. *Larva:* When fully grown creamy-white, 8–10 mm long	Usually two generations a year. Eggs laid in soil surface near carrot in April/June or July/September. After hatching the larvae eventually burrow into root, forming 'mines'; after third moult pupate in soil close to tap root. Some of the second, and possibly a third, generation over-winter in the roots, emerging the following spring	Brown and rusty tunnels (mines often protruding) becoming progressively worse as season proceeds. In severe cases plants may be killed. Makes the mature carrot unmarketable	Rotation and avoid sowing susceptible crops close together. Hygiene round edge of field to cut down shelter for adult. Some varieties are partially resistant. Crop covers are an option on small areas. Sowing late to avoid the first generation or harvesting early help reduce the damage. The number of recommended insecticides is being dramatically reduced. There are a limited number of off-label approvals, e.g. Lambda-cyhalothrin	Celery, celeriac, parsley and parsnips also attacked. A weather-based forecasting system is available to optimise the effectiveness of the insecticide programme

Table 6.1 *(Cont'd)*

Crop attacked	Pest	Description	Life cycle	Symptoms of attack	Control	Notes
Potatoes	**Potato cyst nematode (PCN)**	Two species present in UK –white and yellow	These nematodes spend most of their life cycle below ground in root tissue. They tend not to be very mobile. Cysts in soil hatch and nematodes invade roots. New cysts formed from fertilized females	The biggest pest problem in potatoes. Stunted plants with restricted 'hairy' root system	An integrated approach needed. Where soil analysis indicates, grow resistant varieties. Wide rotations help keep cyst levels low. The effectiveness of nematicides, e.g. fosthiazate, depends on nematode populations and type of product. The use of trap cropping is being investigated	Regular soil sampling for cysts should be undertaken to aid the management and control of this major pest. Seed potatoes cannot be sold in the UK unless land guaranteed free from PCN
	Peach potato aphid (green fly)	See aphids on sugar beet, fodder beet and mangels		A bad infestation (5 aphids per compound leaf) will check the growth of the plant. The main problem is virus spread (PLRV and PVY) in seed crops by aphids	Appling a granular insecticide at planting to control nematodes, e.g. oxamyl, may help to check the aphids. Aphicide sprays can be applied to the growing crop, e.g. pirimicarb and/or lambda-cyhalothrin	Currently there are three types of resistance mechanisms in populations of this aphid to insecticides. Care must be taken in chemical choice and programme

Pest	Description	Damage	Control	Notes
Wireworm	See wireworm on cereals	Maincrop tubers are riddled with tunnel-like holes	Sow early-maturing varieties. Lift the crop in early September if possible	Pre-crop sampling or bait trapping can help identify a potential problem
Slugs	See slugs on cereals	Maincrop potatoes damaged by pests eating holes in the tubers. Crops grown on heavier textured soils are at greater risk	Some varieties are more susceptible than others. If slugs are active in late July/August it may be worthwhile applying pellets on the ridges. Early lifting will help reduce damage	See NIAB recommended list for variety susceptibility ratings
Cutworm	*Adult:* Several moth species but mainly the turnip moth *Larva:* Caterpillars up to 35 mm long; vary in colour–dull/greyish brown to green	Eggs are laid on weeds and hatch in approx. 3 weeks. After early feeding on leaves, caterpillars go into soil and feed (mostly at night) on stem above and below ground level. Most damage is caused in July. Most, but not all, are fully fed in the autumn, and overwinter to pupate in the spring in the soil	Plants cut off at base of stem at soil level, plants wilt and can die. Larvae eat out large hollows in roots. Damage worst on light soils in very dry conditions	Keep land free of weeds to reduce egg laying. Irrigation can reduce the problem in dry years. A number of insecticides are approved, e.g. chlorpyrifos. Treat when the caterpillars are small and still feeding above ground	Also attacks carrots, leeks, turnips, onions, lettuce, parsnips and red beet. A cutworm warning system is available

Table 6.1 *(Cont'd)*

Crop attacked	Pest	Description	Life cycle	Symptoms of attack	Control	Notes
Grass	**Leather-jackets**	See leather-jackets in cereals		Grass dying off in patches, the roots having been eaten away. Larvae found in soil	Sowing new seeds before middle of August and having a consolidated seed bed reduce the problem. Spray with chlorpyrifos if risk of damage high	Field populations can be assessed. The threshold numbers for economic damage is one million/ha
	Frit fly	See frit fly on cereals	The third generation larva can reduce the chances of a successful establishment of some autumn-sown grass seed mixtures		When reseeding, allow at least 4 weeks from the destruction of the old sward to the sowing of the new ley. Chemical control with chlorpyrifos spray	Seedling Italian ryegrass is more susceptible than perennial ryegrass
	Slugs	See slugs on cereals		Seed destruction below the ground is particularly serious	Slug pellets, e.g. methiocarb, broadcast at least 3 days before drilling	Direct drilled grass is most at risk
	Wireworm	See wireworm on cereals		Base of plant chewed just below ground level; plant wilts and	Damage least when time interval between ploughing and	

Red clover	Stem nematode	*Adult*: Slender and colourless, 1.2 mm long	Lives and breeds continuously in plant; passes into soil to infect other plants. Can remain dormant in hay made from an infected crop, becoming active again when conditions are suitable	turns yellow	Thickening at base of stem; some distortion of leaves, petioles and stems. Plants are stunted; infested patches increase in size each year	drilling is small. No chemical treatments are currently available	Some varieties of red clover show resistance. Rotation – several years' break from red clover	Most forage legumes are affected including white clover and lucerne. There are a number of distinct races. Some legumes are host to more than one race

7

Diseases of farm crops

7.1 Introduction to plant disorders

The main agencies of disease are fungi, bacteria and viruses. Mineral deficiencies and physiological disorders are often classified as diseases and are included in this chapter. Diseases, like pests, annually cause millions of pounds worth of damage and loss to the agricultural industry. Diseases can significantly reduce crop yields and quality as well as increasing losses of produce in store.

1 *Parasites*

The organisms that cause plant diseases (pathogens) are called *parasites* as they obtain their food from the infected crop plant. There are several types of parasites:

(a) *Obligate parasites* (or *biotrophs)* are dependent on the living host; they are responsible for causing many plant diseases. They can only grow and reproduce in living hosts. Examples include the rust and mildew fungi, and viruses.

(b) *Non-obligate parasites* can live on either living or dead tissue (most fungi and bacteria).

(c) *Facultative parasites* or *semi-parasites* kill the host tissues and live on the dead cells and are sometimes called *necrotrophs.*

2 *Saprophytes*

These live on dead organic matter and are often present in plants attacked by parasites or plants that have reached maturity and died. They are also found on leaves coated in aphid honeydew. Saprophytes play an important part in helping to break down plant remains into organic matter.

Most pathogens only attack one crop species or family, only a limited number have a wider species range. For a pathogen to be successful it must be able to invade and colonise a plant, grow and reproduce, be able to spread effectively and have some method of survival without the host. This is called the disease life cycle. Each pathogen has a distinct cycle. It is important to understand the life cycle so that the most effective methods of control can be used at the optimum time.

Conditions have to be right for diseases to develop, so not every year is the same disease important. Disease development is affected by:

1   The host plant. The vigour and growth stage can affect infection.
2   The pathogen. Amount and type of infection material and ability to colonise the host plant.
3   The environment. Each pathogen has specific requirements (temperature, moisture/rainfall and humidity) for infection, colonisation, reproduction and spread (see Fig. 7.1 for cereal disease incidence in England and Wales).

7.2   Types of damage

Symptoms of many diseases are characteristic for that disease and crop. Often the disease can be identified without having to grow the disease in the laboratory. Some diseases can be identified visibly, i.e. by seeing the fungal growth or fruiting bodies as with powdery mildew and the rusts. For other diseases it is the type of lesion that helps identification, as follows:

Yellowing or chlorosis: Many of the viruses, such as beet virus yellows, cause leaf chlorosis as do some mineral deficiencies. Where the yellowing is on the leaves, the shape of the infected area may help with identification.
Death of tissues/necrosis: Many diseases can cause the death of the plant tissue. Again, size and shape of the lesions are important. Leaf spots are well-defined lesions and sometimes the dead tissue falls away to leave a shot-hole effect. Leaf blotches tend to be of variable size.
Abnormal growth: Some diseases cause the infected tissue to enlarge either by an increase in cell numbers or size, e.g. potatoes with common scab.
Stunted growth: This is seen in severe cases of barley yellow dwarf virus and with some root diseases such as take-all.
Wilting: There are a number of bacterial and fungal diseases that cause wilting, normally late on in the development of the disease.
Tissue disintegration: This type of damage is associated with many of the root or foot rots and storage rots. The affected cells are broken down and release liquid (wet rot) or become dry and brittle (dry rot).

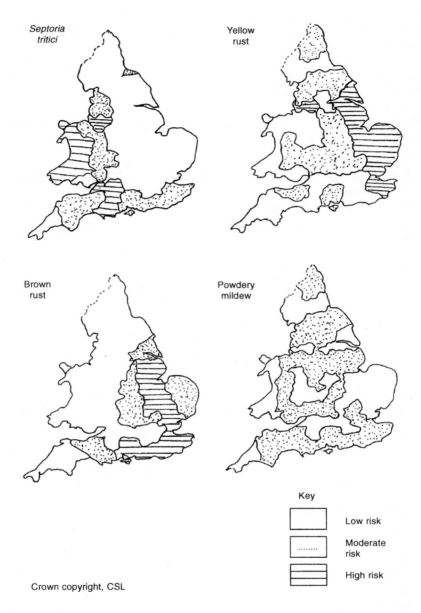

Fig. 7.1   Generalised maps of England and Wales showing variation in risk of foliar diseases in wheat.

7.3   Some important types of pathogens

7.3.1   Fungi

There are many thousands of different species of fungus, the majority of which are invisible to the naked eye and it is only a small number that are important as plant diseases.

A typical fungus is composed of long thin filaments (made up of single cells) termed hyphae. Collectively, these are known as mycelium. Fungi enter the host plant surface either by direct penetration of surface cells (main form of entry) or through wounds or natural openings. Not all fungi that enter plants develop into a disease, sometimes the fungus is killed within the host plant, leaving a visible necrotic spotting or hypersensitive reaction.

Once within a susceptible plant the fungus can grow and spread between cells. Some fungi produce extensions to the hyphae which enter the host cells and act as nutrient absorbing organs (haustoria). In the case of most parasitic fungi, the mycelium is enclosed within the host (only the reproductive parts protruding), although some fungi are only attached to the surface of the host, e.g. powdery mildew. Sometimes, even after a fungus has entered the host, plant symptoms may not be visible at first. This lag time is called the latent infection period and once the symptoms have appeared it can be almost too late to treat with a fungicide.

7.3.1.1   Life cycle

Reproduction

Fungi can reproduce simply by fragments of the hyphae dropping off, but usually reproduction is by production of spores. Spores can be compared to the seeds in ordinary plants, but they are microscopic and occur in vast numbers. Asexual spores are usually important for the spread and increase of a population or epidemic. Sexual spores are important for fungal survival.

Fungal survival and dispersal

Not all fungi have the same mechanisms for spread and survival; this knowledge will help in deciding on disease prevention and control methods.

1   *The seed*. Some fungi are carried from one generation to the next by surviving on the seed coat or inside the seed itself, e.g. smut diseases of cereals.
2   *The soil*. The spores or even resting bodies such as sclerotia (which are a dense mat of hyphae) drop off the host plant and remain in the soil until another susceptible host crop is grown in the field. A suitable rotation will go a long way to check these diseases such as those caused by Sclerotinia sp.
3   *The air*. Spores of many foliar diseases are dispersed on dry air currents. Spores of some of the cereal rusts have been known to travel hundreds of

kilometres in air currents and under favourable conditions can cause widespread epidemics.

4   *Water* is important for the dispersal of some fungal spores over short distances.

5   *Alternate hosts.* Some diseases have the ability to overwinter on alternative hosts as seen with some of the rusts.

6   *Infected plant material.* Many diseases survive from one season to another either on plant debris or volunteer crops. Potato blight often starts from infected potato dumps.

7.3.2   Viruses

The virus was discovered in the nineteenth century. It is a very small organism. Approximately one million viruses could be contained on an average bacterium. Only by using electron microscopes can it be seen that plant viruses have a kind of crystalline form. All viruses are obligate parasites. They are not known to exist as saprophytes. In many, but not all, virus diseases the actual disease is not transmitted through the seed. Examples of exceptions to this are the bean and pea mosaic, and the lettuce mosaic virus disease.

The virus is generally present in every part of the infected plant (apical meristem tissues are often not infected) except the seed. Therefore, if part of that plant, other than the seed, is propagated, the new plant is itself infected, e.g. the potato. The tuber is attached to the stem of the infected plant, and infection is carried forward when the tuber is planted as 'seed'.

In most plant virus diseases, the infection is transmitted from a diseased to a healthy plant by vectors; insects are the main vectors although nematodes and mites are known to transmit some virus diseases. A fungus can also be a vector, e.g. barley yellow mosaic virus (BYMV) transmitted by the soil-borne *Polymyxa graminis;* rhizomania in sugar beet transmitted by *Polymyxa betae.* Viruses can overwinter on perennial infected crops that might not show any visible symptoms and on volunteer crops. One of the main methods of survival is in alternate host plants such as weeds. Viruses tend to have a fairly wide host range. The host species must also be suitable for the vector. The only way viruses can survive in the soil is in a living organism such as a nematode or fungus.

7.3.3   Bacteria

Bacteria are very small organisms, only visible under a microscope. They are of a variety of shapes, but those that cause plant diseases are all rod-shaped. About 12% of the identified bacteria are plant pathogens. Like fungi, bacteria feed on both live and dead material. Although they are responsible for many diseases of humans and livestock, in crop plants they are of minor importance compared with fungi and viruses as causal agents of crop diseases.

Bacteria overwinter in a similar way to fungi except that no resting spores or bodies are produced. They can survive in infected plants, seeds and tubers, in plant debris and, just a few, in the soil. They mainly enter plants through wounds or occasionally through openings like the stomata. Bacteria, causing wet rots, often gain entry into their host root crops after mechanical damage during harvest. Once in the host plant bacteria can reproduce themselves simply by the process of splitting into two. Under favourable conditions this division can take place about every 30 minutes. Thus bacterial diseases can spread very rapidly, once established.

7.4   Other disorders

7.4.1   Lack of essential plant foods (mineral deficiency)
When essential plant foods become unavailable to particular crops, deficiency symptoms will appear. These symptoms, such as chlorosis and necrosis, often look like a foliar disease. Most of the diseases are associated with a lack of trace elements, but shortage of any essential plant food will certainly reduce the yield, affect crop physiological process (depending on where the nutrient is used) and make the crop more vulnerable to pest and disease attack. (See Chapter 3.)

7.4.2   Physiological diseases (stress)
These are often triggered by adverse environmental conditions which can upset the normal physiological processes of the plant. Normally this is only temporary, but there may be occasions when the effect is more permanent.

- Temporary conditions, e.g. a high water table in the early spring. This will cause yellowing of the cereal plant as its root activity is restricted, considerably reducing its oxygen and plant food intake. When the water table falls, the plant is able to grow normally once more, assuming a healthy green colour.
- Permanent conditions, e.g. where the soil has become compacted the root activity of the plant can be restricted. This will result in poor stunted growth with the plant far more vulnerable to pest and disease attack. The yield will certainly be reduced.

Periods of very dry conditions followed by wet weather can cause root cracking and splitting in crops such as carrots and potatoes; it can also cause secondary tuber production in potatoes.

Hail damage can often be confused with a disease. It can cause leaf and flower spotting and even damage the flower parts. In cereals hail and heavy rain at flowering have been known to increase the number of blind grain sites in the ear.

7.5   The control of plant diseases

Before deciding on control measures it is important to know what is causing the disease. Once identified then knowledge of its life cycle and factors affecting the outbreak and spread (*epidemiology*) will help decide on the most appropriate method of control. Depending on the disease (see Table 7.1 at the end of this chapter) there are many non-chemical methods of control that will help reduce the risk before use of any agrochemicals are required. With the increase in number of cases of fungicide resistance, it is more important than ever that alternative control methods are included in the programme in order to maintain the effectiveness of currently available chemicals.

7.5.1   Crop rotation

A good crop rotation can help to avoid an accumulation of the parasite. In many cases the organism cannot exist except when living on the host. If the host plant is not present in the field, in a sense the parasite will be starved to death, but it should be remembered that:

1   Some parasites take years to die, and they may have resting spores in the soil waiting for the susceptible crop to be planted again, e.g. club root of the brassica family.
2   Some parasites have alternative hosts, e.g. the fungus causing take-all of wheat is a parasite on some grasses.

7.5.2   Soil fertility

A crop under stress from low nutrient status can be more prone to disease attack. Too lush a growth can also encourage disease. Early application of nitrogen in winter cereals can cause increased infection from foliar diseases.

7.5.3   Seedbed

Puffy seedbeds can increase the risk of take-all. Over-consolidation or compaction can lead to poor growth and more disease.

7.5.4   Crop hygiene

Diseases should be discouraged by avoiding sources of infection on the farm, e.g. blight from old potato clamps, and virus yellows where sugar beet was stored prior to being sent to the factory. Good stubble cleaning will minimise certain cereal diseases being carried over from one cereal crop to the next year's cereal crop. Some parasites use weeds as alternative hosts. By controlling the weeds, the parasite can be reduced, e.g. cruciferous weeds such as charlock are hosts to the organism responsible for club root.

7.5.5   Clean seed
Seed Certification Schemes ensure that seed lots meet strict standards for presence of seed-borne diseases. These schemes have been very successful at reducing the incidence of some diseases such as bunt or stinking smut in cereals. Home-saved seed should be tested to avoid poor crop establishment and crop growth due to seed-borne disease.

With potatoes it is essential to obtain clean 'seed', free from virus. In some districts where the aphid is very prevalent, potato seed may have to be bought each year. Under the Seed Certification Scheme it is mandatory to have the field inspected and the seed virus tested.

7.5.6   Resistant varieties
In plant breeding, although the breeding of resistant varieties is better understood, it is not by any means simple. For some years plant breeders concentrated on what is called single or major gene resistance. However, with few exceptions, this resistance is overcome by the development of new races of the fungus to which the gene is not resistant.

Breeding programmes are now concentrating on multigene or 'field resistance', which means that a variety has the characteristics to tolerate infection from a wide range of races with little lowering of yield. Emphasis is now on tolerance rather than resistance.

Variety resistance is the main method of control for some fungal-transmitted viruses such as barley yellow mosaic virus.

7.5.7   Varietal diversification
The risk in any year of a serious infection can be reduced in adjacent fields by choosing cereal varieties with resistance to different races of the disease. The National Institute of Agricultural Botany (NIAB) produces variety diversification tables for yellow rust in wheat and mildew in barley.

7.5.8   Variety mixtures (blends)
This is a natural extension of diversification in that varieties from different diversification groups are grown together in the same field. In this way a disease-carrying spore from a susceptible variety, but within a blend of varieties (two or three varieties) making up the crop, has less chance of successfully infecting a neighbouring plant than in a pure crop. Yields of blended crops can be more reliable than pure crops, and this may be achieved with the use of fewer fungicides.

7.5.9   Time of sowing
Early-drilled winter cereals are more likely to be infected by eyespot, take-all and some foliar diseases.

Early-drilled spring barley is more susceptible to *Rhynchosporium*, but late-drilled barley is more susceptible to mildew.

7.5.10   Crop density

Many rain splash diseases, and those that require a high humidity, are more common where there is a high plant population compared with a more open crop with a low population.

7.5.11   The control of insects/virus vectors

Some insects are carriers of parasites, causing serious plant diseases, e.g. control of the green fly (aphid) in sugar beet will reduce the incidence of virus yellows. Other virus vectors such as soil fungi are very difficult to control so that other methods such as use of resistant varieties is the main option for arable farms. (Soil sterilisation will kill both nematode and fungal vectors but is only used on high value horticultural crops.)

7.5.12   Legislation

A number of countries have introduced legislation to ensure that some rare diseases do not spread and become a serious threat. In the UK potato wart disease and rhizomania in beet are two examples. Any outbreaks must be notified to the DEFRA so that spread can be contained. Potato wart disease is now rarely found in commercial potato crops.

7.5.13   Fungicides

Over the last 30 years there has been a large increase in both the number of fungicides available and their subsequent use on crops.

Fungicides are used in the following ways:

1   The dressing of seed with a fungicide; this is carried out to prevent certain soil-borne and seed-borne diseases. In many cases an insecticide is added to help prevent attacks by soil-borne pests. Various fungicides can be used, depending upon the disease to be controlled and the crop.

2   Foliar application to the plant. Different treatment programmes involving the use of fungicides are now considered as an essential part of many crop production programmes.

Fungicides are used when it is considered that a specific disease has developed to a point (the economic threshold) which will actually cause a loss of yield that will pay for the cost of treatment and application. There are now many established thresholds for application of fungicides. Often the rate of the fungicide applied will be adjusted depending on the disease risk and variety resistance. This is called an appropriate dose rate. Some computerised models have been or are being developed to aid decision making, e.g. Decision Support System for Arable Crops (DESSAC).

With some diseases such as potato blight, treatment needs to be applied before symptoms are seen; treatment is based on blight warnings. These warnings rely on weather conditions and forecasts and the likelihood of disease developing.

7.5.13.1 Protectant fungicides

The first chemicals developed for control of fungal diseases were the inorganic compounds such as sulphur and copper compounds. These chemicals do not move in the crop plant (are non-systemic); they simply protect the crop plant from disease infection. They are called protectant fungicides. Good crop coverage is essential for this type of product. The chemicals affect a number of biochemical processes in the fungi so are called multi-site fungicides. Some of the first fungicides produced after these inorganic compounds, such as the dithiocarbamates, have very similar characteristics, e.g. they are protectant, multi-site fungicides. Though these chemicals are not as effective as some of the newer systemic compounds they still have some uses today, particularly in programmes where there is a high risk of disease resistance, e.g. control of potato blight.

7.5.13.2 Systemic fungicides

Systemic fungicides have been developed since the 1960s; they are now the most commonly used fungicides. On entry into the crop plant, they can move to a certain extent within the crop to the site of infection. These fungicides tend to affect a single biochemical pathway within the pathogen and are called site-specific.

Many systemic fungicides can be applied after the initial infection period, before symptoms appear (the latent period); these treatments are called curative. When disease symptoms are visible then an eradicant fungicide is required; these chemicals have the ability to eradicate a disease that is already present and then protect the plant for a certain time after application. Persistency, curative and eradicative activity varies between chemicals.

Some of the first compounds developed were the benzimidazoles (MBCs). Other major groups of systemic fungicides now include the triazoles (EBIs/SBIs and DMIs), morpholines, strobilurins and phenylamides. (See the individual crop chapters for further details on disease control programmes.)

7.5.13.3 Fungicide resistance

Disease resistance to fungicides is now widespread. It is a problem with the systemic products that act on one site only of the fungus. When a fungicide controls a fungal disease effectively, the fungus is 'sensitive' to the chemical. However, other strains of the fungus can and do occur over a period of time, and some of these may be resistant ('insensitive' or 'tolerant') to the fungicide which means that the disease is then not controlled adequately. In some cases once there is resistance it is total and the fungicide is not effective. An example of this type of resistance (single step) is seen in the control of eyespot with the MBC fungicides. In other cases there is a shift in the sensitivity of the fungus population to the fungicide and there is still some control. This has been found with control of mildew in cereals using many of the triazoles fungicides; it is called multi-step resistance.

Resistance builds up through the survival and spread of the resistant strains and it is speeded up by repeated application of the same fungicide treatment. Development of resistance can build up within a year or two of the fungicide being on the market as was found with the introduction of metalaxyl (a phenylamide) for control of potato blight. The more often the same chemical, or chemical in the same group, is used, the greater are the chances of resistant strains developing. There is an increased risk of this happening with fungicides which are site-specific in the fungus compared with multi-site fungicides. Resistance is also more frequent with the systemic rather than the non-systemic or protectant fungicides.

There are several ways of avoiding a build-up of resistance by a fungus or reducing the risk.

1   Combined application of fungicides with non-chemical methods to reduce disease risk.
2   Avoid growing large areas of very susceptible varieties in areas where disease incidence is usually high.
3   Where possible, use fungicides with different modes of action (i.e. from different groups) when more than one has to be used on the same crop.
4   Use approved tank mixtures of fungicides with different modes of action, rather than always relying on single fungicides.
5   Apply fungicides only when necessary; use disease forecasts and thresholds to avoid unnecessary treatment.

Unsatisfactory disease control following the use of fungicides is, at present, not always due to fungicide resistance. There are several other reasons, the main ones being wrong timing and the wrong chemical and the use of too low a dose and poor application.

Table 7.1 indicates the main diseases affecting farm crops and their control.

7.6   Further reading

Agrios G N, *Plant Pathology,* Academic Press, 1997.
Alford D V, *Pest and Disease Management Handbook,* Blackwell Science, 2000.
Bayles R A, Meadway M H and Mitchel A G, *Cereal diseases, potato diseases, peas and bean diseases and diseases of grasses and herbage legumes,* NIAB, 1991.
Bould C, Hewitt E J and Needham P, *Diagnosis of Mineral Disorders,* HMSO, 1983.

Brent K J and Hollomon, D W, *Fungicide Resistance: The Assessment of Risk,* Global Crop Protection Federation, 1998.

Carlile W R, *Control of Crop Diseases*, Edward Arnold, 1995.

Hewitt H G, *Fungicides in Crop Protection,* CABI publishing, 1998.

Lucas J A, *Plant Pathology and Plant Pathogens,* Blackwell Science, 1998.

Mathews G A, *Pesticide Application Methods,* Blackwell Science 2000.

Plant Physiological Disorders, Reference book 223, HMSO, 1985.

Smith I M, Danez J, Phillips D H, Lelliot R A and Archer S A, *European Handbook of Plant Diseases,* Blackwell Scientific Publications, 1988.

Parry D W, *Plant Pathology in Agriculture,* Cambridge University Press, 1990.

Trade guides including: *Cash Crops Diseases; Cereal Diseases; Field Crops Nutrient Disorders.*

Whitehead R, *The UK Pesticide Guide,* CABI publishing, published annually.

Appendix

Table 7.1 Major plant diseases and their control

Crop attacked	Disease	Causal agent	Symptoms of attack	Life cycle	Methods of control
Cereals	(1) **Bunt, covered** or **stinking smut** of wheat (2) **Covered smut** of barley. **Leaf spot** of oats (3) **Covered** and **loose smut** of oats	Fungus	Brown or black spore bodies with distinct smell replace the grain contents	Infected grain is planted; seed and fungus germinate together and thus young shoots become infected. The spores are released when the skin breaks, and so combining contaminates healthy grain	(1) Seed dressing, e.g. products containing guazatine, carboxin or fludioxynil
	(1) **Leaf stripe** of barley (2) **Leaf stripe** of oats	Fungus	The first leaves have narrow brown streaks. Subsequently, brown spots appear on the upper leaves	Infected grain is planted; seed and fungi germinate together and thus young shoots are infected. From the secondary infection, spores are carried to developing grain	(1) Seed dressing containing imazalil
	(1) **Loose smut** of wheat (2) **Loose smut** of barley	Fungus	Infected ears a mass of black spores. They do not remain enclosed within the grain as with the covered smuts	Similar to the covered smuts, but the fungus develops within the grain. The spores are dispersed by the wind to affect healthy ears	(1) Resistant varieties (2) Clean seed (3) The seed can be dressed with a broad-spectrum fungicide, e.g. triadimenol + fuberidazole

Disease	Type	Symptoms	Biology	Control
Yellow rust	Fungus	Yellow-coloured pustules in parallel lines on the leaves, spreading in some cases to the stems and ears. In a severe attack the foliage withers and shrivelled grain results	The fungus mainly attacks wheat though there are distinct races in barley and oats. Infection appears on the plant from May onwards. From the pustules, spores are carried by the wind to infect healthy plants. During winter, spores are normally dormant on autumn-sown crops. Cool, humid conditions favour disease	(1) Resistant varieties, although new races appear against which these varieties soon have no resistance (2) Fungicides, e.g. tebuconazole, cyproconazole, kresoxim-methyl + epoxiconazole (3) Variety diversification
Mildew	Fungus	On winter cereals, grey-white and brown fluffy mycelium on lower leaves in February. Infection spreads to other leaves and plants. Disease common between May and August. Early infected leaves go yellow and shrivelled. Towards the end of season black spore cases formed among brown fungi	From self-sown cereals in stubble, winter and spring cereals can be infected. Specific races attack wheat, barley, oats and rye. Warm, humid (but not wet) weather favours disease	(1) Clean-up old stubble (2) Resistant varieties (3) Fungicides available, although field resistance to some fungicides is now common. Use products such as quinoxyfen, fenpropimorph or cyprodinil (4) Variety diversification
Rhynchosporium – leaf blotch of barley	Fungus	When fully formed lentil shaped blotches (light blue/green/grey with dark brown margins) up to 19 mm long are seen on the leaves. As disease progresses, blotches coalesce	A disease of barley, rye, triticale and oats. Fungus overwinters on self-sown barley plants and on winter barley crops. From here spores are carried to planted barley crops. Disease is spread by rain splash	(1) Clean stubble of all self-sown barley plants (2) Many systemic fungicides are available including epoxiconazole or fenpropimorph

Table 7.1 (Cont'd)

Crop attacked	Disease	Causal agent	Symptoms of attack	Life cycle	Methods of control
	Net blotch of barley	Fungus	Short brown stripes on older leaves; on younger leaves and adjacent plants, in addition to the striping, irregular-shaped dark brown blotches 'spots' or 'netting' symptoms. Can spread to ears	Can be seed-borne but mainly spread from previously infected crop or volunteers. Disease encouraged by cool, wet weather	(1) Seed dressing (2) Crop hygiene to clear stubble of previously infected crop (3) Fungicides, e.g. propiconazole, flusilazole and azoxystrobin
	Halo spot of barley	Fungus	Small pale oval lesions with a dark brown border, scattered across the leaf. Black fruiting bodies (pycnidia) found in lesions	Fungus survives in seed, plant debris and on volunteer cereals. Cool moist conditions favour the disease	Many of currently available fungicide seed treatments and foliar sprays can help suppress the disease
	Brown rust of barley and wheat	Fungus	Numerous, very small individual orange-brown pustules on the leaves. Normally does not develop significantly until the summer. A severe attack causes shrivelled grain	Different species in wheat and barley. The resting spores overwinter on volunteers. From here the pustules are air-borne to infect healthy plants. Encouraged by warm, humid weather; spores can be spread by wind	(1) Crop hygiene to clear stubble of volunteer plants (2) Resistant varieties (3) Many systemic fungicides, e.g. kresoxim-methyl + epoxiconazole
	Crown rust of oats	Fungus	Orange-coloured pustules spread mainly on leaf blade. Later in season black pustules	Only affects oats, a distinct race affects other grasses. Spores are air-borne from	(1) Crop hygiene to clear stubble of volunteer plants

Disease	Type	Symptoms	Life cycle / conditions	Control
		are produced. Severe attack prior to, and including, milk-ripe stage, causes shrivelled grain	overwintering volunteer plants, and winter oats, to the spring crop. Disease encouraged by warm, humid conditions	(2) Keep winter and spring oat crops as far apart as possible (3) Some varieties show reasonable resistance (4) Some conazole and morpholine fungicides give good control
Septoria tritici-leaf spot of wheat	Fungus	Mainly a foliar disease. Oval, bleached/grey discoloured blotches of varying sizes and shapes (on which appear minute black dots or fruiting bodies (pycnidia), seen from late autumn to early summer. Infection of the ear is rare	Leaf spores are liberated in wet weather, and they over-winter on volunteer crops and then transfer to winter cereals. Symptoms can develop 3–4 weeks after the infection period	(1) Crop hygiene to clear stubble of volunteer plants (2) Clean seed (3) Tolerant varieties (4) Rotation (5) Many systemic fungicides, e.g. fluquinconazole or kresoxim-methyl + epoxiconazole
Septoria leaf spot and **glume blotch** of wheat	Fungus	Disease (other than seedling blight) does not develop until midsummer. Leaf lesions start as oval yellow blotches. Fruiting bodies are difficult to see without magnification. Glume blotch becomes prominent in July and August. Irregular, chocolate spots or blotches on glumes, beginning at the tips; later ears become blackened with	Glume blotch has a similar life cycle as *S. tritici* but the fungus can also be carried on the seed. Weather conditions in May which favour *Septoria* are 3 or 4 days of consecutive rain, totalling at least 10mm, or one day with more than 5mm. Symptoms develop 10–14 days after an infection period	(1) Treat when weather conditions favour disease. Too late once symptoms present. Control as for *S. tritici*

Table 7.1 (*Cont'd*)

Crop attacked	Disease	Causal agent	Symptoms of attack	Life cycle	Methods of control
			secondary infection. Shrivelled grain results		
	Barley yellow dwarf virus (BYDV)	Virus	Stunted plants in patches or scattered as single plants. Poor root development. Affected leaves in oats are red, yellow in barley and bronze in wheat. Late heading and reduced yield. Yield losses greatest from autumn infection	Cereal aphid vectors carry the virus, e.g. bird cherry aphid and grain aphid. Volunteers and grasses act as source of infection. Early drilled cereals after grass are most susceptible	(1) Spray off previous crop/green stubble; this reduces aphid carry-over (2) In high-risk areas use seed treated with imidacloprid (3) Use a pyrethroid insecticide end October/early November. High-risk areas may require earlier treatment (4) Oats most susceptible, wheat the least. Some spring barley varieties have different susceptibilities to the disease
	Barley yellow mosaic virus (BYMV) and barley mild mosaic virus (BaMMV)	Virus	Appears in patches in field. Pale green streaks later turning brown, particularly at leaf tip; leaves tend to roll inwards, remain erect to give plant a spiky appearance. Symptoms	Soil-borne fungus. *Polymyxa graminis* is the vector. Disease encouraged by hard winter. Symptoms appear from January. Found only in autumn-sown	(1) No chemical control (2) On affected fields grow resistant varieties (3) Reduce the spread of contaminated soil around the farm

Disease	Type	Symptoms	Cause	Control
		can disappear during warmer weather but the plants are still stunted and late to mature	barley	(4) Widen the interval between barley crops to reduce the risk
Wheat soil-borne mosaic virus (WSBMV)	Virus	Plants are stunted with yellow streaks down the leaf and leaf sheaths. Yield losses can be very high	*Polymyxa graminis* is also the vector for this virus. Disease worst in wet conditions	A new potential threat to winter wheat, rye and triticale in the UK. Control as BYMV
Eyespot	Fungus	Eye-like lesions on stem about 75 mm above ground. Grey mycelium inside stem; straws lodged in all directions. No darkening at base of stem. White heads at harvest. Two strains of eyespot present in the UK	The fungus can remain in the soil, on old stubble and some species of grasses for several years. It usually attacks susceptible crops in the young stages. Early drilled crops most susceptible. Cool moist conditions favour this disease. Disease spread by rain splash	(1) Diagnostic tests available to aid risk assessment (2) Avoid growing continuous cereals (3) Apply a systemic fungicide, e.g. prochloraz at GS 30 or cyprodinil at GS 32
Sharp eyespot	Fungus	Lesions more sharply defined than true eyespot. Brown/purple border followed by cream-coloured area and brownish centre. Lesions more numerous and occur further up stem than true eyespot. Pink/brown mycelium may be found inside the stem. Can cause lodging, 'whiteheads' and shrivelled grain	The fungus is soil-borne but is also found on plant debris. Tends to be more severe on light soils and on early drilled winter crops. Not usually affected by crop rotation	Mainly a problem in winter wheat and barley; though oats and rye can be affected. (1) Dispose of crop residues (2) Delay drilling winter cereals (3) Azoxystrobin and prochloraz may give some disease suppression

Table 7.1 (*Cont'd*)

Crop attacked	Disease	Causal agent	Symptoms of attack	Life cycle	Methods of control
	Fusarium - brown foot rot* and *ear blight	Fungus	Emerging seedlings can be killed. Undefined brown discoloration at base of tillers and lower leaf sheaths, dark brown nodes. Interior of stem and/or grain shows pink fungal growth with some species. Premature ripening and 'whiteheads' or 'blind' ears	A number of *Fusarium* spp. but infection is either seed-borne or from previously infected stubble and crop remains	(1) Stubble hygiene (2) Seed treatment using e.g. fludioxynil or carboxin (3) Foliar fungicides, e.g. metconazole or tebuconazole
	Take-all	Fungus	Black discoloration at base of stem and roots. Infected plants ripen prematurely, in severe cases the ears are bleached and contain little or no grain	Wheat is most affected though barley, rye and triticale can be infected. The fungus survives in the soil in root and stubble residues and on rhizomatous grasses. The host plant is infected when it is grown in the field	(1) Rotation to starve the fungus, but after some years of continuous wheat growing infection appears to lessen (2) Extra nitrogen helps the growth of new roots (3) Bad drainage reduces plant vigour and it is more easily damaged (4) Direct drilling appears to lessen the intensity of the disease
	Pink snow mould (mainly of	Fungus	Patchy crop in autumn. Stunted seedlings occasionally with white/pink mycelium at	A seed-borne fungus, but contamination of crop is also possible from the old	(1) Good stubble hygiene (2) Clean seed (3) Seed treatment, e.g.

wheat and rye)		base. After snow has thawed withered plants in patches temporarily covered by white-pink mould. Thereafter, infected plants stunted, weak root system and shrivelled grain	stubble and other plant debris. Fungus is particularly favoured by low temperatures	using triadimenol with fuberidazole
Snow rot (***Typhula rot***) mainly of barley	Fungus	Thin, poorly tillered crop; plants yellowing and withered in patches. Old leaves often covered by white and pink mycelia and brown resting spores. Young leaves standing erect but eventually yellowed. Weak root system	Soil-borne fungus which can remain dormant for years; when active, infects emerging cereal plants, developing rapidly in dark and humid conditions, i.e. under snow. Continuous winter barley increases the risk of this disease	(1) Sow early in the autumn (2) Some varietal resistance (3) Seed treatment, e.g. triadimenol with fuberidazole (4) Foliar spray with azole fungicide in the autumn
Ergot	Fungus	Hard black curved bodies up to 20 mm long replacing the grain, and protruding from the affected spikelet	Ergots fall to ground and remain until next summer when they germinate and produce short stems with globular heads containing the spores which are then air-borne to the cereal flowers, and certain grasses, depending on the species. Open-flowering cereals most affected, e.g. rye. Infection in oats is rare	Although disease has little effect on yield, ergot is poisonous to mammals (but it does contain medicinal properties). Crop rotation and control of grass weeds (especially blackgrass) in the crop will help. Not considered important enough for special control measures. No ergot is permitted in grain for milling, and one piece only is permitted in Basic seed

Table 7.1 (*Cont'd*)

Crop attacked	Disease	Causal agent	Symptoms of attack	Life cycle	Methods of control
	Manganese deficiency of cereals	Manganese deficiency	Yellowing on leaf veins followed by development of brown lesions. With oats the spots enlarge and can extend across leaf. Thus the leaf can bend right over in the middle. Older leaves wither and die. Can lead to shrivelled grains		Apply manganese sulphate. With a bad attack this may have to be repeated
Oilseed rape	**Phoma– stem canker** and **leaf spot**	Fungus	Beige coloured circular spots with distinct brown margin (0.5–1.0cm diam.) on leaves; spores spread to produce brownish/black canker at base. Stem splits and rots causing lodging; rapid stem elongation and premature ripening	Air-borne spores produced from the stubble carry infection to young crops in the vicinity. Fungus spreads from the leaves to stem where cankers develop. The fungus can affect all parts of the plant including the seed. Leaf symptoms seen from Oct to April	(1) Destroy stubble debris soon after harvest (2) Rotation – grow rape crops in the same field not more than one year in six (3) Use seed treated with iprodione (4) Tolerant varieties (5) Treat with a fungicide, e.g. flusilazole or tebuconazole as soon as leaf spot found in the autumn and/or spring
	Alternaria – dark leaf and **pod spot**	Fungus	Circular small brown-to-black leaf spots sometimes coalescing on leaves and later on pods.	Seed-borne disease although spores can be carried through the air from other	(1) No resistant varieties (2) Use a seed treatment, e.g. iprodione

		Premature ripening and loss of seed	infected brassica crops. Favoured by warm, humid conditions	(3) Foliar treatment from mid- to late-flowering with a fungicide, e.g. iprodione, tebuconazole or difenoconazole
Light leaf spot	Fungus	From January onwards, light green/bleached lesions surrounded by small white spore droplets. Plant population and seed yield can be severely affected	A trash-borne fungus. Disease is spread from plant to plant from rain splash during wet weather. Flower buds can become infected and killed at early extension stage	(1) Dispose of infected crop residues (2) Spray with a fungicide, e.g. cyproconazole or tebuconazole at first sign of the disease in the autumn and/or the spring (3) Some varieties less susceptible than others (4) A forecasting service is available to aid decision making
Stem rot – Sclerotinia	Fungus	Bleached skin lesions which contain black resting bodies (sclerotia)	Sclerotia bodies left in soil after harvest and can survive for 8 years. They germinate in spring and produce spores which infect susceptible crops. Encouraged by warm, wet weather	(1) Wide rotation (2) Fungicides – at early flowering, e.g. prochloraz, vinclozolin or tebuconazole
Linseed **Alternaria**	Fungus	Damage affects seedlings as they emerge. Brick red lesions are found on stems and roots. Can also affect mature plant	A seed-borne disease. Can spread up mature plant during periods of wet weather	(1) Most effective method of control is seed treatment, e.g. with prochloraz

Table 7.1 *(Cont'd)*

Crop attacked	Disease	Causal agent	Symptoms of attack	Life cycle	Methods of control
					(2) Some evidence of varietal resistance (3) Foliar application of a number of fungicides have been effective against late infection
	Grey mould	Fungus	Attacks leaves, stems and seed capsules. Reddish browning on stem base. Plants then become covered in grey mould	A seed-borne disease encouraged by warm, moist conditions	(1) Sow certified seed (2) Seed dressings containing prochloraz can be effective (3) Low nitrogen rates and plant populations tend to reduce problem
	Pasmo	Fungus	Grey/black spots with black pycnidia found on leaves and stems	A disease of winter linseed. Can be seed-borne or spread from plant debris	(1) Sow clean seed 2) Some fungicides are effective when applied at mid-flowering
Peas	**Downy mildew**	Fungus	Greyish-white to a grey/brown mycelium on underside of leaf on young seedlings which usually die. A secondary infection shows as isolated yellow/green spots on the upper surfaces of the leaves.	The disease is soil- and seed-borne from where spores infect the growing point of seedlings. A secondary infection can follow, with spores spreading by air currents and rain splash to	(1) Seed treatment, e.g. using metalaxyl + thiabendazole + thiram (2) Some varieties are more resistant

Disease	Cause	Symptoms	Spread/conditions	Control
		Considerably reduced yield with fewer or no seeds per pod	the developing foliage of other plants. Cool moist conditions favour this disease	
Ascochyta – leaf and **pod spot**	Fungus	Brown spotting on stems, leaves and pods can cause seedling loss. Peas can become stained	Disease is mainly seed-borne. Infected seed germinates and lesions develop on first leaves from where spores spread to rest of foliage, including the pods, and to other plants by air currents and rain splash	(1) Sow healthy seed (2) Dress seed with thiabendazole + thiram
Botrytis – **grey mould**	Fungus	Grey mould develops where petals have fallen onto pods and leaves	Disease spread by damp and humid conditions at end of flowering	(1) Treat with a suitable fungicide, e.g. vinclozolin at mid-flowering
Mycosphaer- **ella – foot** **rot and leaf** **spot**	Fungus	Small brown spots on leaves and pods. Can cause whole plants to die	Spread on seed. Wet weather favours disease	(1) Use healthy seed (2) Chlorothalonil or vinclozolin gives some control at mid flowering
Bacterial **blight**	Bacteria	Dark brown lesions on leaves, stems and pods	Spread on seed. Wet windy weather favours disease	(1) A notifiable disease (2) Only sow healthy seed
Pea wilt – *Fusarium* **wilt**	Fungus	In late May/June, lower leaves tend to turn grey before rolling downwards. All leaves eventually affected. Death of plant either before podding or	The pathogen is soil-borne. It invades the plant and only returns to the soil from the infected dead plant material	(1) Rotation (2) Use resistant varieties *Note:* There are other *Fusarium* species notably root rot

Table 7.1 (*Cont'd*)

Crop attacked	Disease	Causal agent	Symptoms of attack	Life cycle	Methods of control
			before pods have swollen. White mycelium appear on stem after death		
	Pre-emergence **damping-off**	Fungus complex of different species	Poor germination and seedling establishment. The seed rots, or if not the stem is soft and dark brown in colour	Soil-borne pathogens invade seeds and/or plant stems at or just below soil level. Worst when peas sown in cold, wet soil	(1) Seed protected by seed dressing, e.g. thiram
	Marsh spot	Manganese deficiency	Yellowing of leaves between veins which remain green; with severe deficiency; growth is restricted and yield reduced. Brown discoloration (marsh spot) in centre of pea		(1) Manganese sulphate as soon as in full flower. May need repeating 7 days later
Field Beans	**Chocolate spot**	Fungus	Small circular chocolate coloured discolorations on leaves and stems; with bad attack symptoms move to flowers and pods. Spots can coalesce and cause total plant death	Fungus carried over from previous year on debris of old bean haulm and on volunteer plants. Disease favoured by warm, wet weather. Autumn-sown beans, especially early sown, are more liable to attack than spring sown and they suffer more severely	(1) Clean up stubbles containing remains of old crop of bean (2) Fungicide – e.g chlorothalonil + cyproconazole. First treatment in winter beans should be at early flowering stage if disease present on lower leaves

Disease	Type	Symptoms	Notes	Control
Ascochyta – leaf spot	Fungus	Leaves affected by regular brown to black spots, some up to 2 cm in diameter; spots have slightly sunken grey centres (in which can be seen small black spots with brown margins). Pods and seed also affected, the latter covered with brownish-black lesions	Infected seeds when sown may produce seedlings with characteristic disease symptoms on stem at soil level or on lowest leaves. In cold moist conditions the disease will move up the plant and onto other bean plants	(1) Healthy seed – seed can be tested (2) Hygiene – kill any volunteer beans in other fields (3) Seed treatment, e.g. thiabendazole + thiram
Sclerotinia – stem rot	Fungus	Rotting of shoots and roots. Can cause death in severe attacks on winter beans	Soil-borne. The strain that attacks spring beans also attacks peas, red clover and oilseed rape. The sclerotia or resting bodies can persist in the soil for 8 years	(1) No fungicide recommendations (2) Use wide rotations
Bean rust	Fungus	Red/brown rust pustules appear on leaves	Can be spread on seed, trash or volunteers. More of a problem on spring beans and on potassium deficient soils	(1) Hygiene – kill any volunteer beans (2) Fungicides – e.g. fenpropimorph or cyproconazole are very effective
Downy mildew	Fungus	Pale green water-soaked lesions. Grey fungal growth on underside of leaf	Seed- and soil-borne. Spring beans most affected. Disease encouraged by cool wet weather	(1) Rotation (2) Hygiene – destroy debris (3) Fungicides – treat with metalaxyl and chlorothalonil when disease first found

Table 7.1 *(Cont'd)*

Crop attacked	Disease	Causal agent	Symptoms of attack	Life cycle	Methods of control
Sugar beet, fodder beet, mangels	**Virus yellows (Beet mild yellowing virus – BMYV and Beet yellows virus – BYV)**	Virus	First seen in June/early July on single plants scattered throughout the crop – a yellowing of the tips of the plant leaves. This gradually spreads over all but the youngest leaves. Infected leaves thicken and become brittle. The yield is seriously reduced by an early attack	The crop is infected by aphids which have over-wintered in mangel clamps and steckling beds. Several aphids carry the virus, particularly peach potato aphid. After a mild winter and warm spring aphid migration is early and the chances of an epidemic are increased	(1) All beet/mangel clamps should be cleared by the end of March. If not, they should be sprayed to kill any aphids (2) Use a seed treatment, e.g. imidacloprid (3) If the crop has <12 leaves and if > 1 aphid /4 plants use an aphicide. Also if crop > 12 leaves/ plant and > 1 aphid/ plant treat (4) Aphid warnings issued by British Sugar to aid timing of aphicide (5) Foliar aphicide if no seed treatment applied. Care must be taken with insecticide choice (e.g. pirimicarb) as there is resistance to some of the foliar products
	Powdery mildew	Fungus	Powdery greyish-white mycelium on foliage seen in dry weather in late summer,	The spores are air-borne and move from diseased plants, found at loading sites	(1) Spray with sulphur (as soon as disease is seen but before mid-Septem-

Disease	Cause	Symptoms	Notes	Control
		early autumn	and from roots left in the field and also from weed beet, to infect the new crop. Disease favoured by dry, warm weather	ber) or use cyproconazole (2) Some varieties show more resistance than others (3) Use warnings given by Brooms Barn
Rhizomania – root madness (Beet necrotic yellow vein virus – BNYVV)	Virus	Wilted plants (sometimes in patches in field) showing pale yellowing of veins; development of elongated, strap-like leaves often protruding above surrounding plants. Infected roots smaller than healthy roots, constricted below soil level, usually fanging with a proliferation of small lateral roots (bearding). Inside of root shows brown-streaked tissue from tip of tap root upwards	The virus is tranmitted by soil fungus – *Polymyxa betae* which is present in UK soils	A notifiable disease. Widespread in Europe and now spreading in the UK. UK currently has rhizomania free status (1) Tolerant varieties are the main answer (2) Extend the rotation to reduce the risk
Ramularia – leaf spot	Fungus	Pale brown circular spots. Can eventually cause death of all leaves	Mainly soil-borne disease. Wet weather encourages infection and spread	(1) Choose more resistant varieties (2) Apply an approved fungicide, e.g. cyproconazole
Speckled yellows	Manganese deficiency	Small yellowish areas between leaf veins, later fusing to buff-coloured angular, sunken spots (speckled yellows) which	Symptoms often disappear as root system develops. But disease can be a problem on near-alkaline soils or	(1) Foliar spray with manganese sulphate in May and June which may have to be repeated

Table 7.1 *(Cont'd)*

Crop attacked	Disease	Causal agent	Symptoms of attack	Life cycle	Methods of control
				those soils with high organic matter content	
	Heart rot	Boron deficiency	In young plants the youngest leaves turn a black/brown colour and die off. A dry rot attacks the root and spreads from the crown downwards. The growing point is killed, being replaced by a mass of small deformed leaves	This deficiency is more apparent on dry and light soils and can be made worse by heavy liming	(1) Apply borax as soon as the disease is seen. Use a boronated compound fertiliser on suspected soils. Deficiency diagnosed by soil analysis
Potatoes	**Late blight**	Fungus	Brown areas on leaves. Whitish mould on the underside of leaves. Leaves and stems become brown and die off. Affected tubers may rot in store due to secondary infection with rotting bacteria	Infected tubers (either planted, ground keepers, or from dumps) are the main source of blight. Fungal spores are carried by the wind or rain to infect the haulms. Recently resting spores have been found; their importance is unknown. From the haulms the spores are washed into the soil to infect the tubers. Infection can also take place at harvest. The fungus cannot	(1) Ensure that potato dumps do not sprout. Spray off with paraquat or glyphosate or cover with plastic sheeting. (2) Blight spreads rapidly in warm, high humidity, weather and warnings of such conditions are given by various organisations. Early preventative spraying, followed by repeated sprays every 7–14 days, is advisable (3) A wide range of

Disease	Type	Symptoms		Control
			live on dead haulm. Risk of blight encouraged by certain weather conditions – Smith periods – e.g. two days with min temp above 10°C and relative humidity at 90% for at least 11 hours/day. Disease more likely to develop once crop meets between rows. New more virulent strains have been isolated	fungicides are available, both protectants (e.g. dithiocarbamates or organotins) and systemics, e.g. phenylamide group. Due to resistance problems, care should be taken with chemical choice and mixtures (4) The haulms should be burnt off before harvest to prevent the tubers becoming infected whilst being lifted (5) There is some difference in susceptibility to foliage and tuber blight between varieties
Potato leaf roll virus – PLRV	Virus	Lower leaves are rolled upwards and inwards; they feel brittle and crackle when handled. The other leaves are lighter green and more erect than normal. Yield is lowered	The virus is transmitted by aphids from plant to plant. Infected tubers (which show no signs of the disease) are planted and thus the disease is carried forward from year to year	(1) Resistant varieties (2) Use certified 'seed' (3) Ensure no aphids in the chitting house (4) Systemic sprays or granules will control the aphids, and thus reduce the spread of the virus (5) Reduce sources of infection – dumps etc.

Table 7.1 (Cont'd)

Crop attacked	Disease	Causal agent	Symptoms of attack	Life cycle	Methods of control
	Mosaics – Potato virus Y (PVY)	Virus	May range from a faint yellow mottling on leaves to a severe distortion of the leaves and distinct yellow mottling. Yield can be seriously reduced by the severe forms	Peach potato aphid is the main vector of PVY. Symptoms may not appear for 4 weeks after infection	As for leaf roll. Some varieties resistant and control using aphicides can be variable
	Common scab	Bacteria (several species)	Skin-deep irregular-shaped scabs on tuber: these can occur singly or in masses. With a severe attack cracking and pitting takes place with secondary infection by insect larva and millipedes	The soil-borne organism attempts to invade the growing tuber which responds by development of corky tissue to restrict the parasite to the surface layers. Organism re-enters soil when infected seed is planted	(1) Avoid liming just prior to planting potatoes. Disease is particularly prevalent on light sandy, alkaline soils, in dry conditions (2) Irrigation is main method of control. Irrigate when soil moisture deficit reaches 15 mm during the 6 weeks from the start of tuber initiation (3) Some varieties are more resistant than others
	Powdery scab	Fungus	Appearance can be similar to common scab, but the spots are rounder and formed	Spore balls can remain in the soil for many years or may be planted on	(1) Avoid using infected seed or contaminated FYM

Disease	Cause	Symptoms	Infection / spread	Control
Dry rot	Fungus (several species)	as raised pimples under the skin which burst; sometimes cankers and tumours develop. Infected tubers are usually first noticed in January and February. The tuber shrinks and the skin wrinkles in concentric circles. Blue/pink or white pustules appear on the surface	infected tubers and the zoospores attack the new tubers by way of the lenticels, eyes or wounds, resulting in scab development. Usually more troublesome in wet seasons. The soil-borne fungus enters the tuber from adhering soil. Infection can only enter through wounds and bruises caused by rough handling at harvest. Favoured by high storage temperatures	(2) Some varieties are very susceptible (3) Check zinc levels. Soils with a high level of soil zinc may have a lower disease risk. (1) If the potatoes are handled carefully, infection is considerably reduced (2) Use of fungicides at lifting or first grading, e.g. imazalil or thiram
Spraing – Tobacco rattle (TRV) or **potato mop-top virus (PMTV)**	Virus	Foliage – very variable; TRV stem mottling; PMTV yellow blotches and bunching of leaves on short stems – like a mop. Tubers primary (after soil infection): wavy or arc-like brown, corky streaks in flesh of cut tuber. Secondary – from infected tubers: PMTV – badly formed and cracked tubers. TRV – brown spots in flesh	TRV spread by free-living nematodes in soil, especially in light sandy soils. PMTV – spread by powdery scab fungus and can remain in the soil for years in fungal resting bodies	(1) Plant only resistant varieties on infected soils (2) Control free-living nematodes with a nematicide, e.g. oxamyl may sometimes be worthwhile
Blackleg and **tuber soft rots**	Bacteria	Plants stunted and pale green or yellow foliage; easily pulled out of the ground and stem	The bacteria move to tubers via the stolons and in wet soil to healthy tubers to	(1) Rogue or reject seed crops where the disease shows on foliage

Table 7.1 *(Cont'd)*

Crop attacked	Disease	Causal agent	Symptoms of attack	Life cycle	Methods of control
			base is black and rotted. Infected and neighbouring tubers develop a wet rot in the field or in store, especially in damp and badly ventilated (warm) conditions	enter via lenticels or damaged areas. Carried on seed tubers	(2) Do not plant infected tubers (3) No resistant varieties, although some are less susceptible (4) Monitor store regularly. Ensure good store ventilation
	Gangrene	Fungus	A serious tuber rot which develops in storage, usually late; it shows as grey 'thumb-mark' depressions on the tubers and the flesh beneath is rotted; also, pin-head black spore cases	The fungus remains alive in the soil and on trash, and can infect tubers in the soil and from tuber to tuber when handling. Cold wet growing conditions favour this disease. More common-ly found in cool stores	(1) Do not plant diseased seed (2) Reduce mechanical damage (3) Treatment with a fungicide reduces incidence in store, e.g. thiabendazole or imazalil (4) Varieties vary in their susceptibility
	Silver scurf	Fungus	A superficial disease. Causes a silvery skin finish. Can affect quality for pre-pack market	Mainly a seed-borne disease. Symptoms develop in store especially in warm conditions	(1) Plant disease free seed (2) Apply a fungicide, e.g. imazalil or thiabendazole
	Skin spot	Fungus	Tuber symptoms develop during late storage and	Mainly spread by infected tubers. Tuber infection	As for gangrene

Crop/Disease	Type	Symptoms	Conditions	Control
		appear as pimple-like, dark brown, shrunken spots with raised centres. The worst damage is the destruction of the buds in the eyes of seed tubers	occurs at lifting and is worse in cold, wet seasons	
Black dot	Fungus	Similar symptoms on tubers as silver scurf. Small black sclerotia present on the lesions. A problem in potatoes for pre-packing	Soil- and seed-borne disease. Warm and moist conditions in store increase the amount of damage	No resistant varieties or fungicides are available
Brown rot	Bacteria	Symptoms only seen in warm conditions. Initially leaves wilt leading eventually to plant death. Brown streaking on the stem. Tubers have brown vascular discoloration and ooze a bacterial slime	Outbreaks linked to contaminated irrigation water. Woody nightshade is secondary host	A notifiable disease in the EU and UK. Potentially a very serious disease
Brassicae (Brussels sprouts, cabbage, kale, swedes and turnips) Club root or finger and toe	Fungus	Swelling and distortion of the roots. Stunted growth. Leaves pale green in colour	A soil-borne fungus. The fungus grows in the plant roots and causes the typical swellings. Resting spores can pass into the soil, especially if diseased roots are not removed. They can remain alive for several years, becoming active when the host crop is again grown in the field.	(1) Rotation. With a bad attack advisable not to grow the crop for at least 5 years in the field (2) Liming and drainage (3) Resistant crops. Kale is more resistant than swedes or turnips (4) Some varieties of swedes and turnips are more resistant than

Table 7.1 *(Cont'd)*

Crop attacked	Disease	Causal agent	Symptoms of attack	Life cycle	Methods of control
				The spores are more active in acid and wet conditions	others
	Powdery mildew	Fungus	Upper surface of leaves show blue/black discoloration; sprouts turn black	Spores overwinter on infected plants and are carried by air currents to infect the following year's crop. Particularly susceptible crops are Brussels sprouts, swedes and turnips	(1) Varieties differ in their susceptibility to this disease (2) Fungicide at first sign of disease, e.g. metalaxyl or chlorothalonil
	Downy mildew (e.g. Cauliflower)	Fungus	Pale green spots on leaves. Fungal growth/spores can be found on underside of leaves. Badly infected plants may die	A soil-borne problem very important at the seedling stage	(1) Ensure good ventilation of the transplants (2) Apply an approved fungicide at the first sign of disease
	Ring spot (Sprouts, cabbages and cauliflower)	Fungus	Small spots on leaves initially with a well-defined edge. As the disease develops the lesions enlarge to produce several concentric brown rings	Infected debris is the main source of infection	(1) Rotation (2) Destroy crop debris (3) Apply a recommended fungicide, e.g. chlorothalonil
	Brown heart of Swedes	Boron deficiency	No external symptoms, but when the root is cut open a		See Heart rot of sugar

Crop	Disease	Cause	Symptoms	Notes / Control
	and turnips (Raan)		browning or mottling of the flesh is seen. Affected roots are unpalatable	beet
	Stem rot in kale	Boron deficiency	Brown rot in the stem pith followed by stem collapse	See Heart rot of sugar beet
Grass	**Barley yellow dwarf virus (BYDV)**	Virus	Leaves turn yellow, red or brown, discoloration starting at tips and going down leaves. Plants generally stunted, but can produce more tillers. Disease more conspicuous in single plants than in whole sward	Spread by several species of aphids. Ryegrass and the fescues are the most susceptible	No chemical control Some ryegrass varieties more resistant than others
	Ryegrass mosaic (RgMV)	Virus	Yellowish/green mottling or streaking of leaves. Severe infection can show more general browning of leaf	Number of different strains of virus. Spread by wind-borne mite vectors which are favoured by hot, dry weather. Spring-sown Italian ryegrass is most susceptible	No chemical control of vector (1) Some ryegrass varieties are more resistant than others (2) Autumn sowing
	Crown and **brown rust** of perennial ryegrass	Fungus	Crown rust usually seen in late summer. Pale yellow leaf flecking, followed by bright orange/yellow oval pustules. Brown rust is similar in appearance but found in early summer	Spores are air-borne and can quickly infect a clean sward especially in warm dry weather	(1) Frequent grazing (2) Some grass varieties are more resistant than others (3) Propiconazole or triadimefon can be used on silage or seed crops

Table 7.1 *(Cont'd)*

Crop attacked	Disease	Causal agent	Symptoms of attack	Life cycle	Methods of control
	Rhynchosporium – leaf spot	Fungus	Irregular scald-like blotches on leaves. Favoured by cool wet weather and is most apparent in spring and early summer	Spores are air-borne and move from a diseased to a clean sward in the spring. Italian ryegrass most susceptible especially diploids	(1) Choose resistant varieties in the south and south west where disease is always more apparent (2) Propiconazole can be used on silage or seed crops
	Drechslera – leaf spot	Fungus	Causes necrotic spotting or streaking of leaves depending on grass species. Grass yield and quality can be affected	Can be seed-borne. Occurs throughout the year. There are a number of different species	(1) Treat with propiconazole on silage or seed crops
	Powdery mildew	Fungus	Disease found throughout the country especially in Scotland. Present in early spring onwards, particularly in dense swards of short duration ryegrass. Greyish-white mycelium on leaves	Spores are air-borne, and move from an infected to a clean crop. There are many strains of this disease	(1) Resistant varieties (2) Triadimefon or propiconazole can be used on silage or seed crops
Red and white clover, lucerne, sainfoin	**Clover rot (Sclerotinia)**	Fungus	In autumn necrotic spots found on leaves then spreads to stems. Foliage turns olive-green and then black and eventually dies. The root can also die	Resting bodies of fungus produced on affected plants in winter and spring. They are small (size of clover seed), white at first and then turning black. Bodies remain dormant in summer	(1) Clean seed (2) Use resistant varieties where possible (3) Rotation at least an interval of 4–5 years between susceptible crops

Crop	Disease	Type	Symptoms	Spread / Biology	Control
Lucerne	Verticillium wilt	Fungus	Usually seen in fairly isolated patches in first harvest year; in next 2 years spreads to many parts of the field. Normally after first cut, lower leaves turn pale-yellow colour and then white, and eventually shrivel from base upwards. Whole plant finally dies	but in autumn produce seed-borne spores which affect other plants

Disease can be introduced by contaminated seed; spores can also be transported by air, as well as being spread by contaminated fragments of the crop moving from plant to plant and then from field to field by machine | (1) Where suspected use tolerant varieties (2) Use clean seed (3) Harvest healthy crops first before moving onto older infected crops (4) Extend interval between lucerne crops |
Maize	Fusarium – stalk rot	Fungus	Base of plants attacked in August/September; foliage grey/green colour and wilting. Pith brown/pink at base of stem; leads to premature senescence	Soil- and seed-borne disease. Can build up in the soil	Differences in varietal susceptibility. Grain maize more likely to be affected because disease develops most rapidly in mature crops
	Kabatiella – eyespot	Fungus	Round cream spots on leaves surrounded by a dark brown ring. Causes crop to die prematurely	Disease encouraged by cool wet weather and growing successive crops	(1) Crop rotation (2) Plough previous crop residues in the autumn
	Smut	Fungus	Large black galls on any of the above ground parts of the plant, including the cob	Spores from galls can re-infect other maize plants, or they can remain dormant in the soil, surviving for several years	Crop rotation – more serious when maize is cropped frequently.

Part 2

Crop husbandry techniques

8

Cropping techniques

The practical on-farm management of soil, the setting of targets and the monitoring of target achievement and the decisions taken about what crops to grow and when they should appear in the cropping sequence are fundamental to successful crop production.

The management of individual soil types has been dealt with in Chapter 2. Farmers have a responsibility to care for the soil for future generations. This means that, while they are trying to grow profitable crops, they must also take into consideration the long-term effects of their practices on soil health, fertility, structure and stability. This chapter attempts to examine some general management practices which could apply to several different soil types depending on individual situations.

8.1 Drainage

8.1.1 The necessity for good drainage
Normally, the soil can only hold some of the rainwater which falls onto it, particularly during the winter months (see pages 34–5). The remainder either runs off or is evaporated from the surface or soaks through the soil to the subsoil. If surplus water is prevented from moving through the soil and subsoil, it soon fills up the pore space; this will kill or stunt the growing crops as air is driven out of the profile.

The water table is the level in the soil or subsoil, below which the pore space is filled with water. This is not easy to see or measure in clay soils but can be seen in open textured soils (Fig. 8.1).

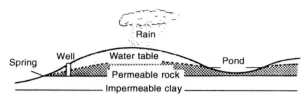

Fig. 8.1 Position of water table and the effect on water levels in wells and ponds.

The water table level fluctuates throughout the year and in the British Isles is usually highest in February and lowest in September. This is because of a higher amount of evaporation, more transpiration from growing crops and usually lower rainfall in the summer.

In chalk and limestone areas, and in most sandy and gravelly soils, water can drain away easily into the porous subsoil. These are free-draining soils. On most other types of farmland some sort of artificial field drainage is necessary to carry away the surplus water and so keep the water table at a reasonable level. For most arable farm crops the water table should be about 0.6 m or more below the surface but for grassland 0.3–0.45 m is sufficient. The withdrawal of grants for drainage work, and the greater protection of some wetland environments have meant a large reduction in the area of land being drained compared with the immediate post-war period and the 1970s and 1980s.

Indications of poor drainage are:

1   Machinery is easily 'bogged down' in wet weather.
2   Stock grazing pastures in wet weather easily poach the sward.
3   Water remains on the surface for many days following heavy rain.
4   Weeds such as rushes, sedges, horsetail, tussock grass and meadowsweet are common in grassland. These weeds usually disappear after drainage. Peat forms in places which have been very wet for a long time.
5   Young plants are pale green or yellow in colour and have generally poor growth.
6   The subsoil is often coloured in various shades of blue or grey, compared with shades of reddish-brown, yellow and orange in well-drained soil.

The benefits of good drainage are:

1   Well-drained land is better aerated and the crops grow better and are less likely to be damaged by root-decaying fungi.
2   The soil dries out better in spring and so warms up more quickly and can be worked earlier.
3   Plants are encouraged to form a deeper and more extensive root system. In this way they can often obtain more plant food and are better able to survive periods of drought.
4   Grassland is firmer, especially after wet periods. Good drainage is essential for high density stocking and where cattle are outwintered if serious poaching of the pasture is to be avoided.

5   Disease risk from parasites is reduced. A good example is the liver fluke. Part of its life cycle is in a water snail found on badly drained land.
6   Inter-row cultivations and harvesting of root crops can be carried out more efficiently.
7   The soil has more potential to support a wider range of crops in the rotation.

The main methods used to remove surplus water and control the water table are:

1   Ditches or open channels.
2   Underground plastic pipe drains.
3   Mole drains.
4   Ridge and furrow (see page 40).

8.1.2   Ditches and open drains

Ditches may be adequate to drain an area by themselves, but they usually serve as outlets for underground drains. They are capable of dealing with large volumes of water in very wet periods. The size and shape of a ditch varies according to the area it serves and the surrounding soil type (Fig. 8.2). Ditches ought to be cleaned out to their original depth once every three or four years. The spoil removed should be spread well clear of the edge of the ditch. Many different types of machines are now available for making new ditches and cleaning neglected ones.

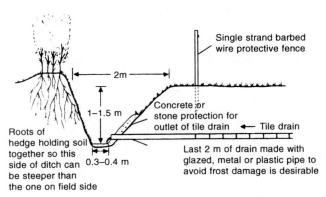

Fig. 8.2   Section through a typical field ditch and tile drain.

Small open channels 10–60m apart are used for draining hill grazing areas. These are either dug by hand using a spade designed for the purpose, or made with a special type of plough drawn by a crawler tractor. Similar open channels are used on low-lying meadow land where underground drainage is not possible.

8.1.3   Underground drains

The distance between drains which is necessary for good drainage depends on the soil texture. In clay soils the small pore spaces restrict the movement of water. Therefore the drains must be spaced much closer together than on the lighter types of soil where water can flow freely through the large pore spaces (Fig. 8.3 and 8.4).

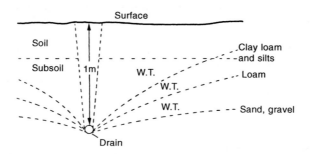

Fig. 8.3   The steepness of a water table (WT) varies with different soil types.

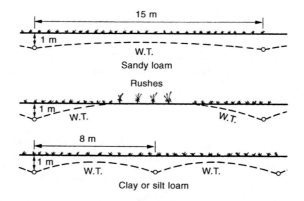

Fig. 8.4   Diagram showing the effect of spacing of drains on the water table (WT) (a) sandy loam (b) and (c) silt or clay loam.

In the past, many different materials have been used in trenches to provide an underground passage for water, e.g. bushes, turf, stones, and flat tiles before cylindrical clay tiles became popular. (Concrete pipes have also been used.) However, the flexible plastic pipe has replaced the clay tile for all new drainage work. Under drainage is very expensive and cost must be related to likely benefits. Consequently, when designing a scheme, the spacing of the drains must be carefully considered. Digging some holes and examining the subsoil can be very helpful in deciding how the field should be drained. If cost is a limiting factor, it would, for example, be advisable to lay the drains about 40 m (two chains) apart in an area where 20 m would be an ideal spacing. This should be quite satisfactory over most of the field and only a few extra drains may be required later in the wetter areas. This should not be difficult if a detailed map of the drainage layout has been made.

The slopes on a field determine the way the drains should run to be efficient in removing water (Fig. 8.5). Water will not run uphill, unless it is pumped, and so the drains must have a gradient. If the slope is more than 2% (1:50) the laterals (side drains) should run across the slope. Ideally, the minimum fall on laterals should be 1:250 and, in mains, 1:400. It is usually preferable that the laterals go into a main drain before entering a ditch; this means fewer outfalls to maintain. In flat areas, and where the soil is silty (causing silting up of the

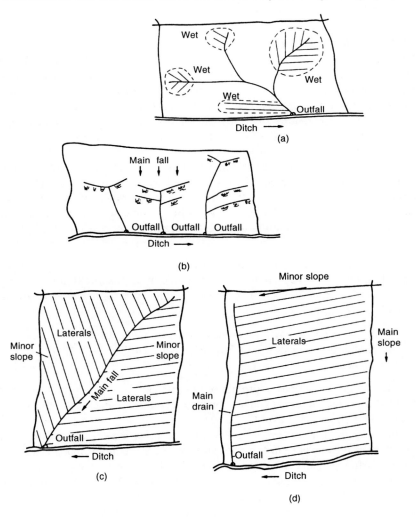

Fig. 8.5   Drain layouts (a) drainage of wet areas (b) intercepting springs (rushy areas) (c) herringbone system in valley (note separation of junctions) (d) grid system.

drains over a long period), it is preferable to let each lateral run straight into the ditch. A high-pressure jetting device can be used to clean them periodically. Iron ochre deposits can also be cleared in this way.

Where the land is low lying and there is no natural fall to a river, the area can be drained in the usual way and the outlet water from the drainage system can lead to a ditch from which it is pumped into a river. Underground drainage is best carried out in reasonably dry weather. It is fairly common practice now to lay drains through a growing crop of winter cereals in a dry period in the spring. With modern machinery the little damage to the crop is more than offset by the beneficial effects of the drains.

8.1.3.1   Underground piped drains

Plastic pipes come in various configurations. Most of them are corrugated with slits cut at intervals to allow water to enter easily. They are usually supplied in 200 metre rolls which are easily transported. Various diameter sizes are available, e.g. 60 or 80 mm for laterals, 100, 125, 150 or 250 mm for mains.

The size of pipe required will depend on the rainfall, the area to be drained by each pipe, gradient and soil structure. In the case of main drains, it will depend on the number of laterals entering.

Various types of machine are available for laying plastic pipes, and the porous backfill (e.g. washed gravel, clinker) required on the heavier soils. The depth of porous fill is more important than the width in the trench. Modern machines can lay pipes very rapidly, especially when a laser beam system is used automatically to adjust the gradient on the drain.

8.1.3.2   Underground drains – mole drainage

Mole drainage is a cheap method which can be used in some fields. Although mole drains are sometimes used on peat soils, the system is normally applied to fields which have:

1   A clay-rich subsoil with no stones, sand or gravel patches.
2   A suitable gradient (a fall of 5–10 cm over 20 m).
3   A reasonably level surface.

A mole plough, which has a torpedo or bullet-shaped 'mole' plus an expander attached to a steel coulter or blade, forms a cylindrical channel in the subsoil along which water can travel (Fig. 8.6).

The best conditions for mole draining occur when the subsoil is damp enough to be plastic and forms a good surface inside the mole channel. It should also be sufficiently dry above to form cracks as the mole plough passes (Fig. 8.7). Furthermore, if the surface is dry the tractor hauling the plough can get a better grip. The plough should be drawn slowly (about 3 km/h) otherwise the vacuum created is likely to spoil the mole. Reasonably dry weather after moling will allow the surface of the mole to harden and so it should last longer.

Mole drains are drawn 3–4 m apart usually at right angles to the direction

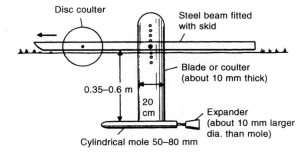

Fig. 8.6 Mole plough.

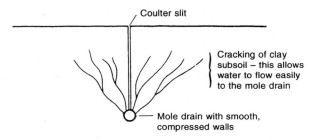

Fig. 8.7 Section through mole drain and surrounding soil.

of the plastic drain pipes. For best results the moles should be drawn through the porous backfill of a plastic main or lateral drain (Fig. 8.8 and 8.9). A new

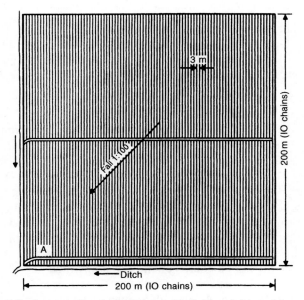

Fig. 8.8 Layout of mole drainage (with field main) in a 4-ha area.

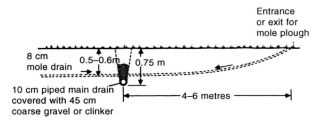

Fig. 8.9 Section through the tile drain shown in Fig. 8.8 to show how water from the mole drain can enter the pipe drain through the porous backing.

set of mole drains can be drawn every three to ten years as required. They will last much longer in a clay soil than in a sandy soil where the sides of the moles collapse quite quickly.

Installing a new drainage system is the first step in removing surplus water from a field. However, the structure of the soil and subsoil must, if necessary, be improved so that water can move easily to the drains (see pages 39–46).

8.2   Irrigation

8.2.1   Soil water availability

Irrigation is the term used when water is applied to crops which are suffering, or about to suffer, from drought.

It is usual to measure irrigation water, like rainfall, in millimetres:

- 1 mm on 1 hectare = $10\,m^3$ = 10 tonnes.
- 25 mm on 1 hectare = $250\,m^3$ (a common application at one time).
- 100 mm on 1 hectare = $1000\,m^3$ = $1\,km^3$ (about a season's requirement).
- 100 mm on 50 ha = $50\,km^3$ (a guide to the amount of water required).

Green plants take up water from the soil and transpire it through their leaves, although a small amount is retained to build up the plant structure. Green crops which cover the ground have a water requirement. This varies from day to day and place to place according to climate and weather conditions. It is called potential transpiration (Table 8.1). When the rainfall is well distributed, crops are less likely to suffer than in wet periods followed by long dry periods. During such dry periods the crops have to survive on the available water held in the soil (Water in the soil, page 35). The difference between the actual amount of water in the soil and the amount held at field capacity is known as the soil moisture deficit (SMD). It is one of the main factors in determining the need for irrigation.

An approximate soil moisture deficit can, at any time, be calculated from the rainfall and the figures given in Table 8.1. For example, if the soil were at field capacity at the end of April, with a green crop covering the ground, and

Table 8.1 Water requirement for green crops, southern UK

Period	per day		per month	
	mm	m³/ha	mm	m³/ha
April and September	1.5	15	50	500
May and August	2.5	25	80	800
June and July	3.5	35	100	1000

with no rain in May, then the SMD would be 80 mm at the end of May allowing for losses due to evaporation and transpiration. However, if 50 mm of rain fell during May, then the deficit would be 30 mm. The amount of water available to a crop at any one time depends on the capacity of the soil per unit depth, and the depth of soil from which the roots take up water. The figures in Table 8.2 are average rooting depths. They would be greater on deep, well-drained soils and less on shallow, badly drained or panned soils. This can be checked in the field by digging a soil profile. Normally, most of the plant roots are concentrated in the top layers and only a small proportion penetrate deeply into the soil.

Table 8.2 Rooting depths of common crops

Crop	Rooting depth (cm)
Cereals, pasture	35–100
Potatoes (varietal differences)	30–70
Beans, peas, conserved grass	45–75
Sugar beet	55–100
Lucerne (alfalfa)	over 120

More than three-quarters of all the farm soils in this country have available water capacities between 60 and 100 mm (12–20% by volume) within the root range of most crops (about 500 mm depth). Sandy, gravelly and shallow stony soils have less than 60 mm (12%), whereas deep silty soils, very fine sands, warpland, organic and peaty soils have well over 100 mm (20%) of available water within root range. The available water capacity of the topsoil is often different from that of the subsoil. This should be taken into consideration when calculating the amount present within the root range of a crop.

It is estimated that the roots of a plant cannot take up much more than half of the available water within the depth of its deepest roots. The top half of the rooting zone contains 70% of roots and they do not penetrate all the pore spaces and so, where available, irrigation should be started when about half the available water has been taken up. This is known as the critical deficit.

8.2.2 Timing and sources of irrigation

Decisions on when to start irrigating must be influenced by the time it will take to apply 25 mm of water over the area involved, usually 5 to 10 days. Some companies and organisations provide helpful computerised predictions on when irrigation should be started. These take into consideration all the relevant factors and are tailored for each field and its crop. The greatest need for irrigation is in the eastern and south-eastern counties where the lower rainfall and higher potential evapo-transpiration means that it would be beneficial about nine years out of ten. The need is much less in the wetter western and northern areas.

Water for irrigation is strictly controlled by licences issued by the National Rivers Authority (NRA). The charge for the water varies between areas and it usually costs up to ten times more for taking water directly from rivers in summer than for taking it to fill reservoirs in winter. Licences are increasingly difficult to obtain and even existing licences are often subject to abstraction restrictions.

Water is generally very scarce in summer in the areas which need it most, but annual rainfall is adequate for most irrigation needs provided the winter surplus is stored in reservoirs or underground aquifers. The main sources of irrigation water are rivers and streams, ponds and lakes, reservoirs and boreholes. Some is taken from mains water but is very expensive. Boreholes are also expensive; they may have to be lined, e.g. those dug through sand or gravel, and require special deep well pumps and sometimes filters. However, they may be the most reliable source in a very dry summer. Above-ground reservoir construction is specialised work and must be designed and supervised by a qualified engineer if over 25 000 m^3 in volume and lie above adjacent land. (A hole in the ground is simpler and may be partly replenished by groundwater.) Lining of a reservoir may be necessary and this could double the cost.

All water used for irrigation should be tested for impurities which may be harmful to crops.

8.2.3   The effect of irrigation on particular crops

Irrigation can be expensive, so it is important to use it as efficiently as possible.

Early **potatoes** will respond with earlier and more profitable production if irrigation is used when the SMD reaches 25 mm. Second earlies and maincrops are not normally irrigated until the tubers are fingernail size, otherwise some varieties may produce too many small tubers and a low saleable yield. On the better soils (with a high available water capacity) irrigation can be delayed until a 40 mm deficit occurs. On light soils where common scab causes problems, irrigation is recommended each time a 15 mm deficit occurs when the tubers are forming. If the soil is very dry and hard at harvest time, some irrigation can make harvesting easier and therefore reduce tuber damage.

With **sugar beet** it may be necessary to irrigate a very dry seedbed to obtain a fine tilth, even germination and more effective use of soil-acting herbicides. After this it is not usually necessary to irrigate until the leaves start meeting across the rows (sugar beet is deep-rooted). It can start with applications of about 25 mm, and increase to 55 mm (in one application) on the high available water capacity soils when the SMD is over 100 mm.

For **peas**, unless the soil is very dry, it is preferable to wait and apply the water at the very responsive stages. At the start of flowering 25 mm should be applied (increases numbers of peas) and 25 mm when the pods are swelling (increases pea size), always provided that the SMD is about 35 mm at these times.

With **cereals**, irrigation can only be justified on very light soils in low rainfall areas and only if higher value crops have had their water need satisfied. The most critical time is about GS 37 when the flag leaf is showing and the ear is developing rapidly. An application of 25 mm should be adequate.

Grass and other **leafy crops** respond throughout the season and about 25 mm of water can be applied every time the deficit reaches 30 mm. Although intensive dairying is the most profitable way to utilise the extra production, it is unusual to irrigate the grass crop.

In the above suggestions, for times and amounts of water to apply, it is assumed that it is a dry period with little or no sign of rain. To irrigate up to field capacity shortly before rain is to risk nitrogen losses by leaching.

8.2.4 Application of irrigation

Different textured soils absorb water at different rates (Table 8.3). When irrigation water or rain falls on a dry soil it saturates the top layers of soil to full field capacity before moving further down. However, if the soil is deeply cracked (e.g. clay), then some of the water will run down the cracks. Very dry soils can absorb water at a faster rate than shown in Table 8.3, but only for a short period.

Table 8.3 Time of absorption of water by various soils

Soil type	Time (hr) required to absorb 25 mm of water
Sandy	1–2
Loams	3
Clay	4–5

Irrigation water is applied by:

1 *Rotary sprinklers.* Each rotating sprinkler covers an area about 21–36 m in diameter. The sprinklers are usually spaced 10.5 m apart along the supply line which is moved about 18 m for each setting. Special nozzles can be used to apply a fine spray at 2.5 mm per hour for frost protection of fruit crops and potatoes. Icicles form on the plants, but the latent heat of freezing protects the plant tissue from damage.

2 *Rain guns.* These are now the most widely used machines. They are used on all types of crops. The droplets are large and the diameter of area covered

may be between 60 and 120 m. Some types are used for irrigation of slurry or dirty water.

3   *Travelling rain gun with hose reel.* This is similar in many respects to a mobile machine, but the hose reel remains stationary – usually on a headland – while the rain gun and its hose (nylon/PVC) are towed out to the start of the run (up to 500 m). It irrigates as it is wound back, using water pressure to turn the hose reel. At the end of the run it switches itself off automatically, and it can then be moved to the next position. About 80% of all irrigation water is now put on using this system.

4   *Centre pivot and linear irrigators.* These are large machines which cover up to 160 hectares. They are mainly centre pivots which move round a pivot (the water source) in a complete circle or part of a circle.

5   *Spraylines.* These apply the water gently and are used mainly for protected horticultural crops.

6   *Trickle.* Water is applied through small bore perforated pipes on or near to the soil surface. They give a much more efficient use of water than sprinklers or rain guns and are often used in perennial horticultural crops or vineyards.

7   *Surface channels.* This method requires almost level or contoured land. It is wasteful of land, water and labour, but capital costs are very low.

8   *Underground pipes.* On level land, water can be dammed in the ditches and allowed to flow up the drainage pipes into the subsoil and lower soil layers. This is drainage in reverse!

On sloping fields, problems can arise with run-off if water is applied too quickly. It may run down between potato ridges, for example, without really wetting the ridges where the crop roots are growing. There is less run-off if a bed system is used (pages 319–20).

Reservoirs with a capacity of over 25 000 m^3 above the natural level of adjoining land must now be registered with the Planning Authorities. Inspection and repairs can be very expensive to ensure that they are safe.

The need for irrigation can be reduced by improving soil structure, breaking pans, correcting acidity (soil and subsoil), increasing organic matter in the soil and selecting crops and varieties with deep-rooting habits.

8.3   Warping

This is a process of soil formation where land, lying between high and low water levels, alongside a tidal river, is deliberately flooded with muddy water. The area to be treated is surrounded by earth banks fitted with sluice gates. At high tide, water is allowed to flood quickly onto the enclosed area and is run off slowly through sluice gates at low tide. The fineness of the material deposited will depend on the length of time allowed for settling. The coarse particles will settle very quickly, but the finer particles may take one or more days to settle. The depth of the deposit may be 0.4 m in one winter. When enough alluvium is deposited, it is then drained and prepared for cropping. These soils are very fertile and are usually intensively cropped with arable crops such as potatoes, sugar beet, peas, wheat and barley.

Part of the land around the Humber estuary is warpland and the best is probably that in the lower Trent valley. Most of this work was done in the nineteenth century. It is too expensive to consider now.

8.4 Claying

The texture of 'blowaway' sandy soils and Black Fen soils can be improved by applying 400–750 t/hectare of clay or marl (a lime-rich clay). If the subsoil of the area is clay, it can be dug out of trenches and roughly scattered by a dragline excavator. In other cases, the clay is dug in pits and transported in special lorry spreaders. Rotary cultivators help to spread the clay. If the work is done in late summer or autumn, the winter frosts help to break down the lumps of clay. As a practice, it is no longer considered economical.

8.5 Tillage and cultivations

8.5.1 The necessity for cultivations

Cultivations are field operations which attempt to alter the soil structure. The main object is to provide a suitable seedbed in which a crop can be planted and will germinate (where applicable) and grow satisfactorily. Cultivations are also used to kill weeds and/or bury the remains of previous crops. The cost of the work can be reduced considerably by good timing and the use of the right implements. Ideally, a good seedbed should be prepared with the minimum amount of working and the least loss of moisture. On heavy soil and in a wet season, some loss of moisture can be desirable. On the medium and heavy soils full advantage should be taken of weathering effects. For example, ploughing in the autumn will allow frost to break the soil into a crumb structure; wetting and drying alternately will have a similar effect. The workability of a soil is dependent on its consistency, which is a reflection of its texture and moisture content.

8.5.2 Seedbed requirements

Listed below are the seedbed requirements for some important crops.

1   *Cereals.*
    (a) *Autumn planted.* The object here is to provide a tilth (seedbed condition) which consists of fine material and egg-sized lumps. It should allow for the seed to be drilled and easily covered with the surface remaining rough after planting. The lumps on the surface prevent the siltier soils from 'capping' in a mild wet winter and they also protect the base of the cereals from the harmful effects of very cold winds. Larger clods of soil will cause problems with residual herbicides if used and could make it easier for slugs to move through the soil and cause damage to seeds and seedlings. Harrowing and/or rolling may be carried out in the spring to break up a soil cap which may have formed. It will also firm the soil around plants which have been heaved by frost action.
    (b) *Spring planted.* A fairly fine seedbed is required in the spring (very fine if grasses and clovers are to be undersown). If the seedbed is dry or very loose after drilling it should be consolidated by rolling. This is especially important if the crop is undersown and where the soil is stony.

2   *Root crops,* e.g. *sugar beet, swedes* and *carrots:* also *kale.* These crops have small seeds and so the seedbed must be as fine as possible, level, moist and firm. This is very important when precision drills and very low seed rates are used. Good, early, ploughing with uniform, well-packed and broken furrow slices will reduce considerably the amount of work required in the spring when, if possible, deep cultivations should be avoided to keep frost mould on top, leaving unweathered soil well below the surface.

3   Direct seeding and/or reseeding of *grasses* and *clovers.* The requirements will be the same as for the root crops.

4   *Beans and peas.* The requirements will generally be similar to the cereal crop, although the tilth need not be so fine. Peas grown on light soils may be drilled into the ploughed surface if the ploughing has been well done. Winter beans are often broadcast onto stubble and ploughed in.

5   *Potatoes.* This crop is usually planted in ridges 90 cm wide or in beds and so deep cultivations are necessary. The fineness of tilth required depends on how the crop will be managed after planting. A fairly rough, damp seedbed is usually preferable to a fine, dry tilth which has possibly been worked too much. Some crops are worked many times after planting – such as harrowing down the ridges, ridging-up again, deep cultivations between the ridges and final earthing-up. The main object of these cultivations is to control weeds, but the implements often damage the roots of the potato plants. Valuable soil moisture can also be lost. Most annual weeds can be controlled by spraying the ridges when the potato sprouts start to appear. This can replace most of the inter-row cultivations and reduce the number of clods produced by the rubber-tyred tractor wheels.

6   *Oil seeds.* Winter oilseed rape is sown in August and moisture conservation at this dry, hot time is of paramount importance. It is usual to carry out minimum cultivations to achieve a fine seedbed for the very small seed.

Spring oilseed rape and linseed also need fine seedbed conditions to give intimate contact between soil and seed. It is usual to plough in late autumn and then work the resultant frost tilth carefully with straight-tined shallow implements.

8.5.3 Non-ploughing techniques

Direct drilling (slit-seeding, zero tillage) and minimal cultivation techniques have now become alternatives to conventional cultivations on many farms and on a wide range of soil types, including difficult clays.

Several types of special drills are available for direct drilling, using such developments as heavily weighted discs for cutting slits, strong cultivator tines or modified rotary cultivators. For it to be successful, direct drilling has relied on achieving a good straw and stubble burn to remove surface trash. Now that burning is banned, the practice has declined. With cereals it should only be used on very clean stubbles where the previous crop has been cut and the straw removed from the field. Even in these circumstances, there will be a small risk of disease spread from the stubble to the new crop. It can still be seriously considered for seeding grass and for crops such as kale.

Minimal cultivation systems are those using various types of cultivators instead of the plough, and in such a way that only the minimum depth of soil is moved to allow drilling to take place.

8.5.4 Tillage implements

The main implements used for tillage are described below.

Ploughs

Ploughing is the first operation in seedbed preparation on most farms and is likely to remain so for some time yet. However, many farmers are now using rotary cultivators, heavy cultivators with fixed or spring tines, and mechanically-driven digging or pulverising machines, as alternatives to the plough. Good ploughing is still the best method of burying weeds and the remains of previous crops. It can also set up the soil so that good frost penetration is possible. Fast ploughing produces a more broken furrow slice than slow steady work. The mounted or semi-mounted plough has replaced the trailed type on most farms because of ease of handling. General-purpose mouldboards are commonly used. The shorter digger types (concave mouldboards) break the furrow slices better and are often used on the lighter soils. Deep digger ploughs are used where deep ploughing is required, e.g. for roots or potatoes. The one way (reversible) type of plough has become popular for crops such as roots and peas. It has right-hand and left-hand mouldboards and so no openings or finishes have to be made when ploughing; the seedbed should therefore be at least slightly more level. Round-and-round ploughing with the ordinary plough has almost the same effect, although this is not a suitable method on all fields.

The proper use of skim and disc coulters and careful setting of the plough for depth, width and pitch can greatly improve the quality of the ploughing. The furrow slice can only be turned over satisfactorily if the depth is less than about two-thirds the width. The usual widths of ordinary plough bodies vary from 20 to 35cm. If possible, it is desirable to vary the depth of ploughing from year to year to avoid the formation of a plough pan. Very deep ploughing, which brings up several centimetres of poorly weathered subsoil, should only be undertaken with care: the long-term effects will probably be worthwhile but, for a few years afterwards, the soil may be rather sticky and difficult to work. Buried weed seeds, such as wild oats which have fallen down cracks, may be brought to the surface and may spoil the following crops. 'Chisel ploughing' is a term used to describe the work done by a heavy duty cultivator with special spring or fixed tines; unlike the ordinary plough, it does not move or invert all the soil. Disc ploughs have large saucer-shaped discs instead of shares and mouldboards. Compared with the ordinary mouldboard plough, they do not cut all the ground or invert the soil so well, but they can work in harder and stickier soil conditions. They are more popular in dry countries. Double mouldboard ridging ploughs are sometimes used for potatoes and some root crops in the wetter areas.

Cultivators

These are tined implements which are used to break up the soil clods (to ploughing depth). Some have tines which are rigid or are held by very strong springs which only give when an obstruction, such as a strong tree root, is struck. Others have spring tines which are constantly moving according to the resistance of the soil. They have a very good pulverising effect and can often be pulled at a high speed. The shares on the tines are of various widths. The pitch of the tines draws the implement into the soil. Depth can be controlled by tractor linkage or wheels. The timing of cultivations is very important if the operation is to be effective.

Harrows

There are many types of harrow. Harrow tines are usually straight, but they may vary in length and strength, depending on the type of harrow. These implements are often used to complete the work of the cultivator. Besides breaking the soil down to a fine tilth, they can have a useful consolidating effect due to shaking the soil about and rearranging particle distribution. 'Dutch' harrows have spikes fitted in a heavy frame and are useful for levelling a seedbed as well as breaking clods. Some harrows, e.g. the chain type, consist of flexible links joined together to form a rectangle. These follow an uneven surface better and are useful on grassland to remove dead, matted grass at the end of the season. Most chain harrows have spikes fitted on one side.

Rotary power harrows can result in much better movement of the soil in one pass. They are most valuable on the heavier soils when preparing fine seedbeds for potatoes, sugar beet and other crops. Power harrow/drill combinations are used by some farmers as one-pass operations to prepare seedbeds for cereals. They can be used in conditions where ordinary drills may not be appropriate.

Hoes

These are implements used for controlling weeds between the rows in root crops. Various shaped blades and discs may be fitted to them. Most types are either front-, mid- or rear-mounted on a tractor. The front- and mid-mounted types are controlled by the steering of the tractor drive. The rear-mounted types usually require a second person for steering the hoe.

Disc harrows

These consist of 'gangs' of saucer-shaped discs between 30 and 60cm in diameter. They have a cutting and consolidating effect on the soil. This is particularly useful when working a seedbed on ploughed-out grassland; some discs have scalloped edges to improve the cutting effect. The more the discs are angled, the greater will be the depth of penetration, the cutting and breaking effect on the clods and the draught.

Disc harrows are widely used for preparing all kinds of seedbeds, but it should be remembered that they are expensive implements to use. They have a heavy draught and many wearing parts, such as discs, bearings and linkages, and so should only be used when harrows would not be suitable. They tend to cut the rhizomes of weeds such as couch and creeping thistle into short pieces which are easily spread. Discing of old grassland before ploughing will usually allow the plough to get a better bury of the turf. Heavy discs, and especially those with scalloped edges, are very useful for working in chopped straw after combining.

Rotary cultivators (e.g. rotavator)

This type of implement consists of curved blades which rotate around a horizontal shaft set at right angles to the direction of travel. The shaft is driven from the power take-off of the tractor; depth is controlled by a land wheel or skid. This implement can produce a good tilth in difficult conditions and, in many cases, it can replace all other implements in seedbed preparation. A light fluffy tilth is sometimes produced which may 'cap' easily if wet weather follows. The fineness of tilth can be controlled by the forward speed of the tractor; for example, a fast speed can produce a coarse tilth. It is a very useful implement for mixing crop remains into the soil. The rotating action of the blades helps to drive the implement forward and so extra care must be taken when going down steep slopes.

In wet heavy soils the rotating action of the blades may have a smearing effect on the soil. This can usually be avoided by having the blades properly angled. Rotavating of ploughed or cultivated land when the surface is frozen in winter can produce a good seedbed for cereals in the spring without any further working or loss of moisture. Narrow rotary cultivator units are available for working between rows of root crops.

Rolls

These are used to consolidate the top few centimetres of the soil so that plant roots can keep in contact with the soil particles, and the soil can hold more moisture. They are also used for crushing clods and breaking surface crusts. Rolls should not be used when the soil is wet; this is especially important on the heavier soils. The two main types of rolls are the flat roll, which has a smooth surface, and the Cambridge or ring roll, which has a ribbed surface and consists of a number of heavy iron wheels or rings (about 7 cm wide) each of which has a ridge about 4 cm high. The rings are free to move independently and this helps to keep the surface clean. The ribbed or corrugated surface left by the Cambridge roll provides an excellent seedbed on which to sow grass and clover seeds or roots. Also, it is less likely to 'cap' than a flat rolled surface.

A *furrow press* is a special type of very heavy ring roller (usually with three or four wheels) used for compressing the furrow slices after ploughing. It is usually attached to and pulled alongside the plough. On light soil it can be used to prepare a surface suitable for drilling.

8.5.5 Pans and soil loosening

A pan is a hard, cement-like layer in the soil or subsoil which can be very harmful because it prevents surplus water draining away freely and restricts root growth. Such a layer may be caused by ploughing at the same depth every year. Such ploughing produces a plough pan and is partly caused by the base of the plough sliding along the furrow. It is more likely to occur if rubber-tyred tractors are used when the soil is wet, and there is some wheelslip which has a smearing effect on the bottom of the furrow. Plough pans are more likely to form on the heavier types of soil. They can be broken up by using a subsoiler or by deeper ploughing.

Pans may also be formed by the deposition of iron compounds, and sometimes humus, in layers in the soil or subsoil. These are often called chemical or iron pans and may be destroyed in the same way as plough pans. Clay pans form in some soil formation processes.

Soil loosening (which is justified only when necessary) is a term now used to cover many different types of operation, including subsoiling. This aims to improve the structure of the soil and subsoil by breaking pans and having a general loosening effect. An ordinary subsoiler can be used. It is very effective when worked at the correct depth. A subsoiler is a very strong tine (usually two or more are fitted on a toolbar) which can be drawn through the soil and subsoil (about 0.5 m deep and 1 m apart) to break pans and produce a heaving and cracking effect. This will only produce satisfactory results when the subsoil is reasonably dry and drainage is good. The modern types with 'wings' fitted near the base of the tine produce a very good shattering effect.

Other implements which may be used include the Paraplow. The angled mainframe carries three or four sloping tines which have adjustable rear flaps; there are also adjustable sloping discs. The effect is to give a fairly uniform lifting and cracking to about 35 cm. The Shakaerator has five strong tines which are made to vibrate in the ground by a mechanism driven from the tractor. Modified rotary cultivators with fixed tines attached can also be described as soil looseners.

The soil loosening process is very necessary in situations where shallow cultivations – up to 150 mm – have been used for many years and the soil below has become consolidated and impermeable to water and plant roots. Rain water collects on this layer and causes plant death by waterlogging and/or soil erosion. When attempting to loosen a compacted soil/subsoil, it is advisable to start with shallow cultivations, then to work deeper with stronger tines and then, if necessary, with a winged subsoiler. This can be achieved by several passes over the field. However, the one-pass method can also be used. The unit has a toolbar fitted with shallow tines in front, strong spring tines in the middle and winged subsoilers attached to the rear. It obviously requires greater tractor power, but it is an energy-efficient method. This work should only be carried out when the soil crumbles and does not smear.

8.5.6 Control of soil erosion

Soil erosion by water is an increasing problem on many soils, especially on sandy, silty and chalk soils which are in continuous arable cultivation, and where the organic matter is below 2%. Rain splash causes capping of such soils and heavy rain readily runs off instead of soaking into the soil. There may also be panning problems. Erosion can be serious on sloping fields (especially large fields) where the cultivation lines, crop rows, tramlines, etc., run in the direction of the slope, and the wheelings are compacted. In these situations, in a wet time or a thunderstorm, deep rills and gullies, up to a metre deep, can be cut in the fields and up to 150 tonnes/ha of soil washed away. Sometimes crops can be covered at the lower end of a field with soil washed from the upper slopes. Fitting tines behind tractor wheels to rectify compaction, e.g. when spraying, can reduce erosion damage.

Erosion by wind can cause very serious damage on Black Fen soils and on some sandy soils. Various controls are being used, e.g. straw planting by special machines, sowing nursecrops such as mustard or cereals which are selectively killed by herbicides when the crop (e.g. sugar beet) is established, claying or marling and applying sewage sludge.

8.5.7 Soil capping and its prevention

A soil cap is a hard crust, often only about 2–3 cm thick, which sometimes forms on the surface of a soil. It is most likely to form on soils which are low in organic matter. Heavy rain or large droplets of water from rain guns, following secondary cultivations, may cause soil capping. Tractor wheels (especially if slipping), trailers and heavy machinery can also cause capping in wet soils.

Although a soil cap is easily destroyed by weathering (e.g. frost, or wetting and drying) or by cultivations, it may do harm while it lasts because it prevents water moving into the soil, as well as preventing air moving into and out of the soil in wet weather. It also hinders the development of seedlings from small seeds such as grasses and clovers, roots and vegetables.

Chemicals such as cellulose xanthate can be sprayed on seedbeds to prevent soil capping. They enhance seedling establishment and do not affect herbicidal activity in the soil, but are very expensive.

8.6 Control of weeds by cultivation

Historically, the introduction of herbicides reduced the importance of cultivation as a means of controlling weeds. Cereal crops were regarded by many farmers as the cleaning crops instead of the roots and potato crops, mainly because chemical spraying of weeds in cereals was very effective, if expensive. Now that crop prices have fallen dramatically, farmers are looking to return to cultural methods of controlling weeds in order to save on variable costs. Cultivations are now an important part of an integrated crop-management system, with weeds controlled between crops by the production of stale or false seedbeds and inter-row cultivations important in row crops such as potatoes and sugar beet. In organic farming the use of in-crop weeders is an important control measure (see below). The use of rotational set-aside is also an important management tool for the control of noxious weeds such as creeping thistle and couch grass by cultivations.

Annual weeds can be tackled by:

1   Working the stubble after harvest (e.g. discing, cultivating or rotavating) to encourage seeds to germinate. These young weeds can later be destroyed by harrowing or ploughing. Unfortunately, this allows wild oats to increase.
2   Preparing a 'false' seedbed in spring to allow the weed seeds to germinate. These can be killed by cultivation before sowing the root crop.
3   Inter-row hoeing of root crops which can destroy many annual weeds and some perennials.

It should be noted that in organic farming mechanical weed control is the only option and various fine-tined implements have been developed for use in the growing crop.

Perennial weeds such as couch grass, creeping thistle, docks, field bindweed and coltsfoot can usually be controlled satisfactorily by the fallow (i.e. cultivating the soil periodically through the growing season instead of cropping). However, this is very expensive. A fair amount of control can be obtained by short-term working in dry weather.

The traditional method of killing couch has been to drag the rhizomes to the surface in late summer. This is then followed by rotary cultivation – three or four times at two to three week intervals – to chop up the rhizomes. To get a good kill of the plant, adequate growing conditions are necessary for the couch rhizome to respond to being chopped up by sending out more green shoots and thus to hasten its exhaustion. A much more reliable method is to use glyphosate.

The deeper-rooted bindweed, docks, thistles and coltsfoot cannot be controlled satisfactorily just by cultivations, although periodic hoeing and cultivating between the rows of root crops can generally reduce the problem.

Thorough cultivations which provide the most suitable conditions for rapid healthy growth of the crop can often result in the crop outgrowing and smothering the weeds.

8.7   Crop management: key issues

Any good management technique should follow the management cycle:

1   Set targets.
2   Assess progress.
3   Adjust inputs.
4   Monitor success.

This applies equally both to car production and to crop production. The main difference will be the effect of the natural environment on *crop* production, with factors such as the weather being largely outside the farmer's control. However, the cycle can still be useful.

8.7.1   Setting targets

Most farmers set targets of one kind or another. Arable farmers set targets of profitability, of cash flow and of capital investment. From an agronomy point of view it is possible to set targets for crop structure so that maximum profit is obtained. For instance they might set a target of 600 fertile ears/m^2 for a wheat crop at harvest. This can be achieved by sowing the correct number of seeds allowing for targeted losses in the field and manipulating tiller numbers by the judicious use of nitrogen. They would set targets for the amount of fertiliser and agrochemicals to be used to *maintain* this desired crop structure. They would set drilling date and harvest date targets as well as cultivation targets for the next year's crops. These targets should be realistic and achievable. They should involve an element of risk management as well as trying to satisfy the *personal* goals of the farmer or farm manager.

8.7.2 Assessing progress

This is often the hardest thing to do. It requires a quantitative as well as a qualitative analysis of the crop's progress, and this in itself requires good monitoring techniques. It is similar to the actual figures on a forward cash flow being compared to the budgeted figures. It needs frequent crop walking and a good recording system for factors such as leaf emergence, pod formation, ear size, tiller numbers, disease, weed and pest incidence, rainfall and soil moisture deficit. It also needs a knowledge of the science which underlies the decision-making process. Very often, farmers are too busy to do this successfully and will employ people such as crop consultants and agronomists to do it for them.

8.7.3 Adjusting inputs

This is reactive management to problems that have occurred during the growing season. Inputs can be raised or lowered or changed according to crop progress. This is where the external factors such as weather can have a big part to play. There is no such thing as an average season in agriculture so no management blueprints can ever be 100% successful without some kind of modification. Heavy summer rainfall will mean an increase in disease on most crops; a drought may cause problems of tiller survival; late frosts may mean loss of grain sites in early-sown cereals. The farmer must react to these influences by adjustment of inputs in order to maintain his targeted margin.

8.7.4 Monitoring success

Ultimately the bottom line measurement of success is the profit of the farm as a whole, but it is possible to evaluate the contribution given by each enterprise by the use of gross margins. A comparative analysis can look at the margins between different crops on the farm, between different varieties of those crops, between different fields on the farm, between those crops and standard data and between those crops and potentially new crop introductions to the farm system. A farmer can then see which crops are contributing positively or negatively to the business, how much they are contributing and which elements within the margin could be improved. Decisions for the following seasons can then be made. Management programmes for the farm computer can now make this analysis speedy and accurate, although it must be remembered that the output is only as good as the input, and a system of data collection and management should be in place on most farms these days.

8.8 Break crops and crop rotations

8.8.1 Break crops

Since the mid 1980s, when the price of cereals began to fall, farmers have been looking for ways of diversifying their cropping systems. During the 1970s and early 1980s, good prices for wheat and barley meant that high inputs such as sprays and fertilisers could be afforded because the extra yield obtained would easily cover the material and application costs. Cereals were grown year after year on the same land. As margins were squeezed alternative crops were introduced onto farms. Collectively, these were known as break crops and they fall into two categories: a) combinable break crops such as oilseed rape, peas and beans and b) non-combinable break crops such as sugar beet, potatoes, field scale vegetables and grass.

Alternative crops in a break crop system should do several important things:

1   They should break the cycle of weeds, diseases and pests found in the cereal crops.
2   They should improve the soil condition in terms of structure, organic matter levels and nutrition, especially nitrogen.
3   They should make money; certainly, they should make more money than the alternative continuous cereal system of production.
4   They should spread risk. This could be a marketing risk or the risk of crop failure or another type of risk.
5   They should spread the labour and machinery workload throughout the year.
6   They should take the pressure off the storage, cleaning, drying and handling system used on the farm.
7   They should allow the farmer to carry out more effective cultural control of weeds, diseases and pests, thus saving on pesticide costs.
8   They should allow the use of different active ingredients in pesticides to try and overcome the problems of resistance build-up, and to provide cheaper control of the serious grass weeds which can build up in continuous cereal production.

All these factors should give advantages to arable farmers. They should be more profitable, they should have a more interesting system and it should bring environmental advantages to the farm as a whole. However, there are some disadvantages. The cropping system can become too complicated to the point where the crops are not managed well. Small amounts of different crop commodities do not put farmers in a strong marketing position and there can be a problem with broad-leaved weed and volunteer build-up over the course of the rotation.

Some break crops are better at fulfilling the above requirements than others. Oilseed rape is a very good break crop. It breaks the cycles of pests, diseases and weeds, it leaves the soil in good condition, it is drilled and harvested at different times than cereals and there are approved herbicides for use on rape which are effective graminicides. Linseed, on the other hand, is a poor break crop. It is not very competitive against weeds, there are few approved herbicides for the crop, it leaves little nitrogen in the soil and is often harvested late and in difficult conditions so that drilling of the following crop is delayed. Legumes such as peas and beans fix atmospheric nitrogen and so provide nutrition for the following crop. However, both species can leave weed problems in the field and can be difficult crops to predict in terms of yield and profit. However, they do attract a premium arable area payment, unlike oilseeds, so their popularity as break crops could increase in the future.

As arable area payments have fallen, and oil and protein prices have been reduced nearer to those on the world market, so farmers are beginning to think about returning to a simpler system of cereal production with, maybe, one break crop as part of the rotation. During the previous incarnation of long runs of cereals, one of the most difficult problems to overcome was that of the root disease Take-All. This disease builds up in second, third and fourth wheats and then declines to manageable levels thereafter. Because it was soil-borne it was impossible to control other than by rotation and other cultural methods. A seed dressing is now available (and others are in the pipeline) which can help to control Take-All and this may lead to a revival of continuous wheat growing in the main arable areas.

8.8.2 Crop rotations

The rotational use of crops in a fixed sequence around the farm is now the norm on most arable farms. Rotations can be flexible in terms of crops used and length of the sequence. There is nothing new about the use of break crops in rotations with white straw crops. Farmers have been using fodder crops, roots, grass, legumes and cereals since organised agriculture began. Fine-tuning of the rotations has taken place over the centuries as we have developed better cultivation and harvest equipment, better varieties, better fertilisers and better crop-protection materials. The principles remain the same: you need a mixture of exhaustive and restorative crops in an attempt to introduce as much sustainability into the cropping system as possible.

There are nearly as many rotations as there are farmers. A typical combinable crop rotation might be:

Year 1 – oilseed rape
Year 2 – first wheat
Year 3 – winter beans
Year 4 – first wheat
Year 5 – peas
Year 6 – first wheat
Year 7 – winter barley

and then back to oilseed rape and the cycle begins again.

This rotation can be lengthened by the inclusion of some second wheats or three-year leys, or can be shortened by the removal of a break crop and one of the first wheats.

A typical root crop rotation might be:

Year 1 – potatoes
Year 2 – first wheat
Year 3 – second wheat
Year 4 – sugar beet
Year 5 – first wheat

and back to potatoes.

The important factors are that the rotation should be agronomically sound and should not cause management problems with such issues as timeliness of operations, clashes of harvest, lack of separate storage facilities.

8.9   Further reading

Bailey R, *Irrigated Crops and their Management*, Farming Press, 1990.
British Agrochemical Association, *Integrated Crop Management*, BAA, 1996.
Castle L, McCunnall R and Tring B, *Field Drainage – Principles and Practices*, Batsford, 1984.
Clarke J H, *Rotations and Cropping Systems*, AAB, 1996.
Davies B, Eagle L and Finney S, *Resource Management*-Soil, Farming Press, 2001.
Davies B, *Reduced Cultivations for Cereals*, HGCA, 1998.

9

Integrated crop management

9.1 Introduction

Integrated crop management or ICM was first introduced in 1991 in an attempt to improve the public's perception of farming. A new organisation was formed, called Linking Environment and Farming (LEAF), with the aim of promoting good agricultural practice and reassuring the consumer that produce from British agriculture was safe to eat by informing them about both the reasoning and methods of food production. This was a direct response to the criticisms that agriculture was facing following adverse media coverage of several issues. In the ten years since then, there have been many such events and these have included nitrates in drinking water, organophosphate residues in food, salmonella and *E. coli* contamination of food, the BSE crisis and the Foot and Mouth outbreak in 2001. Ten years after the initial conception, the objective remains valid.

A second objective of integrated crop management was to introduce a standard for crop production, which would publicise the high standards of farming in the United Kingdom. This would encompass the principles of good farming practised by producers for many years, and also highlight aspects of environmental management to counteract criticisms about the damaging effects of intensive agriculture on the countryside.

However, it was apparent that although these aims were correct and laudable, it was not sufficient just to demonstrate and promote the principles of ICM. The fresh produce sector realised that further action was required to protect its industry and meet the demands of its customers – the supermarkets. The introduction of Assured Fresh Produce in 1996 gave the field vegetables and salad crops sector an externally verified assurance scheme that covered the health and safety aspects of growing these crops. The Assured Combinable Crops Scheme followed in 1997 to cover the small grain crops such as cereals, oilseed rape, peas and beans.

This chapter outlines the principles of ICM and, with the use of some examples, illustrates how ICM can be implemented on farms.

9.2   Definition

Before defining ICM, it is very important to understand that ICM is a concept and a holistic or whole farm approach. Each farm is unique, not only in respect of the physical landscape, features and climate, but also in the farming system and management. Thus, the implementation of ICM principles on each farm will be different and there cannot be a blueprint.

Whilst there is no single definition of ICM, the following encompasses three major issues: 'The economic production of safe, wholesome food in an environmentally sensitive way' (LEAF, 1993). This has been updated to a 'Whole farm policy aiming to provide efficient and profitable production which is economical, viable and environmentally responsible' (LEAF, 2000).

The farming system adopted must be profitable to be sustainable; and there is no virtue in introducing all the beneficial features demanded by pressure groups if the business is not viable in the long term. It draws attention to the importance of food safety in agriculture, and that producing the food on the farm is an integral part of the food chain, together with the processors, distributors and retailers. In addition, food production does impinge on the environment and water quality, soil erosion, wildlife and conservation are included as key aspects of farm management plans.

Other issues could be highlighted which are important components of the overall crop production system. For example, ICM involves the use of traditional farming practices alongside new techniques. Some commentators argue that there is nothing new in ICM, but when the use of Geographic Information Systems (GIS) for yield and nutrient mapping or disease and insect forecasting systems are considered, it is clear that modern technology has been embraced in current farming systems. The use of cultural control methods and the reduction in agrochemical use is a key part of ICM philosophy. This may not always be the case, but alternative strategies of pest control are considered before pesticides are used, which introduces the idea that ICM requires informed decision making. There is no doubt that the approach requires a high management capability and a commitment in both time and effort.

9.3   Economics

Comparisons of the profitability of ICM and conventional systems have been carried out in farm scale studies by Focus on Farming Practice (FOFP) at Stoughton, Leicestershire, the Rhone Poulenc Management Study (RPMS) at Boarded Barns Farm, Ongar, Essex (Aventis, 2001) and in large scale plot trials in Hampshire, Herefordshire, Cambridgeshire, North Yorkshire and Midlothian as part of a LINK funded programme of research studying Integrated Farming Systems (IFS) (MAFF, 1998a). These studies have shown that, in general, farm income from ICM systems is similar to that from conventionally managed farms. The reduction in inputs achieved with more targeted inputs in an ICM system generally results in a slightly lower yield. However, the reduction in value of the output is compensated for by the reduction in inputs and lower variable costs, giving the overall effect of similar levels of profitability for conventional and ICM systems. Similar findings have been published by Lloyds TSB from calculations they have made on farms following the two systems, which include the fixed costs and not simply comparisons of gross margins. It should be noted that one exception to these findings was at High Mowthorpe, North Yorkshire where the ICM rotation included potatoes, and this gave a much lower yield and hence significantly lower profitability than when potatoes were grown in a conventionally managed rotation. The work concluded that the selection of a suitable break crop was essential to ensure the economic viability of the ICM or IFS rotation.

When ICM was introduced it was hoped that farmers implementing the philosophy would gain commercial advantage by selling their produce at a premium. To date this has not been possible. The commercial advantage is that produce grown to assured standards can be traded, and that grown with no quality assurance cannot. LEAF is attempting to launch its own logo on food sold in the supermarket.

9.4   Crop rotations

Growing a sequence of crop species has been practised in Britain for many years with the Norfolk Four Course Rotation of 'Turnip' Viscount Townshend of Raynham the first documented example dating from 1730, which included roots (originally turnips), spring barley, red clover for hay or grazing and winter wheat. Sound rotations are a cornerstone of the concept of ICM, because they provide the opportunity to control pests, diseases and weeds by cultural means rather than relying solely on chemical control. They provide a mixture of exhaustive crops such as cereals, which remove valuable nitrogen and potassium from the soil and restorative crops such as legumes, which fix nitrogen from the air into organic compounds. These are released to the following crop after the residues are incorporated into the soil. This reduces the requirement for inorganic nitrogen fertiliser that is not only expensive to produce, but is also a major energy input to growing crops because the manufacturing process involves using high temperature and pressure to produce nitric acid and then ammonium nitrate.

9.4.1 Examples of rotations
The two rotations given below are examples taken from farms practising ICM that show these principles.

Winter Beans → Winter Wheat → Winter Wheat →

Winter Oilseed Rape → Winter Wheat

The arable rotation was taken from a heavy land farm that had a five year rotation with two break crops, two first wheats and a second wheat crop. This provides two high yielding, lower input first wheats and does not allow the build-up of take-all in the cereal crops. Although both break crops are affected by *Sclerotinia* stem rot, the winter beans are attacked by *S. trifoliorum* and the oilseed rape by *S. sclerotiorum*, which means that a five year rotation provides sufficient break between susceptible crops. Winter beans are normally sown by ploughing down in late October or early November, resulting in the early flush of autumn-germinating weeds, especially problem ones such as cleavers, blackgrass and sterile brome, being buried. The winter beans fix nitrogen and leave the residues for the following wheat crop, reducing the spring fertiliser recommendation by around 30 kg of nitrogen per hectare. There may also be the possibility of using minimal cultivation techniques to establish the first wheats following beans or oilseed rape provided the soil structure is good, and no compacted layers are present to impede root growth and drainage. There are also the added benefits of reducing energy costs and more timely drilling because work rates are higher.

Grass Ley → Grass Ley → Winter Wheat → Winter Wheat → Oats

The second rotation was employed on a mixed farm, where the ley was utilised by the dairy herd and followers and the winter wheat and oats were grown as feed for the cattle. The two-year ley is a fertility-building phase, where organic matter accumulates and improves the structure of the soil. The nutrient status of the soil depends on the use of the ley, with grazing allowing the return of animal wastes, whereas taking silage cuts or hay crops has the potential to deplete soil potassium levels because of luxury uptake in grass removed for feed. This rotation also has a first wheat crop and does not allow the build-up of take-all (*Gauemannomyces graminis*), because the oat crop is affected by a different species to the winter wheat. Winter wheat is attacked by *G. graminis* var, *tritici* and oats are immune, whereas oats are attacked by *G. graminis* var, *avenae*. There are several potential pest problems – wireworms, leatherjackets, slugs, frit fly, volunteer ryegrass, broad-leaved weeds and aphids – for the first wheat following ploughing up of the ryegrass ley. Cultural control of the invertebrate pests would involve ploughing the grass out and leaving as long a period as possible before the cereal crop. However, on the farm where this rotation was practised, frit fly was still a major problem for the first wheat sown in late September or early October. The only option for this economically damaging pest is to apply an organophosphate insecticide, such as chlorpyrifos, for control and significant yield loss is not suffered. This illustrates that ICM involves looking at alternative methods of control by examining the cultural options available, considering alternative pesticides which may be less harmful in the environment, but in the absence of alternatives that give effective control, the only option left is to recommend an organophosphate which is not only potentially damaging to other invertebrate species, but also the populations take time to recover following application.

9.4.2 Set-aside

One part of arable cropping systems that has not been considered in the two rotations described above is set-aside, which the European Union has set at a default level of 10% until 2006 under the Agenda 2000 proposals (MAFF, 2001a). There is great flexibility in how set-aside can be incorporated into arable cropping systems. Rotational set-aside is still an option, with farmers generally selecting the least profitable crop, such as a second wheat or winter barley, in a rotation for the area to be set-aside. This land is generally managed as a low cost option, allowing natural regeneration of weeds and volunteers with a spray of glyphosate applied after 15 April to leave a weed-free entry for the following crop. Leaving stubbles through the autumn is very valuable as it provides a food source of seeds, both weed and crop, as well as invertebrates on volunteers and other weeds, for foraging birds. However, there are now other options, such as multiannual set-aside, which allow specific habitats to be created for species of farmland bird. One example would be to create a spring and summer fallow to provide suitable nesting sites for skylarks and lapwings. Set-aside can also be used on headlands provided it is at least 20 m wide and 0.3 ha in area, adjacent to field margins. Recently this was reduced to a ten m margin along water courses which is compatible with the Local Environment Risk Assessments for Pesticides (LERAPS), and fulfils the need for buffer strips adjacent to water courses to prevent pesticide drift from boom-spraying operations (Pesticide Safety Directorate, 1999). This shows that farmers implementing the philosophy of ICM will manage the arable area of the farm to ensure profitable crops are produced, and will consider carefully the options that they have to manage the non-cropped areas on their farms, such as set-aside.

9.5 Soil management

Good soil management has always been recognised as an essential element of growing high yielding, profitable crops. The aim is to produce a seedbed which is firm enough to provide good soil–seed contact but has spaces between aggregates that allow free movement of water, air and no impedance to root growth. Establishing the desired plant populations and conditions that facilitate uptake of nutrients and water will provide the potential for a high yielding crop provided disease is carefully managed.

Traditionally, ploughing has been the primary method of cultivation, inverting the soil to bury weeds and alleviate surface compaction. This is a costly operation, both in terms of the time taken but also the amount of energy required, with contractors currently charging around £30 per ha. In the 1990s combination drills, incorporating a Roterra to produce a seedbed and a pneumatic seed drill, have been used successfully as a one-pass system. However, with the reduction in the grain price from above £120 to approximately £70/t (2001) severely affecting the profitability of winter wheat, interest in minimal cultivation and direct drilling has increased again.

These crop establishment techniques fit comfortably with the philosophy behind ICM. The use of minimum cultivation or 'Lo till', as it has been referred to recently, has the benefit of reducing the amount of energy used to establish a crop and therefore meets the aim of reducing farm energy inputs. There are also additional benefits to invertebrates and wildlife. Trials carried out comparing direct drilling, minimal cultivation techniques and ploughing in the 1970s and 1980s showed that earthworm populations were inversely related to the amount of cultivation carried out. The slicing and inversion associated with ploughing kills earthworms resulting in fewer earthworms/m^2 than minimum cultivation and direct drilling. In trials carried out on the LIFE project at Long Ashton, there was circumstantial evidence to suggest that a higher incidence of Barley Yellow Dwarf Virus (BYDV) on winter cereals, established by ploughing compared with minimum cultivation, was linked to lower numbers of predatory invertebrates, such as ground or rove beetles, which were killed by the soil inversion. Currently research is being carried out to discover if the reduced soil disturbance associated with minimum cultivation leaves more seeds and insects that are potential food sources for birds foraging in autumn and winter.

Although these benefits indicate that minimum tillage could be regarded as a more environmentally friendly production system than ploughing, there are potential risks associated with the system that need to be considered. Firstly, the system will only be successful if the soil structure is good, emphasising the need for growers to take a spade and dig inspection pits to assess if compaction is a problem and if deep cultivation is required. Failure to establish the desired plant population in autumn is an indication of impeded drainage and restricted rooting. The second area of concern highlighted by some agronomists is the potential of autumn-germinating grass weeds, e.g. black-grass and sterile brome, to thrive with minimum cultivations and cause serious problems. Again, there was plenty of evidence collected from earlier minimum cultivation trials to show that weed numbers built up quickly and affected yield when seed was left on the surface or incorporated into the top few cm.

Therefore, there must be a compromise situation to ensure the economic viability and sustainability of the arable cropping system. Taking the arable rotation presented previously as the example, minimum cultivation could be used to establish the first wheats following oilseed rape and beans, with ploughing used to establish the break crops and second wheat. This would provide some environmental benefit in terms of lower establishment costs, a reduction in fossil fuel use, improved foraging for farmland birds in autumn, whilst ensuring that autumn-germinating grass weeds are kept under control over the rotation and the threat of herbicide resistance is not increased.

9.6   Crop nutrition

While good crop establishment is essential for successful and profitable crop management, the provision of nutrients at the correct rate and timing allows a crop to grow to its full potential. However, the inappropriate use of fertilisers by applying at too high a rate or at times of the year when they are not utilised, can lead to losses to water and air causing pollution problems.

In an ICM system this starts with an understanding of the soil and the inherent properties associated with sand, silt, clay and organic matter particles that influence soil fertility. This will provide information on the ability of a soil to hold nutrients or, conversely, the risk of nutrient leaching, and the likely contribution of nitrogen or phosphorus from mineralisation of organic matter. Levels of the major nutrients phosphorus, potassium and magnesium together with soil pH should be checked periodically every three to five years by soil analysis to form the basis of crop nutrition planning.

Nutrient planning can be carried out on a field by field basis, tailoring the crop's requirement based on recommendations given in the *Fertiliser Recommendations for Agricultural and Horticultural Crops:* RB209. Nitrogen recommendations have been adjusted according to previous cropping, reducing the amount of inorganic fertiliser nitrogen applied based on estimates of the release of the residues from soil sampling, and crop trials data. The updated fertiliser recommendations concentrate on soil mineral nitrogen (SMN), which is a measure of the amount of ammonium and nitrate nitrogen available to the crop in spring, which is influenced by soil type, climate and previous crop.

Where there is a fixed cropping rotation on a farm, the nutrient management plan can be improved by carrying out nutrient balance calculations. From records of crop yields, the nutrient removal of phosphorus in the form of phosphate and of potassium as potash can be calculated from standard values. These are then totalled over the rotation and compared with the recommendations calculated from RB209. In many situations there will be imbalances between offtakes and inputs that can be corrected. This should only be carried out by trained agronomists, such as those holding the Fertiliser Advisers' Certification and Training Scheme (FACTS), who can make adjustments according to individual crop needs.

ICM places great emphasis on the utilisation of organic manures in cropping systems and assessment of their nutrient contribution when calculating inorganic fertiliser requirements. For an individual crop this means preferably analysing the manure to assess the nutrient content, and subtracting the quantity of available nitrogen, phosphate and potash from the fertiliser recommendation obtained from RB209. The second option is to use standard values provided by MAFF in RB209 for the available nutrients, which in the case of nitrogen is adjusted according to the time of application.

Records should be kept of all operations, including primary cultivations and applications with dates, operator name, field, crop, soil and weather conditions. These should also include the type of fertiliser or organic manure applied and storage details. In Nitrate Vulnerable Zones (NVZs) records of all organic and inorganic nitrogen applications must be kept. All staff carrying out field operations should have access to records, be fully trained and have access to the Nutrient Management Plan and the Codes of Good Agricultural Practice for Air, Soil and Water. Careful planning of crop nutrition is essential for profitable crop production, but can only be achieved through correct timing of application, machinery maintenance and calibration, which will avoid waste and pollution.

9.7 Crop protection

The official government position regarding the use of crop protection chemicals is to minimise their use (Pesticide Forum, 2000). The concept of ICM, which looks to utilise cultural control methods where the opportunities arise and reduce the reliance on chemical control is very much in line with this policy. Therefore, many publications, leaflets and the MAFF (now DEFRA) website promote ICM as a farming system that will deliver many of their objectives and ICM is included as one of the sustainability indices.

The history of agricultural production has been inextricably linked to the control of weeds, pests and diseases in crop production practices. There is no doubt that the production of consistently high yielding, high quality crops requires the use of pesticides. Although organic production methods provide a viable alternative, the yields achieved are lower and the quality of the produce is generally much more variable. The aim, therefore, of an ICM system is to utilise the cultural control methods that are known to reduce the incidence of pests on crops, and then use crop protection chemicals as a final resort. An excellent example of this approach is the production of a fine, firm seedbed for cereal and other small seeded crops, which restricts the movement of slugs in the surface soil horizons where the seed is placed. However, the point must be made that cultural control techniques are not always effective in providing sufficient protection to the crop. In the grass rotation discussed a major problem pest was the attack of frit fly on the cereal crop established immediately following the grass ley. Whilst ploughing early and leaving a period of six weeks before drilling the winter wheat reduced the attack of frit fly on the emerging plants, there was still a major reduction in plant numbers and final yield. In this situation the only other option would have been to spray with a crop protection chemical to prevent frit fly damage.

Whilst it has been stated that rotations are a cornerstone of ICM, there can be little doubt that the choice of variety selected to be grown is an integral part of the philosophy. If the aim is to reduce the application of crop protection chemicals the starting point must be to choose varieties that have resistance to disease from the NIAB recommended lists. The next important step in the implementation of a crop protection policy involves the correct identification and evaluation of a pest problem. This has been recognised by the requirement of all advisers, recommending crop protection products and fertilisers, to hold the BASIS Certificate of Competence and the FACTS Fertiliser Certificate respectively, in order to meet the criteria of the Assured Fresh Produce and Assured Combinable Crops Schemes. From December 2000 Integrated Crop Management was incorporated as an integral part of the BASIS Certificate to ensure that all advisers are competent to implement ICM policies on farm.

Following correct identification of a particular pest, the next step is to try to evaluate the seriousness of the problem and the need for corrective action or prevention. Threshold values for pests and diseases have been successfully used. For example, the threshold for pollen beetle on oilseed rape is 15 per plant at flower bud formation. Although it may be considered by some to be prophylactic, the use of prediction models to justify the application of certain products can still be recommended in an ICM system. The forecasting system, based on aphid catches that is co-ordinated by Brooms Barn Experimental Station, is a good example of how simple monitoring and forecasting can be invaluable tools for predicting the likely invasion of aphids on sugar beet, and then by controlling them reduce the impact of Virus Yellows.

Once the need for a crop protection product has been justified the choice of product, rate and timing are all very important decisions to make. The crop protection chemical selected will not only depend on its ability to control the pest, but also on its environmental profile. Properties such as the ease with which the chemical leaches to ground water, persistence in the environment, volatility and effect on non-target organisms can be taken into account. The best known example is the use of pirimicarb to control aphids, because it is not toxic to other beneficial invertebrates, such as ground beetles, ladybirds and bees, whereas the cheaper alternative pyrethroids are toxic to these organisms.

Integrated crop management also places great emphasis on operator training for all aspects of work carried out on the farm. The area of pesticide application provides an excellent illustration, with the spray operator being required to hold a PA2A certificate, issued by the National Proficiency Tests Council to show professional competence in calibrating and operating a boom spray applicator. Further training will be required to understand and follow the code of practice for the safe use of pesticides on farms, which will involve carrying out a Control of Substances Hazardous to Health (COSHH) assessment of the risks to health from using a pesticide before work starts. This will involve assessing the level of personal protective equipment which the spray operator is required to wear.

The spray operator will also need to be familiar with the Health and Safety Executive (HSE) Information Sheet No 16, *Guidance on Storing Pesticides for Farmers and other Professional Users*. There are also issues involved in the disposal of pesticides, such as the EU Groundwater Directive, which requires farmers to have designated areas for disposal of pesticides. Knowledge of the sprayer and how it operates can reduce risk to the operator and environment through the use of closed transfer systems for filling or low-drift nozzles to reduce spray drift. The field of crop protection demonstrates the requirement for a well-trained workforce, conversant with the relevant legislation and codes of practice, as well as the practical skills necessary to carry out certain tasks in order to conform to ICM principles.

9.8   Food quality and safety in the food chain

The farmer is the first link in the food chain and determines the way crops are planted, the inputs applied to the growing crop and the method of harvesting and storage. Therefore, the farmer has a significant impact on the safety and quality of a food product, especially where fruit and vegetables reach the consumer with very little further processing. The farm assurance schemes set standards for good agricultural practice, and require farmers to provide evidence of the methods used in producing their crops. A further quality control imposed is the maximum residue level (MRL) of a pesticide in food on the supermarket shelf. This is monitored and tested for by DEFRA.

The management of food safety in the food chain is being addressed by assurance schemes, with the use of Hazard Analysis of Critical Control Points (HACCP) systems being seen as the preferred approach to management at all stages of food production, including agricultural production. The HACCP system identifies hazards, which can cause the consumer temporary or permanent injury, and risk as the probability of a hazard occurring. Current EU hygiene regulations include reference to HACCP, proposing that it should be extended to primary production of food on farm, which provide effective control of pathogens and their hazards, fulfilling the key aspects of the Assured Produce or EUREP Schemes.

9.9   Wildlife and conservation

Farmers have been seen as the custodians of the countryside because of the way they have managed and developed landscapes and habitats over the centuries. However, recent surveys have shown that the numbers of breeding birds in farmland habitats have declined substantially. This has focused attention on the farming practices in an attempt to find the reasons and halt or reverse these declines. Loss of habitats, such as hedgerows, woodland and rough grazing, drainage of wetlands, change of cropping practice from spring-sown crops to predominantly winter planting, intensive cropping systems dependent on the use of pesticides, fertilisers and increased mechanisation are considered to be the main factors causing farmland birds to decline.

Integrated crop management acknowledges the countryside is our heritage and that it is incumbent on everyone, including farmers and the farming industry, to work towards conserving it. The first step in any on-farm improvement is to carry out a survey of the site, to determine what landscape features are present, record the habitats and their extent, and make some assessment of the species present, possibly quantifying the numbers of selected indicator species. This would then form the basis of a management plan for both the cropped and uncropped areas on the farm.

On many farms in the UK this would focus on field margins and particularly hedgerows as a habitat that is a prominent component of the rural landscape. The management of field margins and Beetle Banks has been well documented. The same publications also detail how the cropped area adjacent to field margins can be managed in the form of Conservation Headlands, which are selectively sprayed to encourage less troublesome weeds and thus insects that are essential chick food.

Field margins have been viewed by some farmers as a source of pernicious weeds, such as cleavers, couch and sterile brome. The previous answer was to prevent spread into the cropped area by spraying into the hedge bottom with a broad spectrum herbicide. This was not only expensive and time-consuming because it was carried out each year, but also destroyed the field margin vegetation and frequently reduced the growth of the hedge. ICM advocates leaving a minimum one metre grass strip of non-invasive perennial grass species (e.g. mixtures of Cocksfoot, Yorkshire Fog, Timothy and Red Fescue) adjacent to the hedge or field margin. On many farms this grass margin has been extended to widths varying from 2 to 20m using options from the Countryside Stewardship Scheme or the strategic use of set-aside. Some conservationists would like to see the biodiversity of these sown strips improved by the addition of native wild flowers to encourage beneficial insects, such as bees, hoverflies and lacewings.

Conservation Headlands have been promoted to increase further the biodiversity adjacent to the field margins. A grassy strip adjacent to the hedgerow could provide suitable nesting habitat for grey partridge. However, in cereal crops treated with residual herbicides very few weeds survive to provide habitat for insects and particularly sawfly larvae, which are a preferred food source for partridge chicks. A conservation headland is typically the outer six m of a field that is selectively sprayed to control serious grass weeds (wild oats and blackgrass) and cleavers, but allows other broad-leaved weeds that are hosts for invertebrates to survive. This concept has been shown to be successful in promoting biodiversity and increased partridge chick survival, but does increase the weed seed burden in the harvested crop from the field headland.

Beetle banks are grassy strips created in the centre of large arable fields to provide suitable habitat for predatory insects as well as nesting habitat for partridge or skylark. They are designed to provide a habitat of tussocky grasses (e.g. cocksfoot) where ground beetles can overwinter in relatively dry conditions. The predation of aphids is more effective because the beetles move into the crop from the beetle bank in the centre of the field and the field margin round the outside.

These measures together with the sympathetic management of existing field margins, such as hedgerows, ditches and associated ponds can have a major impact on wildlife. For example, the introduction of a management plan for cutting hedges would advocate rotational cutting on a two- to three-year cycle to allow berry development. The actual cutting would be carried out in late winter, where possible, to allow birds to forage through the autumn and winter when food is scarce. Similarly, cereal stubbles could be left undisturbed until spring before sugar beet or maize is sown to provide valuable feed for a wide variety of farmland birds. These are options that all farmers can consider undertaking to enhance the biodiversity of arable farming landscapes.

9.10   Crop assurance schemes

Following the introduction of ICM as an attempt to improve the public image of agriculture, some sectors realised that the industry needed to take positive action to improve consumer confidence. The growers supplying vegetable and salad crops were the first to act by launching the Assured Fresh Produce Scheme in 1996. This was an initiative by the producers themselves, putting in place a system which not only records farming operations and inputs but also allows external scrutiny or verification of the production systems. There had already been poor publicity in the arable sector with newspaper reports linking cancer clusters in Lincolnshire with organophosphate residues in carrots. The voluntary introduction of the scheme would demonstrate to supermarkets that field vegetables and salad crops were grown to professionally agreed protocols and provide traceability. This has given rise to several quotes, such as 'from Plough to Plate' and 'from Field to Fork' indicating that the whole food supply chain is included.

More recently the Assured Combinable Crops Scheme was launched in 1998 to address the same issues for arable crops. Whilst not placed directly onto supermarket shelves like many vegetables and salad crops, wheat grown for bread making is processed and baked at the mills prior to being sold in retail outlets. It is therefore just as important for these to be subjected to the same quality assurance standards as other food. Both the Assured Fresh Produce and Assured Combinable Crops Schemes concentrate on ensuring that good farming practice is followed when growing crops. Although it may address some environmental issues, there is no requirement for farmers to undertake positive habitat management that will benefit wildlife. Sainsbury Plc. did address this issue by offering preferred growers the opportunity to carry out a Farming and Wildlife Advisory Group (FWAG) Landwise Assessment of their farms. This assesses the range of habitats and wildlife present, and then provides a plan for the development and improvement of the farm in the future. This has been taken further by developing a Biodiversity Action Plan (BAP) for species that are present on the farm and are highlighted as important species for conservation in the locality.

Both of the assurance schemes have a cost to the farmer, which depends on the area of land being used for crop production. Typically, this starts at around £100 for less than 30ha which can rise to £350 for more than 250ha. The fee pays for the administration of the scheme, and includes a visit by the external verifier – organised by Checkmate International Ltd.

The EUREP GAP Protocol was introduced in November 1999 to standardise the requirements of retailers in the EU and bring benefits to the food supply industry. The Euro-Retailer Produce Working Group (EUREP) is a technical working party aimed at promoting and encouraging good agricultural practice (GAP) in the fruit and vegetable production industry. This scheme aims to fulfil the same role as the Assured Fresh Produce and Assured Combinable Crops Schemes established in the UK, but actually goes further in requiring the producer to have an environmental management plan that addresses wildlife and conservation as well as health and safety issues.

9.11   Further reading

Aventis, *Food for Thought. Sustainable Food Production for the 21st Century Consumer,* Ongar, Essex, Aventis Crop Science UK Ltd, 2001.

British Agrochemicals Association, *Arable Wildlife: Protecting Non-target Species,* Peterborough, BAA, 1997.

Campbell L H, Avery M I, Donald P F, Green R E and Wilson J D, *A Review of the Indirect Effects of Pesticides on Birds,* Joint Nature Conservation Committee Report 227, Peterborough, JNCC, 1997.

Euro-Retailer Produce Working Group, *EUREP GAP Verification 2000,* EUREP EHI-EuroHandelsinstitut, 50829 Köln, 1999.

Game Conservancy Trust, The, *Guidelines for the Management of Field Margins,* Fordingbridge, Hants, The Game Conservancy Trust, 1995.

LEAF, *Guidelines for Integrated Crop Management*, Stoneleigh, Warwickshire, LEAF, 1993.

LEAF, *The LEAF Handbook for Integrated Crop Management*, Stoneleigh, Warwickshire, LEAF, 2000.

MAFF, *Integrated Farming. A Report from the Integrated Arable Crop Production Alliance for Farmers, Agronomists and Advisers*, MAFF Publications, 1998a.

MAFF, *Code of Practice for the Safe Use of Pesticides on Farms*, MAFF Publications, 1998b.

MAFF, *Arable Area Payments Scheme. Explanatory Guide: 2001 Update*, MAFF Publications, 2001a (www.defra.gov.uk).

MAFF, *Fertiliser Recommendations for Agricultural and Horticultural Crops*: RB209. MAFF Publications, 2001b.

Pesticide Forum, 2000: See website www.defra.gov.uk

Pesticide Safety Directorate, *Local Environmental Risk Assessments for Pesticides. A Practical Guide*, MAFF Publications, 1999 and 2002.

10

Organic crop husbandry

10.1   Organic farming

Organic farming is also known as biological or ecological farming.

The ideas on organic farming started early in the twentieth century. In 1946 a group of like-minded farmers and scientists, including Lady Eve Balfour and Sir Albert Howard, founded the first UK organisation, The Soil Association. It was not until the 1990s that there was a large increase in interest in organic farming both in the United Kingdom and in many parts of Europe and the world. About 3% of land in the European Union is now farmed organically.

With the many recent food safety scares the demand for organic produce has increased both for fresh and processed food. There are still good premiums to cover some of the extra costs, although competition is increasing due to increased availability. Commodity prices for conventional growers have been poor and combined with the grants available to support farmers during the conversion period; more farmers have been encouraged to convert to organic farming. In the past some farmers wanted to convert as they agreed with the ideals of organic farming, but without grant aid or a market were forced economically to abstain from converting.

Organic farming is a system which avoids the routine use of readily soluble fertilisers and or agrochemicals whether naturally occurring or not. It is a system which aims to use renewable resources where possible and thus it is considered to be more sustainable than conventional methods. Organic farming is not a return to pre-1940s farming. It uses current technology and relies heavily on good husbandry. Organic farming aims to maintain soil fertility and organic matter as well as encouraging biological cycles. There is a large reliance on legumes for supplying nitrogen requirements. Weeds, pests and diseases are controlled mainly by non-chemical methods. Animal welfare, pollution and the wider social and environmental issues are also very important.

Some organic crops such as cereals are cheaper to grow than conventional crops, but to be viable they cannot be grown so frequently in the rotation. Many field vegetables are much more expensive to grow than conventionally grown crops especially if much hand weeding is required, as in carrots. Yields from organic crops can be very variable and are often only 60–70% of normal average yields.

For organic farming to work, the grower has to be committed to the system. There has to be a change in the approach to the husbandry of growing the crops as agrochemicals and fertilisers cannot be applied if a serious problem arises. The risk of crop loss is much higher in organic than in conventionally grown crops.

10.2   Achieving organic status

All products sold as organic have to be certified. All countries have their own certifying bodies.

10.2.1   UK Standards

In the UK there are now a number of certifying bodies including the Soil Association, Organic Farmers and Growers, Organic Food Federation, United Kingdom Register of Organic Food Standards (UKROFS), Scottish Organic Producers Association and Irish Organic Farmers and Growers. UKROFS oversees and checks that these organisations conform to the UK and EU regulations. Each farm is annually inspected by its registered body to confirm that the Approved Guidelines are being followed. UKROFS then carries out random check inspections.

Standards are constantly being amended; it is important that farmers maintain good farm records and keep up to date with the changes. Current recommendations include: no Genetically Modified Organisms (GMOs) can be fed to stock or grown within a 6 mile radius of an organic farm; farmyard manure can only come from stock that has not been fed GMOs; written proof is necessary for all these stipulations. Bought-in farmyard manure has to be composted. As organic seed stocks are limited, farmers can currently obtain derogations from their certifying body to plant non-organic seed. In the future the target is for all seed sown to be organic. A limited number of fertilisers and crop protection products are permitted to be applied to organic crops. Some products have restricted use such as meadow salt (a potassium source) and copper for control of blight. Farmers must have a very good reason to apply these restricted products, and they have to obtain a derogation from their certification body before applying these products. The permitted and restricted lists are constantly being amended according to UK and EU regulations. There are very strict livestock regulations relating to BSE, stocking rates and inclusion of conventionally produced feed stuffs in diets.

During the annual inspection, input and output records as well as accounts have to be available. Crop records required include: treatments applied for the last 3 years for land in-conversion; crop rotation plans and cropping areas; cropping history including yields; application of farmyard manures and other fertiliser, composting treatments, rates and source; source and types of products used for pest and disease control and source and types of seeds and transplants.

After inspection a compliance form is sent to the farmer with details of the certification decision for each enterprise on the farm, and details of any action to be taken. Each farm must provide certification details when selling organic produce.

10.2.2 The conversion period

When a farmer decides to convert to organic production methods a conversion plan is produced in order to get certified by an approved body. Depending on the enterprises on the farm the plan includes details of land areas, dates of conversion, cropping and rotations, stocking rates for livestock, forage management and details of stock housing and produce storage. Conversion can start from the date of application of the last non-permitted product. The ground is not fully organic until two years after that date. Some farms convert all their land in one go; others find it easier for rotational purposes to convert over a few years. At present, in the UK, a farmer does not have to convert the whole farm.

Currently DEFRA runs an organic conversion information service (OCIS). This is a free service. On initial contact the farmer will receive information on organic farming methods, marketing and certification. In addition, the farmer can have one-and-a-half days free advisory visits. These visits are with organic advisors from a number of advisory organisations such as ADAS (Agricultural Development and Advisory Service) and The Organic Advisory Service (OAS) at Elm Farm Research Centre, Berkshire.

During the first year of conversion, crops can only be sold as conventionally grown non-organic. Second year in-conversion produce can sometimes be sold at a premium which is usually lower than for full organic products. It is common practice to grow fertility-building crops rather than cash crops during the two-year conversion period. Set-aside sown with an approved quantity of clover, cut and mulched, can be a useful in-conversion crop.

There is usually a reduction in farm returns during the in-conversion period; this can be partially offset by the current UK Organic Farming Scheme grant. Payments are for five years but are greatest in the two in-conversion years. Farmers who receive these payments are still able to claim other standard grants such as the arable area payments. Some of the Countryside Stewardship and Environmentally Sensitive Area (ESA) payments may be reduced.

10.3 Rotations

Rotation is the basis of a good organic system. Initially, when changing to this farming system, during the in-conversion period fertility-building crops such as clover are usually grown. A good rotation will aim to balance nutrient requirements, minimise nutrient losses, maintain soil organic matter and structure and help eliminate or reduce weed, pest and/or disease problems. Organic rotations use more legumes than conventional cropping. Red and white clover, lucerne and even vetches are grown to boost soil nitrogen supply.

Green manures such as mustard, vetches or rye are commonly grown after a summer-harvested crop and before a spring-planted crop. Green manures reduce the losses of soil mineral nitrogen during the winter. They also provide plant cover during the winter so that the soil is not prone or liable to suffer from soil water erosion. When ploughed-in in the spring the green manures release the readily available nitrogenous compounds for the following crop.

Normally no more than two years of cereals are grown in succession. Milling wheat commands the highest cereal premium, often more than double the conventional price. With the increase in organic livestock enterprises there is growing demand for organic feed grain. Other arable crops grown include grain legumes, potatoes and sugar beet. There is no one perfect rotation; each farm will have a rotation to suit the enterprises of that specific farm. Continuous cereals are prohibited and there must be at least a two year break before cropping again with alliums, brassicas or potatoes on the same land.

An example of a rotation used on mixed organic units is the following:

- Two to three years grass clover.
- Potatoes.
- Cereal undersown (red clover).
- Red clover.
- Wheat.
- Spring cereal (undersown).

An alternative version might be:

- Two to three years grass/clover ley.
- Wheat.
- Rye or oats.
- Grain legume or red clover.
- Barley or oats undersown.

Most of the organic manure available would be applied to the leys or potatoes.

A number of farms practise a stockless system. In stockless rotation, usually two years of cereals will follow a legume. The leguminous crop could either be field beans, a forage legume for seed or set-aside. Straw is incorporated and any spring crops grown could follow a green manure. This system has been successfully run for many years on some farms. Other farms have found that livestock needs to be introduced. This is only feasible on farms where there are adequate livestock facilities or where there are local organic livestock units looking for grass keep.

A rotation for a stockless system:

- Set-aside (containing clover) cut and mulched.
- Potatoes.
- Winter wheat.
- A green manure crop.
- Spring cereal (undersown).

This rotation can be adjusted to include vegetables. For example, two years of vegetables could follow the set-aside. A *Brassica* such as cabbage would be planted first, followed by a less nitrogen-demanding crop such as carrots or leeks.

10.4   Soil and plant nutrition

An important aim in organic farming is to maintain and, if possible, improve the soil – the structure, flora and fauna. The ideal is to balance nutrients removed from the farm, in harvested crops or livestock, with inputs from bought-in feed, straw and seed. There is limited data to suggest that organic farming increases soil bacteria and fungal activity and that these are very important in mobilising some of the soil's nutrient reserves.

Soil structure can be affected by crops grown, cultivations and application of organic manures. Some crops, which have deep tap roots such as the vetches and field beans have been shown to improve soil structure. Shallow cultivations tend to be favoured to try and maintain the level of available nutrients in the topsoil. It is not always possible to cultivate shallowly when ploughing out a ley or after a vegetable crop with large crop residues. When ploughing out grass leys there can be an initial flush of available nitrogen released, and it can be difficult to time cultivations and crop drilling to reduce nitrate leaching. Quite often the use of the stale seedbed technique for weed control and avoidance of some pests and diseases leads to late drilling for some crops, and this means that the soil can be liable to nutrient leaching before the crop is able to use the available nutrients.

Grazed leys and incorporation of crop remains such as straw and green manure crops help to maintain organic matter.

10.4.1   Inorganic fertilisers

Nutrient off-takes should be balanced by nutrient inputs. There is only a limited range of fertilisers that are permitted for use on an organically grown crop. Most of these products are not very soluble and tend to become available to the crop over a long period. Production should be planned to minimise the need for bought-in nutrients. Lime, e.g. ground limestone or chalk; phosphorus compounds, e.g. natural rock phosphate; magnesium, e.g. magnesium rock or kieserite; and trace elements, e.g. liquid seaweed are examples of permitted products.

There are no permitted potassium sources other than farm yard manure (FYM), bought-in straw and animal feeds. Under certain circumstances some restricted products can be applied. Potassium, e.g. meadow salt or Silvinite, is an example of a restricted fertiliser. In the case of Silvinite, the grower has to submit a recent soil analysis result for the field in question. The Index must be 0 or 1 for derogation to be accepted. This product is prohibited for use in some other European countries.

Because of the restrictions on fertiliser and manure sources several stockless organic farms have now introduced a livestock system. There are, however, a number of successful stockless farms, which often have started with high levels of soil potassium or have soils with a reasonable clay content. These soils will have sufficient amounts of potassium mineralised each year for crop production.

10.4.2 Organic manures

As the use of inorganic fertilisers is limited, farmers rely on the application of organic manures including approved sources of amenity waste for both soil improvement and provision of crop nutrients. Some organic manures can be brought onto a holding as long as they come from an approved livestock system. The use of plant wastes and animal manures from non-organic sources are restricted and must be approved by the certifying body before use. FYM from non-organic sources has to be composted for three months or stacked for six months. As noted earlier, no GM feed should have been fed to the stock producing the manure.

There are several ways of managing compost. The aim is to limit nutrient loss. FYM should be composted to reduce soil-borne diseases and weed seeds. Turning is encouraged to ensure an adequate breakdown process. Turning usually increases aeration and speeds up the microbial activity, but the more the manure is turned the greater the gaseous loss of nitrogenous compounds. The nitrogen in well-composted FYM is usually fairly stable and not prone to soil leaching. During composting the manure will reduce in volume by 50% so reducing spreading costs. There is a lot of debate about whether to cover the manure pile. Covering can help lower the moisture content and leachate losses, but more commonly the heap is outside, uncovered. If so, the aim is to build the heap so that excess water runs off and, if possible, the leachate is collected.

Manure management is important in order to minimise nutrient losses during the composting period. Care must also be taken to ensure that there is no risk of run-off and pollution of water courses during application. In an organic system, when the manure has been composted, there can be a loss of nitrogen and potassium. As organic feed can be lower in potassium than in conventional sources, normally organic manure contains less available potassium than standard data would suggest. Use of sewage sludge, effluents and sludge-based composts is prohibited. Manure from some livestock systems such as intensive poultry battery systems is also prohibited.

10.4.3   Nitrogen sources

Most of the nitrogen in well-composted FYM is in an organic form and will be slowly released over the following years. Only a relatively small amount will be available in the first season after application. Organic farmers are restricted to applying no more than the equivalent of 250 kg/ha of total nitrogen in FYM per year. Timing of application of manure can affect the amount of soluble nitrogen compounds that are present. Losses are usually lowest from a late winter/early spring application.

The major nitrogen source for organic crop production is from nitrogen-fixing legumes. Residues of nitrogen are usually highest after lucerne or red clover. Ploughing-out a legume and planting the following crop can lead to high losses of soluble nitrogen. It is very difficult for an organic farmer to minimise these nitrogen losses as planting date is also governed by weed, pest or disease control needs.

There are a number of pelleted-approved organic fertilisers. These usually contain a very low percentage of nutrients (N, P and K) and are very expensive. Their main use is on specialist horticultural holdings.

10.5   Weed control

In an organic rotation a much broader range of weed species is commonly found. Some of the most difficult weeds to control in organic crops are perennial weeds such as creeping thistle and docks. The importance of different weeds in organic compared with conventionally grown crops is partly due to the different rotations and the soil nutrient status. Many of the most serious conventional arable weeds, such as cleavers and blackgrass, tend to be very responsive to applied nitrogen and continuous winter cropping and so are less important in an organic system.

Many growers converting to organic production are often seriously worried about potential weed problems. In practice, there can be problems but there are some very effective methods of control available.

10.5.1   Rotations and cropping

A mixed rotation, rather than mono-cropping, can have a major effect on weed species. A rotation of winter- and spring-sown crops will tend not to favour any weed species in particular, and the inclusion of good cleaning crops such as potatoes will also help to reduce weed populations. Some crops themselves have an effect on weeds (allelopathy). Commonly, different weeds grow in different crops planted in the same field at the same time. Oats and red clover are two crops that appear to be very good at suppressing weeds. Growing a mixture of crops such as beans and wheat has also been shown to reduce weed biomass compared with the amount found in the individual crops. Weed problems tend to be reduced in new grass leys if they are established in the spring as an undersown crop rather than after a cereal crop in the autumn.

Green manures can reduce weeds by ensuring good ground cover during the autumn and winter. Ploughed-in green manure crops such as rye and mustard can also suppress weed (and occasionally crop) germination.

10.5.2   Variety selection
Variety should not be underestimated as a method of weed control. Many varieties of the same crop have very different growth and/or leaf habits. Varieties should be chosen that produce a vigorous leaf canopy, such as *Cara* maincrop potato, which will help suppress weeds. Varieties with horizontal (lax) rather than vertical leaves will also have an effect on weeds. It has been found that the very old variety of wheat, *Maris Widgeon* which has a tall straw and lax leaves can reduce weed biomass by around half that found in short strawed varieties such as *Mercia*.

10.5.3   Time of sowing
Delayed sowing is often practised on organic farms usually due to the avoidance of pest and disease problems; it can also aid weed control. Delayed drilling can give time for some weeds to germinate and be killed by cultivations before drilling takes place – the 'stale' seedbed technique.

When growing organic *Brassica* crops, transplants are usually used. This gives scope to kill weeds before planting the crop. Once the crop is planted it is also able to compete very quickly with most weeds. The combination of transplants, late drilling and in-crop cultivations, means that these crops can be virtually weed free, so making them excellent cleaning crops.

10.5.4   Seed rates
High seed rates can help reduce weed competition. In winter cereals drilling 500 seeds/m^2 compared with 200 seeds/m^2 can half the weed biomass. In horticultural crops it is not always possible to plant high populations as this technique can affect crop quality and lead to poor marketable yields.

10.5.5   Cultivations

This is one of the major tools that an organic farmer can use. Often on organic farms more cultivations are required than on conventional farms due to the requirements for weed control. It starts as soon as the first cultivations take place. Most organic farms use the plough more than many conventional farmers. When cropping after grass/clover leys the ground is often rotavated before ploughing as this can help kill the grass. Ploughing is a very good method for controlling some of the grass weeds and some perennial weeds. The 'stale' seedbed technique is commonly used before sowing the next crop. Once sown or planted there is now quite a range of in-crop weeders that can be very effective if the timing and weather conditions are right. The most commonly used weeder is the tine or harrow comb weeders. These can be used on a wide range of crops from cereals to many field vegetables such as the *Brassica*. In cereals these harrows can be very effective especially if the weeds are small; results have ranged from 0 to 80% weed control. Mat-forming weeds such as ivy-leaved speedwell and chickweed can be easily pulled up, even late season, compared with weeds with tap roots such as charlock which are not very well controlled. In winter cereals autumn weeding, when the crop is not competitive, can lead to more weeds emerging. To get best results with these weeders there will be some crop damage, but usually there is little effect on crop yield. An effective harrowing may cover the crop plants with up to 30% of soil. Depending on the season, crops may be treated several times with the harrow comb.

As well as whole-field treatments with the harrow comb some crops are also cultivated with an inter-row hoe. Inter-row cultivators are very important for weed control in organic field-vegetables and root crops. The use of some of these implements is very effective at minimising the amount of hand weeding required. Some of these weeders are self propelled, others are tractor mounted. A wide variety of hoes is used, from goose-foot shares to brushes or even finger tines. Brush hoes do appear to work better than conventional hoes in wet conditions. The hoes can have an effect on weeds within the crop row by throwing soil into the rows and so smothering the weeds.

As in conventional farming it is important to start the weed control programme in time before the weeds start competing. In many crops such as beans, carrots, swedes and onions the first in-crop weeding is the important one. It should be timed when the weeds are small, usually only 3 to 4 weeks after 50% crop emergence. This is when the rows are just visible and the crop plants have their first true leaves. In transplanted crops, the first weeding can take place within a week or two of planting as long as the transplants are not easily uprooted. Weeding with a tine weeder is usually done as soon as the next flush of weeds are at the cotyledon stage. Dry conditions after weeding will ensure better control.

10.5.6   Flame or thermal weeding

This method of weed control is very important in some horticultural crops. The use of propane gas burners can be effective at controlling a range of weeds.

The heat from the burner causes the plant cell walls to burst and the weed normally dries up within two to three days. These burners do not control all species. Weeds such as chickweed, redshank and cleavers are very susceptible even past the two true leaf stage. Charlock is only susceptible at the cotyledon stage and many grass and perennial broad-leaved species are only checked. In crops such as carrots, onions and parsnips, flame weeding is carried out just before the crop emerges. This gives these small seeded crops a chance to get established before the weeds start to compete. Flame weeders can also be useful to burn off potatoes before lifting.

There are several tractor-mounted models available. The problem with flame weeding is the high energy consumption and slow work rates, but it can reduce hand-weeding costs by half in some horticultural crops.

10.5.7 Other methods of weed control

Hand weeding cannot be ruled out on organic farms as it is important to avoid the development of potential weed problems. In some horticultural crops such as carrots and onions, even after flame weeding and inter-row cultivations, hand weeding is still needed. At least 200 man hours/ha are commonly required. To make weeding easier some farms use bed weeders. These are tractor drawn or self-propelled frames on which the weeding gang lie so that both their hands are free for pulling weeds. This ensures that the task is not back-breaking.

Some perennial weeds such as couch and docks can be difficult to control. One option is to cultivate the set-aside ground within the permitted time limits. If weather conditions are dry then repeated cultivations can be very effective at dragging the couch rhizomes to the soil surface where they will dry out.

A mulch such as black polythene is another method of weed control that can be very effective; the problem is cost. Mulches are only economically viable for use on high value horticultural crops. Black polythene is usually laid using tractor-mounted equipment. Polythene mulches have to be removed after harvest; occasionally they can be reused. Currently a number of biodegradable mulches, such as those made out of recycled paper, are being developed.

10.6 Disease control

Some diseases common in conventional crops, such as the cereal foliar diseases, are not so serious in organic crops. This is partly because many organic crops are thinner than the conventionally grown crops and this reduces the risk of the spread of rain splash diseases. The thin crops also lead to more air movement and less of a microclimate within the crop canopy. Lush crops tend to be those most prone to damage by diseases. It is suggested that a balanced soil and crop nutrient status will help to reduce the incidence of some diseases. Application of manure has been shown to reduce disease possibly due to affecting the soil micro-flora.

Choice of crop and species is very important. In an organic rotation some crop species are grown less frequently than on many conventional farms; this can significantly reduce disease pressure. A limited number of organic variety trials are now undertaken in a range of crops and some recommended lists are being produced. However, recommended lists for conventionally grown crops can still be useful when deciding on the most appropriate variety. Growing mixtures of cereal varieties can help to lower disease incidence and lead to small yield increases. Mixtures are only feasible if growing cereals for the feed market, rather than for a specific quality market.

Good crop hygiene is very important for organic farmers to help reduce disease incidence. Seed and storage diseases can be serious; organic growers should only sow, or plant, disease-free seed or transplants and only disease-free crops should be stored.

Even with using all these disease-control methods there are still some diseases that can be very serious in organic crops. One example is potato blight; with potato blight there is only a limited difference in susceptibility between varieties, although there are a number of eastern European varieties that are showing promise as regards resistance. Currently in the UK growers can apply for derogation to apply products such as copper oxychloride once blight appears. These copper products have restricted use and are likely to become prohibited in the near future.

10.7 Pest control

As with diseases there are a number of pest problems that can be reduced by good husbandry practices such as wide rotations. Increasing the natural predator population can reduce some pests, such as aphids. Organic farmers are encouraged to manage field margins and boundaries sympathetically. On some organic farms higher predatory insect numbers have been found compared with conventional farms. There has been some success from reduced pest damage using companion planting and mixed cropping.

Time of sowing can be very important for reducing pest problems. Cereal growers in high-risk areas normally drill winter cereals after the middle of October to reduce aphid attack and transfer of barley yellow dwarf virus (BYDV). Maincrop carrots are sown at the end of May or in early June to avoid the first generation of carrot fly.

On a limited scale crop covers are used over susceptible crops when there is likely to be damage. These crop covers form very effective pest barriers against pests such as carrot and cabbage root fly and caterpillars. Other than cost the problem with crop covers is that they have to be removed before weeding can be carried out.

Pheromone traps are sometimes used as an aid to warn of possible pest problems. There are also a limited number of permitted treatments including soap sprays against aphids and the bacteria treatment, *Bacillus thuringiensis* for control of caterpillars such as the cabbage white. There are several other permitted, naturally occurring, biological control agents mainly for use in glasshouses.

10.8   Husbandry examples

As an organic farmer cannot always control the problems that appear it is important to get the husbandry correct at the beginning and to realise any potential problems. The following is an outline of standard husbandry for a range of organic crops.

10.8.1   Winter wheat

Organic winter wheat crop usually follows a grass/clover ley or a crop that has been manured such as potatoes. After grass, at least four weeks are normally left with bare ground to reduce the risk of frit fly. Ploughing is as shallow as possible, but adequate grass burial is required.

A disease-resistant variety should be chosen for the required market. Feed varieties are higher yielding than the quality milling varieties but the end price is lower as in conventional farming. If the farm is growing any conventional cereal crops it is important that a different variety is grown on the organic area. The crop should not be drilled too early, after mid-October is best to reduce the BYDV risk. A fairly high seed rate should be used, 400–500 seeds /m^2; this seed rate will make up for bird and other pest and disease losses as the seed is untreated. Bird problems can be much more serious when using untreated, compared with standard dressed, seed. In practice, crop establishment can be as low as 40 to 50%.

Weed control will depend on species present and soil conditions. Up to three treatments may be required overall with a harrow comb between November and April. Wild oats and docks will need hand roguing. If the crop is drilled on wide rows then the weeds can be controlled using an inter-row cultivator.

The crop is harvested at a similar time as conventional crops in August. Yields can vary between 3 and 7 t/ha, depending on the fertility of the site. Recently, the major market has been for grain for human consumption, although there is now an expanding feed market. Grain quality can be a problem, particularly low protein levels. Prices for milling quality have been double the conventional price.

10.8.2   Potatoes

Potatoes are usually grown after a fertility-building crop such as a grass/clover ley. If the potatoes follow a long-term ley then wireworm could be a potential problem. Potatoes are a very responsive crop to composted manure which can be ploughed into the seedbed. Cultivation systems are similar to those for conventionally grown crops, the only difference being time of ploughing. Autumn ploughing of grass/clover leys can lead to high nitrate losses. In an organic system the aim is to try to limit leaching losses, so cultivations are sometimes delayed.

Varieties should be selected that are suited to the market requirements and with good resistance to foliage and tuber blight. Choose slug-resistant varieties if slugs are a problem. Good quality seed should be used; and organic seed is now readily available; it is important that planted seed is as free of disease as possible. Seed-rate calculations are as for conventional crops, but adjustments should be made for more expensive seed and lower yields. To reduce the effect of blight there is an advantage to chitting the seed so that the crop is earlier to mature.

Potatoes are very competitive against weeds. Weed control is usually completely mechanised, and specialised potato ridge cultivators are available. The other option is to chain harrow the ridges, then re-ridge afterwards. These cultivations are usually done just before crop emergence.

Blight is the biggest problem when growing organic potatoes especially in high-risk areas. Currently growers can obtain derogation to apply copper compounds once the disease is found. These treatments can help to slow the progress of the disease, but if infection starts early in the summer yields can still be significantly reduced. The potato haulm is usually mown off once blight becomes very serious. Removing the haulm when there is less than 25% infection can reduce yields. The haulm is best burnt off with a flame weeder straight after flailing to reduce the risk of blight spores being washed down to the tubers. (Rounded rather than steep edged ridges can help reduce the number of blight spores washed through the ridges.) Flame burning will also clean the ridges of many weeds and ease harvesting a few weeks later. Only disease-free tubers should be stored.

Potato prices have been two to three times higher than conventional prices, but yields are very variable. Average marketable yields of organic maincrop potatoes are about 26 t/ha, but can vary between 15 and 40 t/ha.

10.8.3   Field vegetables, e.g. winter cabbage

Organic vegetables are grown on a range of farm types. There are a number of specialist organic vegetable holdings, although more generally they are included in the rotation on mixed farms where there is a supply of compost and nitrogen-supplying legumes. Some small-scale producers concentrate on local markets and vegetable box schemes. Other growers market their produce through co-operatives or packhouses, which supply the supermarkets. Marketing needs to be organised before growing the crops.

Meeting the quality requirements for organic vegetables can be much more difficult than for conventionally grown crops, both in relation to size and damage from pest and disease. Storage losses can also be higher than for conventional crops. The procedures used for growing and harvesting are very similar to those for conventionally grown crops except for rotation, crop protection and fertilisers.

With winter cabbages for example, rotations are no more than one year in four as this helps to reduce the incidence of some diseases. Varieties are often chosen not just for yield, disease resistance and competitiveness against weeds, but also for taste. Transplants are used so that there is plenty of time for seedbed cultivations and weed control. Care has to be taken with plant population; this should be adjusted depending on the fertility of the site so that marketable yield can be maximised. Weed control is best carried out using a harrow comb weeder. Harrow combing (up to three times) should take place at each flush of new weeds followed by an inter-row cultivation; little hand weeding is required. Cabbage caterpillars are effectively controlled using *Bacillus thuringiensis*. Occasionally aphids can be a problem; they can be controlled using soft soap. Crop harvest is sometimes later, depending on soil fertility, than for the same conventional crop. Marketable yields for winter cabbages can be slightly lower than those for the standard crops. Recently prices for organic winter cabbages have not been much higher than for non-organic crops.

10.9   Other systems

Another ecological farming system which is practised by a few growers in the UK, but is more popular in other parts of Europe, is Biodynamic Farming. This is very similar in many ways to the standard organic systems. It was started in the 1920s following the lectures of Rudolf Steiner. It is a whole system involving organic ideals as well as the spiritual approach. Special preparations are used when composting manure. Lunar study can be important when planting crops. Biodynamic produce is sold under the Demeter label.

10.10   Further reading

Blake F, *Organic Farming and Growing,* Crowood Press, 1994.
DEFRA, *Organic conversion information service,* 2000.
Howard A, *An Agricultural Testament* – (Reprint), The Other India Press, 1998.
Koepf H H, Pettersson B D and Schaumann W, *Biodynamic Agriculture: An Introduction,* The Anthroposophic Press, 1976.
Lampkin N, *Organic Farming,* Farming Press, 1990.
Lampkin N and Measures M, *Organic Farm Management Handbook,* Aberystwyth, University of Wales Elm Farm Research Centre, 2001.
NIAB, *Organic Vegetable Handbook,* 1999.
Newton J, *Organic grassland,* Chalcombe Publishing, 1993.
Newton J, *Profitable Organic Farming,* Blackwell Science, 1995.
Organic Farmers & Growers publications including *Organics Today.*
Soil Association publications including *Organic Farming, Living Earth* and *Growers leaflets.*
Younie D, Taylor B R, Welch J P, Wilkinson, Organic Cereals and Pulses, Chalcombe Publishing, 2001.

11

Plant breeding and seed production

11.1 Introduction

The objective of agricultural plant breeding is to develop crop plant varieties which are well adapted to human needs. Most of the crops grown in the United Kingdom today were originally introductions from other parts of the world: potatoes from South America, wheat and barley from the Middle East, oilseed rape from China and maize from Central America are some examples. Early farmers achieved improvements in yields by propagating the most desirable looking plants. These are now sometimes referred to as 'landraces' and the selections made by these farmers were largely based on appearance. The processes of modern plant breeding have enabled further improvements to be made such as enhanced yield, crop quality and other desirable traits which give some present-day crop varieties only a passing resemblance to their predecessors.

11.1.1 Gregor Mendel
The basis of modern plant breeding lies in the work of a Moravian monk called Gregor Mendel who lived between 1822 and 1884. His work with a variety of organisms established the principles of the heritability of specific characteristics. Mendel demonstrated that plants and animals are inherently variable and from that variability selection can take place. Using peas he showed that by cross-pollinating selected parents it was possible to combine desired characteristics in the offspring in a predictable way. This was the first step in refining the natural process of plant evolution.

11.1.2 Plant breeding in the twentieth century

It was not until the twentieth century and the isolation of the specific parts of the deoxyribonucleic acid (DNA) in the nuclei of plant cells (the 'chromosomes' and the 'genes' which they contain), which predictably convey characteristics from one generation to another, that the significance of Mendel's work was appreciated. It now forms the basis of all conventional modern plant breeding.

Much of the UK plant breeding efforts of the twentieth century were undertaken by publicly funded organisations such as the Plant Breeding Institute (PBI) at Cambridge, the Welsh Plant Breeding Station (WPBS) at Aberystwyth and the Scottish Plant Breeding Station. The Plant Varieties and Seeds Act (1964) introduced Plant Breeders' Rights and a system whereby the breeders of commercial crop plant varieties became eligible for royalty payments. This development helped to fund an increase in the commercial breeding of crops by private sector companies.

11.1.3   The National Institute of Agricultural Botany (now NIAB) and the Official Seed Testing Station (OSTS)

During the First World War it became apparent that increases in home food production were essential and legislation was introduced in 1917 to establish the Official Seed Testing Station and, significantly, the testing of seed before sale. The NIAB was founded at Cambridge two years later in 1919 and has performed an important function right through to the present day in encouraging farmers to use improved varieties and high-quality seeds. The NIAB Recommended Lists and Descriptive Lists of crop varieties have become important benchmarks for information to farmers. With the increasing move towards the use of farm-saved seed in recent years the previously MAFF-funded Official Seed Testing Station (OSTS) can still perform an important function for individual farmers, testing seed samples for viability, the presence of weed seeds and for important seed-borne diseases.

11.2   Plant breeding methods

Techniques of plant breeding vary according to the particular crop. The production of a new variety is a slow, painstaking and costly operation involving a time-scale of up to 12 years or even longer. There is, even then, no guarantee that the variety will become a commercial success.

11.2.1 Conventional plant breeding

Conventional plant breeding usually involves the cross breeding of specifically chosen parent plants to form first (F1) and second (F2) generation populations.

Selections are then made, usually from the second generation of plants, which exhibit very large amounts of genetic variation, according to the specific combinations of genes which are present. The breeder selects promising plants from this large pool and progressively, generation by generation, selects lines which conform to more specific requirements. These lines are then 'purified' by growing on for several more generations until the new variety is ready for entry into the official National List trials.

11.2.2   Breeding hybrids

A particular phenomenon sometimes observed in the first (F1) generation cross is known as hybrid vigour or heterosis. This is usually exhibited for one year as improved growth and ultimately yield but may also affect the uniformity or other quality characteristics of the crop. Vegetable (especially members of the *Brassica* family), oilseed rape, forage maize and sugar beet hybrids are commonly grown by farmers. The introduction of hybrid wheat and rye varieties so far has not met with great success in the UK. However, trials with recent introductions of wheat have produced very good yields and hybrid barley varieties too are reported to be showing promise.

The technique for the production of hybrids usually involves the production of self-fertilised 'inbred' lines which are subsequently crossed to form the F1 hybrid which exhibits improved characteristics. A major disadvantage of the use of hybrids is that farmers are precluded from saving their own seed and have to purchase a fresh supply each year.

Oilseed rape has received most attention regarding hybridisation in recent years. In 2001 about half of the NIAB recommended and provisionally recommended varieties were hybrids of one sort or another and many show substantial yield advantages over conventional varieties. Some are 'varietal associations', i.e. two distinct varieties sown together in a predetermined ratio (usually 80: 20). Others, known as 'restored hybrids' or 'top crosses', are F1 hybrids produced in a variety of ways, usually from crossing inbred lines but sometimes from more complex breeding operations involving open pollinated varieties.

11.2.3   Enhanced plant breeding techniques

Plant breeders are always seeking ways of reducing the interval between the initial cross breeding and the final introduction of a commercially viable variety. Accelerated development in environmentally controlled growth rooms, micropropagation and even cross breeding by the fusion of plant cells (so-called protoplast fusion) are ways in which the process can be speeded up. Another way is by the maintenance of common shuttle breeding and multiplication programmes in the northern and southern hemispheres to reduce production time.

Increasing the chromosome number or 'ploidy' of plants is a phenomenon which occurs naturally at a low level, often as a response to physical conditions such as extreme heat or physical damage. Applied to normal diploid (two sets of chromosomes) seeds or small seedlings, the chemical colchicine (a naturally occurring substance extracted from autumn crocus plants) will readily cause an increase in the chromosome number and the creation of 'polyploid' plants. In comparison with diploids, these are normally larger and more lush in growth but often suffer from reduced fertility levels. Some breeding programmes have been devoted to the production of tetraploids (four sets of chromosomes), particularly in the cases of grasses and forage plants but they have by no means replaced diploids. Many modern sugar beet varieties are 'triploid' hybrids formed as a result of crossing diploid with tetraploid varieties. Triticale is the result of a breeding programme involving crosses between durum wheat and rye which has created a new polyploid crop.

Recent advances in plant biotechnology have enabled plant breeders better to identify the genes that determine the characteristics of an individual plant. This is known as the science of 'genomics', DNA profiling or, more colloquially, 'genetic fingerprinting'. It enables breeding programmes to be undertaken with a much greater degree of precision using marker genes for faster selection. It can sometimes be exploited to facilitate the processes of genetic modification (GM) which are covered in the following sections.

11.2.4 Genetic modification
11.2.4.1 Applications and principles
There is no doubting the greater progress which can be made in plant breeding programmes by genetic modification. These techniques have an almost infinite range of potential applications in the development of new varieties. Enhanced yields, quality, pest and disease resistance and the better tolerance of drought and salinity are just a few of the many examples of what may be achieved in future commercial varieties. New varieties can and have been engineered to be resistant to specific non-selective herbicides (e.g. glyphosate). This enables farmers to control all weeds very easily and substantially to reduce the volume and range of their herbicide applications. Most importantly, of course, it also enables them to reduce their costs. Enhanced nutritional qualities and flavour, extended shelf-life and improved appearance are all characteristics which may make foods prepared from GM crops more attractive to the consumer, and some (e.g. genetically modified tomatoes, soyabeans and maize) are already being used. An important recent advance has been the creation of a new strain of tomato which is capable of growing in saline conditions. The commercial exploitation of these developments may, however, be many years away.

The principles involved in genetic modification of crop plants have already been alluded to. Having identified from the genome (gene map) of the donor organism the gene which expresses a particular trait, the plant breeder is able to remove it from the DNA using enzymes and then to transfer it into the host plant, often using another organism (e.g. a virus) as a vector. The expectation is then that the same characteristic will continue to be expressed in the new host plant. A specific example concerns genetic material from a bacterium (*Bacillus thuringiensis*) which causes it to produce a toxin called Bt. This has been successfully inserted into the cells of a number of crop species (e.g. maize) as a way of reducing attacks from important insect pests. This can preclude the need for routine applications of broad spectrum insecticides. Although such applications are closely regulated and rigorously tested for environmental impact, the need for reduced treatment is an environmental bonus.

There are also mechanical techniques for introducing DNA containing specific genes into plant cells. One of them concerns the use of DNA 'bullets' whereby metal beads combined with DNA fragments are mechanically propelled into the host cell where the donor DNA combines with that of the host plant.

11.2.4.2   Risks and consumer attitudes

In spite of the potential benefits conferred by accelerated breeding programmes incorporating genetic modification there are widespread concerns about the risks associated with these so-called 'transgenic' plants. Such concerns have prompted the UK government to restrict (for the time being) the growth of GM crop varieties to a small number of farm-scale trials designed to asses the likely environmental impact of their more widespread introduction. Furthermore, products containing GM maize or soya have to be appropriately labelled.

There is no doubt that genetically modified crop plants can create difficulties. Organic growers in particular are concerned about the potential of their crops becoming contaminated by pollen from GM crops or pollinating insects in the neighbourhood. Another example is the almost certainty of 'volunteer' GM plants occurring as weeds in subsequent crops (non-GM oilseed rape is already an important arable weed in the UK). The suggestion is that cross-pollination with wild plants of the same family may give rise to unique and herbicide resistant 'super-weeds'. Further concerns have been expressed about the potential of insect-resistant GM crops to reduce populations of other attractive or beneficial species (e.g. Monarch butterflies in the USA) although these experiments have been shown to be flawed. The use of antibiotic resistance as a marker gene in transgenic selection has also been queried and this practice is now ceasing.

However, there is widespread public ignorance and mistrust about GM technology and extremist environmental organisations have not been slow to capitalise upon this. There have been several very well publicised incidents in recent years where GM crops have been destroyed by activists and many more where there have been confrontations resulting in farmers being forced to destroy the offending crops as a result of public pressure. However, it must also be said that farmers in other parts of the world are starting to benefit from the use of GM varieties. About 40 million hectares were grown in 2000 (mostly in North America) and there is little, if any, hard evidence of either environmental damage or of health problems among consumers.

11.3 Target traits in breeding

11.3.1 Improvements in crop yields

The huge improvements in crop yields observed during the latter half of the twentieth century were due to many factors, including the increased use of fertilisers and pesticides and improved mechanisation techniques to allow planting at optimum times. In addition to these factors, plant breeding has played a major part, not only influencing the harvestable yields (the 'harvest index') of crops, but also, and very importantly, their physical characteristics, pest- and disease-resistance, maturity times and quality.

The harvest index of wheat has been substantially increased by the breeding of semi-dwarf varieties, and a closely related trait, improved resistance to lodging, has resulted in improvements both to yield and crop quality. Improved yields have been an important trait in all of the other major crops. In some cases improvements have been obtained by plant breeders selecting for winter hardy types which can be autumn sown. Oilseed rape, linseed, peas and lupins are examples.

Crop quality has become increasingly important as farmers attempt to maximise revenue at times of depressed prices. Milling and malting quality of cereals, oil percentage and fatty acid constituents of oilseed rape and juice quality in sugar beet have all been important targets for plant breeders. In the potato crop the suitability of specific varieties for the processing industry (taste, texture, cooking quality, and colorations) are extremely important and related to the levels of enzymes and sugars, or to the storage characteristics of the variety.

Genetic resistance to pests and diseases has also provided important targets for the plant-breeding industry; farmers are able to take advantage of the results of this effort to an increasing extent in their quest to minimise the use of all pesticides. Some diseases of great economic importance, such as *Rhizomania* in sugar beet are the subject of specific programmes and partially resistant varieties, suitable for UK conditions, are currently in trial. However, some diseases remain perennial problems and can only be controlled by strict adherence to plant health precautions.

11.3.2 Grasses and clovers

One of the main outcomes of the bovine spongiform encephalopathy (BSE) and Foot and Mouth Disease epidemics, in terms of livestock production methods, has been a re-evaluation of the systems of production in favour of grass fed animals and an expansion of organic production.

One of the immediate results of these important developments has been an increase in the sales of clover seed. The recent products of the Institute of Grassland and Environmental Research (IGER) white clover breeding programme such as *AberHerald* and *AberDai* are by now well known. Future introductions will concentrate on persistent varieties for harsh upland conditions, and varieties for lowlands which are able to yield well in spite of moderately heavy (< 250 kg N/ha) applications of fertiliser. In a further breeding programme involving the use of Caucasian clover it is hoped to introduce the valuable characteristic of drought resistance to the range of white clover varieties.

There has recently been a large increase in interest in the use of red clover and a much needed and very welcome reinstatement of the breeding programme for this legume at IGER Aberystwyth. Important objectives for the plant breeders will be increased longevity, coupled with improved resistance to clover rot and stem eelworm.

Bloat remains one of the major inhibitions among stock farmers to the widespread uptake of the use of clovers. Whereas the introduction of a 'non-bloating' clover still seems some way off, there is increased interest in the non-bloating characteristics of legumes such as birdsfoot trefoil and sainfoin.

Although of limited interest to farmers in the UK, there is increased interest in the USA in varieties of lucerne (alfalfa) which are being bred for freedom from bloat and suitability for grazing.

Developments in the breeding of grasses have concentrated mainly on the ryegrasses, which constitute more than 80% of the grass seeds sown in the UK. The main thrust, in recent years, has been breeding for improved quality. The introduction of 'high sugar' varieties such as *AberDart* should lead to much-enhanced palatability and intake at grazing, as well as improved silage fermentation. Following on from this should be better performance from stock at grass and a reduced reliance on concentrates.

With increasing concern about nitrate leaching, improving the way grasses utilise nitrogen, both from fertilisers and clover, will also be important in the future; ryegrass/fescue hybrids are being investigated for this purpose. Specific grass/legume variety combinations are also starting to be evaluated. These can make substantial differences both to dry matter yield and to the performance of livestock.

11.4   Choosing the right variety

11.4.1   National Lists and Recommended Lists

When a farmer purchases seed of a particular variety within the UK (or any other member country of the EU) it is possible to guarantee its performance potential as a result of the extremely rigorous testing that all varieties are subjected to under the EU Seeds (National Lists of Varieties) Regulations. Varieties which do not appear on the National List cannot be marketed legally. All varieties are assessed for their 'Distinctness, Uniformity and Stability' (DUS). These trials also assess the variety's 'Value for Cultivation and Use' (VCU). Many of the varieties submitted by plant breeders do not pass. National List varieties also appear in the EU Common Catalogue of varieties and, subject to plant health requirements, can be traded in other member countries.

The second tier of variety testing in the UK involves the selection of the best varieties, independently recommended for commercial production. These are the NIAB recommended lists for England and Wales and the corresponding lists prepared by the Department of Agriculture and Rural Development (DARD) and the Scottish Agricultural College (SAC) for Northern Ireland and Scotland. These lists are updated annually and varieties will only gain entry to them if they have consistently outperformed existing varieties in one or a number of ways over a three-year period. In recent years these national recommended lists have been further enhanced by the addition of English regional recommendations to take account of differential performance of varieties in specific regions.

Local and regionally based commercial and farmer-owned organisations such as Agricultural Research Centres undertake a certain amount of variety trialling as well and give valuable advice on variety choice.

11.5 Seed quality

Seed quality is a term which encompasses a number of important criteria, in particular genetic purity, seed physical quality, viability, the presence of pests or diseases and moisture content. The following section describes the various factors which can impact upon seed quality.

11.5.1 Classes of seed, generation control and seed certification

The production of marketable quantities of seed on a commercial scale requires a series of multiplication steps which take place over a number of years. Each time a crop is grown for multiplication there is a danger of some deterioration in the genetic quality of the variety. This may take place through contamination or unplanned cross-pollination. Cross-fertilised species such as beans and sugar beet are more likely to exhibit variation than self-fertilised crops such as wheat. It is necessary therefore to limit the number of generations over which seeds of a particular species can be multiplied in order to minimise the level of potential genetic deterioration.

The categories of certified seed which are available for most crops are as follows (each represents a single year's step in the multiplication process): 'Breeders Seed', 'Pre-Basic Seed', 'Basic Seed', 'Certified Seed 1st Generation (C1)' and 'Certified Seed 2nd Generation (C2)'. Breeders and Pre-Basic seed are normally produced by the plant breeder or agent whereas Basic seed is produced by the maintainer of the variety. Farmers specialising in seed production would normally only be involved in the growing of Basic seed to produce C1 seed or of C1 seed to produce a crop of C2 seed. Basic seed is normally too expensive to justify its use for growing commercial crops, and C1 and C2 seed are the categories of seed normally purchased and grown commercially by farmers. After the C2 generation, no further multiplication for commercial seed production is permitted.

The labels of seed bags are important documents for seed growers. They contain the evidence of the variety sown and the reference number of the seed lot. It is important that they are retained and that one example label is displayed in the crop (although these are often vandalised) and that at least one is retained in the farm office with evidence of the name and Ordnance Survey (OS) number of the field where the seed is sown. There is a colour coding of labels which identifies the generation of a particular seed lot: Pre-Basic seed – white with a purple stripe; Basic seed – white; C1 seed – blue; C2 seed – red.

Hybrid varieties are produced from the first generation (F1) of a cross between two lines of Basic seed. The harvest from a hybrid crop is unlikely to exhibit the same characteristics as the original cross and so should never be used for farm saved seed.

11.5.2 Varietal identification and crop inspection

It is very important for farmers and those engaged in the plant breeding and seed-production industries to be able accurately to identify specific varieties. This is particularly important when stocks are being multiplied for bulk seed production. All crops must be carefully inspected to ensure a very high degree of varietal purity. If this is not the case the crop may be rejected for seed.

Each individual variety of any crop has specific distinguishing characteristics which, to the trained eye, enables any significant deviation to be spotted easily. These characteristics may include such factors as the size, shape and colour of leaves, flowers or even parts of flowers. Hairiness or waxiness of leaves and factors such as the number of seeds per pod or the numbers of tillers or stem branches per plant may also be important. And finally, evidence of rate of maturity (e.g. date of flowering) and pest- or disease-resistance may also be taken into account.

All of these identifying features, relevant to the individual species, will be well-known to the seed crop inspector who will have received training and updating on the characteristics of new varieties. It is important that crop inspections are made at a time when meaningful observations can be made which, for many species, is likely to occur around the time of flowering. The seed crop inspector makes a walk through the crop which should enable him or her to see as much of it as possible. Sample areas of the crop are then selected at random and the numbers of plants that do not conform to the description of the variety being grown are counted and recorded. Other factors, such as the presence of important weeds (e.g. wild oats), diseases (e.g. virus diseases in potato crops) or the proportion of the crop which has lodged are also recorded. On the basis of one or more inspections the crop will be accepted or rejected for seed. The penalties of having a crop rejected can be substantial since the grower will have undoubtedly incurred the very high cost of the seed (usually Basic or C1) for multiplication, and a crop rejection will deprive him of the extra income which the seed crop can command. It is common therefore to observe the very highest levels of attention to the detail of crop husbandry on farms growing crops for seed.

11.5.3 Identifying seed of particular varieties
In cases involving confusion, dispute, contamination or, increasingly, for routine confirmation, it may become necessary to identify the variety of a particular seed lot. Although seeds of individual varieties have some visual distinguishing characteristics it is very difficult to identify them by eye with any degree of certainty. In such cases the normal procedures which can be carried out are known as electrophoresis tests. These tests involve the extraction of the proteins from the seed under the influence of an electric field and production of a characteristic banding pattern when the proteins are separated out on a gel. These patterns provide the equivalent of a fingerprint for a specific variety.

DNA profiling can also be used to identify varieties but is not used at present to the same extent as electrophoresis.

11.5.4 Maintaining the physical purity of seeds
Farmers, when purchasing seed lots, will not expect them to contain weed seeds, seeds of other crop species, seeds of other varieties of the same species or any extraneous material (e.g. stones, dirt or plant vegetation).

Choice of the field in which seed crops are grown is of the utmost importance since the probable presence of important weeds or diseases is likely to be well-known by the seed grower. Furthermore, the likely presence of volunteer plants or groundkeepers from previous crops of the same species is highly predictable from a knowledge of the previous cropping on the farm. Fields where livestock have been fed on hay, straw or coarse rations should be avoided as they often contain cereal or pulse seed, wild oats or grass weeds. There are in fact, for each species and generation of seed, specific requirements about previous cropping laid down in DEFRA regulations for the production of seed.

Isolation is also an important requirement and the regulations always require seed crops to be separated physically from non-seed crops to prevent physical contamination. Another reason why isolation is important concerns the potential for pollen contamination in the case of cross-pollinated species. An added complication is that such contamination may potentially occur from crops grown on other farms. For each species where cross-pollination to any degree is likely to be of concern, specific recommendations are given concerning the minimum isolation gap which is acceptable for the various generations of seed. A simple way in which contamination of seed crops can be minimised and isolation improved is to discard the produce of the headlands. Checking that appropriate isolation distances have been observed is an important function of crop inspection.

Farm equipment, e.g. seed drills, combine harvesters, trailers and barn machinery are all important potential sources of contamination and should be rigorously cleaned before drilling, or harvesting seed crops. Furthermore, mistakes in handling seed in store can occur and seed lots should always be carefully identified and staff made aware of the need to avoid accidental contamination. Inevitably some contamination of seed with weeds or other materials will occur. A variety of techniques involving screening and gravity separation can be used to remove most of the extraneous material.

There are specific tolerances allowed in seed samples for the incidence of such items as seeds of other species, weeds, ergot pieces and other contaminants. Certified seed may not legally be sold if it contains more than the specified amounts of contaminants.

11.5.5 Seed viability

Viability refers to the proportion of the seed that is alive. However, this fundamental requirement also concerns the ability of seeds to germinate normally, and to produce a vigorous and healthy seedling plant population which in turn is capable of producing an acceptable crop yield.

All seeds marketed must be tested for germination by law and the percentage declared. There are minimum statutory germination percentage levels for each species which vary between 75% and 90%. Germination tests are normally undertaken in satellite laboratories certified by the OSTS on seed merchants' premises. Seed samples (usually 100 at a time) are germinated under standard conditions involving specific growth mediums and in environmentally controlled growth rooms. In some cases alternative and more rapid viability testing can be carried out satisfactorily using the chemical 2,3,5,tri-phenyl tetrazolium chloride which, when administered to a seed cut into two, will produce a red stain in a viable embryo.

The fact that seed is alive does not necessarily mean that it will germinate and seed dormancy is sometimes an important consideration. In some cases it is necessary to submit seeds to specific conditions (e.g. cooling) in order to break dormancy and to encourage germination. Another important factor which may affect seed germination percentage concerns the drying operation. If excessive temperature is used during the drying process this can damage the seed and may reduce the germination percentage. Also, in seasons when high temperatures occur during harvest, it may be necessary to cool seeds by ventilation in order to avoid damage.

The use of desiccants, especially diquat, to aid the harvest of seed crops is common. Particular note should be taken of the adverse effect that desiccants, such as glyphosate and glufosinate-ammonium, can have on subsequent seed germination in some crops and the label recommendations and manufacturer's advice should be strictly adhered to.

Seed vigour refers to the total performance of a seed lot throughout the process of germination and seedling emergence. High-vigour seed is beneficial in practice for field establishment in poor conditions. The use of low-vigour seed is likely to result in reduced levels of field emergence and the need to increase seed rates. Electroconductivity is a test that is often carried out on vining pea seed lots. This establishes the degree of physical damage which has occurred to the seed during harvest and storage and which may affect seedling vigour and disease incidence in cold or wet conditions.

11.5.6 Seed size
Seed size has little effect on viability but it is an important factor to take into account when deciding upon seed rates. An important and frequently used aid is the 'thousand seed weight' which can be used accurately to assess the weight of seed required in order to achieve a particular plant population. The following formula can be used to calculate an accurate seed rate:

$$\text{Seed rate (kg/ha)} = \frac{\text{Target plants per m}^2 \times 1000 \text{ seed weight in grammes}}{\text{Predicted percentage establishment}}$$

[11.1]

11.5.7 Seed health
Specific diseases such as leaf and pod spot (*Ascochyta fabae*) in field beans and pea bacterial blight (*Pseudomonas syringae*) and a whole range of well-known diseases of cereals and oilseeds, can be carried by seeds. It is important therefore that all seed crops are carefully assessed for them in order to avoid a high level of contamination of the crop. Chemical seed treatments are capable of controlling many seed-borne diseases, but not all, and it is sometimes necessary to undertake seed testing in order to establish the levels of infection. Seed potato tubers too can be tested to establish the levels of infection with specific viruses. Standards are set out in the seeds regulations for all crops regarding the proportions of infected seed which are acceptable in commercial certified seed lots.

11.6   Seed production

11.6.1   Cereals
Breeders seed, Pre-Basic, Basic, C1 and C2 generations are produced for wheat, barley and oats. There is no C2 generation for rye or triticale. Basic, C1 and C2 seed can at present be certified at two levels, EU minimum standard and HVS (higher voluntary standard) although the extra costs involved with the production of HVS seed have led to questions about its cost-effectiveness.

The minimum requirements for previous cropping are outlined in Table 11.1. Certain cultural practices, such as early autumn drilling and minimal cultivation create greater risks of volunteers being carried over to a seed crop and in such circumstances a longer break will be advisable.

Table 11.1   Previous cropping requirements for fields growing cereal seed

Crop	Grade of seed	Previous cropping requirements
All except rye	Pre-Basic and Basic	No cereals for the previous 2 years
All except rye	C1	2 years clear of the same species 1 year clear of other cereal species
All except rye and triticale	C2	1 year clear of the same species or the same variety grown from certified seed may be grown without a break
Rye	All except C2	1 year clear of rye and triticale

Important weeds are wild oats, cleavers, sterile brome and blackgrass. Seed crops should not be grown in fields where these are likely to be a major problem. Light populations of wild oats and other cereal species (e.g. barley in wheat) may be rogued out of the seed crop and should then (in the case of wild oats) be removed from the field in a plastic bag and burned.

Isolation is not normally a problem for the self-pollinating wheat, barley and oats and a 2 m gap between crops is all that is statutorily required. However, winter barley (especially 6 row varieties) is prone to out-crossing and so a voluntary 50 m isolation gap is often observed. Rye and triticale are both cross-fertilising species and isolation gaps of 250 m for C1 and 300 m for Basic seed are required.

All cereal seed crops are inspected twice for varietal authenticity and for the presence of other cereal species, weeds and diseases. Wild oats are not permitted at more than 7 plants/ha. An important seed-borne disease is loose smut of barley, which, in the case of heavy infestations, can be the cause of the rejection of a crop for seed. Lodging in wheat may also cause problems with seed crops in particular as it is likely that the grain on lodged plants may start to sprout before harvest. Crops which are badly lodged will almost certainly be rejected for seed.

Harvesting, transport and storage should all be carried out in such a way as to minimise the chances of contamination of the seed crop with other species or varieties. Seed grain should not be dried at a temperature of more than 49 °C since high temperatures will adversely affect its germination.

The minimum standard at which certified seed can be marketed is 98% analytical purity (99% for HVS), 85% germination and a maximum of 17% moisture content. There are specific standards for the content of wild oats and ergot pieces and also seeds of the weeds corn cockle, couch, sterile brome and wild radish.

11.6.2 Peas and field beans

Certified seed of the Breeders, Pre-Basic, Basic, C1 and C2 generations are produced for varieties grown for harvesting dry for incorporation into livestock rations. Vegetable (vining pea) varieties are marketed as 'standard seed' although there is a voluntary certification scheme for Pre-Basic, Basic and C1 generations. The regulation governing previous cropping in fields intended for seed production, however, is very straightforward; no pea or faba bean crops (including legume/cereal mixtures for ensiling or dredge corn) in the previous two years are permitted.

So far as isolation is concerned, there should be at least a 2 m gap (or a substantial physical barrier) between pea or bean seed crops and any other crops. In the case of field beans an isolation gap is required between crops of different varieties. For crops of over 2 ha this has to be 50 m for C2 and 100 m for the C1 and Basic generations. For small crops of less than 2 ha these distances rise to 100 and 200 m respectively.

Crop inspections take place at least twice for both pea and bean crops to assess for varietal purity standards and freedom from other crop species and weeds. Bean varieties especially, because of the high degree of cross-pollination (between 30 and 70%) which takes place, are often quite variable in terms of plant height, time of flowering and maturity. Only 1% of off-types are permitted in certified seed crops of peas and beans and the standards for vegetable varieties are even higher. Wild bees undertake much of the pollination in bean crops but hives of honey-bees are also sometimes introduced and are generally thought to be beneficial.

Pea bacterial blight (*Pseudomonae syringae*) is an important notifiable disease of peas. In order to minimise the likelihood of contamination all pea seed crops must be isolated from other pea crops by at least 50 m.

Leaf and pod spot (*Ascochyta fabae*) is probably the most important seed-borne disease of beans although it affects peas as well and, in fact, seed treatment with fungicides appears to be more effective in the latter. Most disease stems from infected seed, but volunteer plants and the movement of tractors and other machinery from crop to crop can also spread it. The testing of seed stocks for *Ascochyta* is routine and maximum levels for seed infection are prescribed. There is no effective chemical control.

Stem eelworm (*Ditylenchus dipsaci*) is an important seed-borne pest of field beans. Home saved seed should always be tested for it and the status of certified seed should be ascertained as well.

An increasingly important pest of bean seed crops is the bean beetle or bruchid beetle (*Bruchus rufimanus*), the larva of which burrows neat circular holes into the near-ripe seed which, after combining, can provide a refuge for weed seeds such as wild oats and, if infestation is greater than 10% of seeds, adversely affect the germination percentage. Insecticide applications to the crop when adult beetles can be found have become common in order to reduce the levels of this pest.

Combining pea and bean seed in excessively dry conditions should be avoided, if possible, because of the increase in damage to the seed which occurs. Drying, if necessary, should be at relatively low temperatures with peas not exceeding 38–43 °C; for beans the most satisfactory system is to ventilate in a bin or on the floor with ambient air or with a very small amount of extra heat. Minimising the damage to both pea and bean seed should ensure adequate germination. Certified C1 and C2 pea and field bean seed has an analytical purity of at least 98 % and a minimum germination of 85 % (in the case of broad bean seed the minimum germination is 80 %). There is a maximum content of seeds of other species of just 0.5 % and there are specific standards set for wild oats, docks and dodder.

11.6.3 Oilseed crops

Oilseed rape and turnip oilseed rape are two very similar species. Breeders, Pre-Basic, Basic and certified (C1) seed generations are available. There is no second generation (C2) of certified seed permitted.

The minimum requirement in terms of previous cropping is that the field should not have grown a cruciferous seed crop in the previous five years. In practice, seed growers are encouraged to choose fields (if available) which have never grown a cruciferous seed crop, and where the incidence of weeds such as charlock and wild radish is at a minimum.

Oilseed rape and turnip rape are 60–70 % self-fertile but they will also out-cross readily with other species, if in flower, such as swede, fodder turnip, fodder rape, black and brown mustards and Chinese cabbage. Volunteer oilseed rape plants in neighbouring fields can be an important source of pollen contamination (especially the high erucic acid varieties). Isolation requirements for seed crops are therefore quite stringent. There should be a physical barrier or at least 2 m fallow between the crop and any other crop likely to cause contamination. In addition, there should be an isolation gap of at least 200 m between a certified seed crop and any source of pollen contamination such as other varieties of oilseed rape or any of the crops mentioned above. For the production of Basic seed this minimum isolation gap should be doubled to 400 m.

Field certification standards are also stringent and for the production of certified seed only 0.3 % of off-types are allowed and for Pre-Basic and Basic seed this is decreased to 0.1 %.

The disease *Sclerotinia sclerotiorum* is important in oilseed rape and a number of other crops as well. It infects the stems of crops during flowering and has a substantial effect on yields. During combining, fragments of the sclerotia of this fungus often break away from infected stems and become mixed in with the harvested seed. There are no field standards for the incidence of this disease but the permissible number of sclerotia in a 100 g certified seed sample is 10.

Crop inspections may take place up to three times, firstly in the vegetative stage when it should be possible to identify the variety through its leaf characteristics. A further inspection at the stem-elongation stage should allow for the identification of any off-types in the crop. The final inspection is made around the time of early flowering when the characteristics of the flowers can be confirmed, and the surrounding fields checked for potential sources of pollen contamination.

Certified, Pre-Basic and Basic seed of oilseed and turnip rape has a minimum analytical purity of 98% and a minimum germination of 85%. The maximum content by weight of other seed species is 0.3%. There are specific standards applied to wild oats, dodder, dock and wild radish contamination.

11.6.4 Linseed and flax

Linseed and flax have been heavily subsidised by the Common Agricultural Policy (CAP) and until recent years additional funding in the form of Seed Production Aid was available for certified seed production. Breeders, Pre-Basic, Basic, C1 and C2 generations are available.

Fields which are selected for linseed seed production should have had no linseed or flax crops for the past five years and, because diseases such as *Sclerotinia* can affect both oilseed rape and linseed, no cruciferous seed crops should be grown during the previous two seasons. Because linseed and flax are self-pollinating only a 2 m fallow strip or a physical barrier is required between a seed crop and any other crops likely to cause contamination. Crop inspections usually take place twice. The first is when the crop is in the tillering stage and the second around the time of flowering when varietal off-types may be readily identified. Field standards for varietal purity allow just 2.5 off-types/ha for C2 seed, 2/ha for C1 and 0.3/ha for Basic seed.

Linseed and flax are susceptible to some important seed-borne diseases. No more than 5% of seeds should be infected with the following: *Botrytis* spp., *Alternaria* spp., *Fusarium* spp., *Colletotrichum lini* and *Phoma exigua* var *linicola*. Seed crops are normally sprayed with an approved fungicide before the end of flowering in order to try to achieve these standards.

At harvesting, particular attention should be paid to the cleaning of the combine trailers and any barn machinery. Oilseed rape and linseed can be dried easily to about 8% moisture but if drying on floor it should not be in layers more than 1.25 m deep. Drying air temperature should not exceed 50°C. Ventilation should be applied to cool the seed down to less than 15°C. At higher temperatures or moisture contents infestation with insect pests or grain mites can become a problem.

11.6.5 Sugar beet

Basic and certified generations of sugar beet seed are grown but very little in the UK. British Sugar at present controls the distribution of all seed in the country and growers place an order for their preferred varieties when they sign the annual contract.

High quality seed is one of the most important factors in the production of sugar beet. Since the introduction of monogerm varieties in the 1960s the majority of commercial crops have been drilled to a stand without the need for thinning. Many modern varieties are hybrids and more than half of the current varieties offered by British Sugar are triploids, the result of crossing diploid and tetraploid parent lines. Special climatic conditions are required for high-quality seed production especially during the periods of flowering, maturing and harvesting. This has now largely precluded the growing of the crop in the UK or indeed anywhere else in northern Europe.

11.6.6   Potatoes

Crop production from true potato seed, taken from the fruits of the plant is rare, but it can be done. There are two main advantages: one is that in parts of the world where food is scarce an extra supply is secured by saving the tubers which would normally have to be replanted; the other is that some of the important potato viruses are not transmitted by true seed.

The majority of potato crops, however, are raised from replanted tubers known as seed potatoes. Seed production has traditionally been carried out in Scotland, Northern Ireland and the hill areas of England and Wales. The main advantages of these areas are that the low temperatures and strong winds keep aphid populations in check. This means that the severe virus diseases (leaf-roll and the mosaics) which are spread from diseased to healthy plants by aphids are less likely to occur. However, recent advances in aphid control and concerns over the quality of seed from some traditional areas have seen successful seed production extended to some of the English arable areas as a profitable break in predominantly cereal and break crop rotations.

The choice of field for seed potato production mainly reflects the need to minimise the possible carry over of groundkeepers from previous crops. Seed potatoes should not be grown in the same field for more than one year in five (one in seven is preferable). Fields for seed production must be certified free from potato cyst nematodes (PCN) and seed crops must never be grown on land where potato wart disease has occurred. The main thrust of the practice and the regulations surrounding potato seed production concern the minimisation of the incidence of virus diseases. All seed crops must be certified during the growing season. This ensures that they are true to variety but also that they are as free as possible from virus and other diseases. Most growers inspect crops carefully prior to official inspection and any diseased plants are rogued out of the crop.

The main certification grades of seed potato are set out in Table 11.2 together with the field tolerances allowed for impurities. Virus tested stem cuttings (VTSC) is the highest grade of seed potato and is produced by rooting stem cuttings from virus-free plants in sterilised compost in isolated and aphid-free glasshouses. Other micropropagation techniques are now used for producing disease-free plants even more quickly than the VTSC method. Subsequent generations ('super-elite' and 'elite') are grown on in isolation from other stocks to ensure, as far as possible, freedom from viruses and the other important tuber-borne diseases such as gangrene, skin spot and blackleg.

Table 11.2   Tolerances: percentage of plants affected (final field inspection)

Disease and defects	Basic seed				Certified seed CC
	VTSC	SE	E	AA	
Rogues and variations	0	0.05	0.05	0.1	0.2 ⎱ 0.3 ⎰ 0.5
Leaf roll	0	0.01 ⎱	0.1	0.25	2.0
Severe mosaic	0	0.00 ⎰			
Mild mosaic	0	0.05	0.5	2.0	10.0
Blackleg	0	0.25	0.5	1.0	4.0

Notes: VTSC = virus tested stem cutting; SE = super elite (generation 1–3); E = elite. 0.01% = 5 plants/ha; 1.00% = 500 plants/ha.

Ideally, seed potato crops should be planted with sprouted tubers and at a high seed rate to produce a good yield of seed-size tubers (35–55 mm). The haulm should be destroyed chemically, to reduce the risk of blight or virus transmission, when most of the tubers are of seed size. This reduces not only the risk of spread of virus diseases by aphids but also the spread of blight to the tubers. It is an undesirable practice to let the crop grow to maturity so that additional income can be obtained from sale of large ware-size tubers. Seed crops should be kept free from weeds and blight infection throughout the growing season and this attention to detail should continue through to storage. Wounds should be allowed to heal and wet loads kept out of the store. Sprout-suppressant treatment should never be used on seed potatoes in store.

Only Basic seed can be used for further commercial seed production, although AA grade seed is sometimes used for ware production. Certified seed (CC) grade is healthy commercial seed, also for ware production. Crops grown from Basic seed should contain less than 4% of plants with virus infection and those grown from certified seed, less than 10%. Growers should not expect virus infection levels in purchased seed to be limited to the tolerances set for crop inspection. Extra infection may occur during the growing season when the inspection takes place, or by re-infection by aphids during chitting.

11.6.7   Herbage seed

All herbage seed production in the UK is overseen by NIAB. Only Basic seed is used for C1 generation production. In some cases it is permitted for seed crops to be taken from the same crop for two seasons, but no C2 generation multiplication is permitted. All seed crops are inspected at around the time of ear emergence to establish variety type and to check for off-types, weeds and isolation.

In total about 8000 hectares of grass are grown for seed in the UK at present. Of this over 6500 hectares is ryegrass of various types. European Union seed production aid is available at the time of writing for Italian, hybrid and perennial ryegrasses. The reader is advised to check with DEFRA the current status of this aid. Whereas it is possible to achieve good gross margins from seed production, the vagaries of the weather, the difficulties of combining seed crops at high moisture levels and very low prices in the late 1990s have brought about substantial reductions in the area grown.

Seed production from grass requires considerable planning as well as skill and perseverance. Good yields are dependent on good growing conditions and dry weather during the critical harvesting period. Grass seed is grown most successfully in low rainfall areas such as in the south and east of England, and usually on chalk or limestone based soils. The Italian and hybrid ryegrasses are harvested for one season only, but most of the perennials will produce seed crops for two seasons. Grass for seed production will, in addition, provide threshed hay and some grazing.

It is very important that there are no grass weeds in the field – especially blackgrass, couch, rough stalked meadow-grass, sterile brome and other grasses with similar sized seed. Docks too are an important contaminant of grass seed. Every effort should be made to rid the field of these weeds prior to growing a seed crop. High weed infestations can lead to a failure of the crop to pass field inspection. It is also important that the field is not contaminated with volunteer plants of the same species; NIAB stipulate up to four years of arable cropping prior to sowing a grass seed crop. The herbicide ethofumesate is regularly used on establishing seed crops to reduce the incidence of grass weeds. Broad-leaved weeds can be easily controlled by a range of post-emergence treatments.

Where there is a danger of cross-pollination (i.e. when neighbouring grass crops contain the same species), their flowering period overlaps and they are of the same ploidy (e.g. both diploids), then a gap of at least 50 m between the seed crop and its neighbour should be established before flowering begins. If this is not possible then, when flowering is over, a discard strip must be cut out of the seed crop adjacent to the neighbouring field. This material must be discarded and no seed may be used from it. It is important to remember that grass seed crops can be laid down for several years and careful planning is required to avoid cross contamination. Even areas such as waste ground and motorway verges can be sources of pollen.

The ryegrasses and fescues are usually undersown in spring cereals, either in narrow rows or broadcast, whereas cocksfoot and Timothy are usually sown direct in wide rows (about 50 cm). Seed rates are often substantially below (as little as 12.5 kg/ha for Italian ryegrass) those which would be considered normal for forage production. Italian and hybrid ryegrasses are usually cut early for a high quality silage crop or grazed prior to being laid up for the seed crop. This must be accomplished before the end of April. Moderate levels (about

70 kg/ha) of nitrogen fertilisers are then applied and the seed crop harvested usually in July or August. Perennial ryegrasses receive nitrogen at similar times to the Italians and hybrids, but no cutting or grazing can be carried out. A good ryegrass crop will usually 'lodge' about a fortnight before harvest and this will reduce losses by shedding. The crop may be combined direct or from windrows. The seed must be carefully dried (usually by prolonged blowing with ambient air) and cleaned. Yields of cleaned seed range from 1 to 1.75 t/ha but can be extremely variable and considerable skill is needed to achieve high yields.

Red clover. The USA and Canada are the main world centres of production of red clover seed. Only very small areas have been sown for seed in the UK in recent years, but the substantial increase in interest in this legume creates a new opportunity for herbage seed producers.

White clover. New Zealand, the USA and some South American countries are the primary producers of white clover for seed. In Europe, the largest producer is Denmark. At present, the only UK white clover crops for seed are very small areas of the *Kent wild white* local variety.

11.6.8 Organic seed

It is important that seed used by organic producers is, wherever possible, of organic origin and this is stipulated by EU regulation 2092/91. Organic seed crops, if entered for UK certification, are currently subject to exactly the same standards as non-organic. However, a lack of organic seeds has brought about a derogation across Europe which allows organic producers to use non-organic seed where no appropriate organic seed is available. This derogation will expire on 1 January 2004 after which date organic seed or other approved plant propagation material must be used for all organic production.

11.6.9 Farm-saved seed

As a way of saving costs many farmers now save seed from a number of species for replanting in the following season. There is relatively little risk associated with this provided that some basic rules are obeyed. The first is to grow the crop as a seed crop with exactly the same precautions as have already been indicated. Many farmers intending to use farm-saved seed will purchase a relatively small quantity of C1 generation seed to grow on. It is important to have the seed cleaned from weed seeds as far as possible and treated professionally with the required approved seed treatment by a reputable mobile seed treatment company. In some cases, if seed-borne disease levels have been confirmed by testing to be absent, it may even be possible to omit the fungicidal seed treatment. The other important rule is to have the seed tested professionally for germination percentage.

When seed is to be tested it is vital that really thorough and rigorous sampling is carried out throughout the seed lot. In the case of oilseed rape, where farmers are intending to use farm-saved seed it must be tested for glucosinolate levels to conform with EU regulations. Potato growers are allowed to grow on their own 'once grown' seed from a crop grown from CC grade but a sample of tubers should be tested to ensure that they are as free as possible from viruses and fungal diseases. Farm-saved non-certified seed of any species may not legally be sold. Farmers using farm-saved seed are now required to pay royalties to the breeders of the specific varieties that they are growing. These royalties are collected and distributed on behalf of breeding companies by the British Society of Plant Breeders.

11.7   Further reading

Gooding M J and Davies W P, *Wheat Production and Utilization*, CAB International, 1997.

Kelly A F and George R A T, *Encylopaedia of Seed Production of World Crops*, John Wiley and Sons, 1998.

MAFF, *Guidance Notes for Growers of Seed Crops in England and Wales*, MAFF, 1997.

RASE, *Old Crops in New Bottles?*, Royal Agricultural Society of England, 1998.

Simmonds N W and Smartt J, *Principles of Crop Improvement*, Blackwell Science, 1999.

Wellington P S and Silvey V, *Crop and Seed Improvement – a History of the National Institute for Agricultural Botany*, Henry Ling Ltd, 1997.

Part 3

The management of individual crops

12

Cereals

12.1 Introduction

Cereals are the most widely grown arable crops in Europe (Table 12.1) Over the last 40 years there have been large changes in the relative proportion of the main cereals. Yields have increased more rapidly in wheat than in barley. In the UK, for example, the average wheat yield is now 8 t/ha, compared with 6.3 t/ha for winter barley and 5.1 t/ha for spring barley. This increase in yield is largely due to improvements in varieties and husbandry techniques. Fertiliser use – particularly nitrogen – and application of crop protection chemicals for the control of pests, diseases and weeds, have also been important reasons for the increase. The majority of wheat crops in the UK are autumn-sown, compared with approximately 60% of the barley crop and 75% of the oat crop.

In the EU wheat now accounts for around 50% of the cereals produced followed by barley and grain maize (southern Europe). Other cereals grown to a limited extent include oats, durum wheat, rye and triticale.

Table 12.1 UK and EU cropping and yields 2000

Crop	UK	EU – 15 member states
Wheat area (Mha)	2.09	14.4
Yield (t/ha)	7.99	6.63
All barley area (Mha)	1.13	10.58
Yield (t/ha)	5.74	4.88
Spring barley area (Mha)	0.54	5.81
Yield (t/ha)	5.13	4.41
All grain area (Mha)	3.36	36.8
Yield (t/ha)	7.14	5.76

As well as changes in production of cereals, there have been changes in their usage. Over 80 % of the wheat used by millers in the UK is now home grown due to improvements in wheat quality and milling technology. Very little Canadian hard wheat is now imported for bread making. Wheat, compared with barley, is also used in greater quantities by the feed compounders. Since the early 1980s, with the improvements in cereal yields, Europe has become a major exporter of cereals on the world market. Surplus grain is expensive either to store (Intervention) or support export subsidies. The Agenda 2000 reform package, using set-aside and a reduction in Intervention prices is aimed at reducing the surplus of cereals and making EU grain competitive at world prices. Farmers receive arable area aid payments, which are paid on area of crop grown.

Wheat, rye and maize grains consist of the seed enclosed in a fruit coat (the pericarp) and are referred to as 'naked' caryopses (kernels). In barley and oats, the kernels are enclosed in husks formed by the fusing of the glumes (palea and lemma) and are referred to as 'covered' caryopses (Fig. 12.1). The cereals are easily recognised by their well-known grains, ears or flowering heads (Fig. 12.2–12.6) and in the early leafy stages (Fig. 12.7).

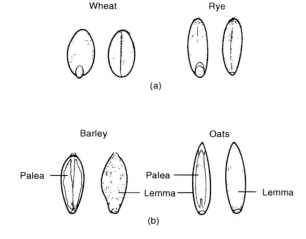

Fig. 12.1   Kernels (a) 'naked' kernels (b) 'covered' kernels.

12.2   Grain quality in cereals

There are several markets for cereals, including milling, animal feed, malting, seed, export, industrial uses and Intervention (Table 12.2). Grain quality requirements will be affected by the proposed market.

Standard tests when selling grain include the following:

1   *Moisture content.* This is very important when storing grain. For long-term storage it should be about 14 %. Too high a moisture content will be penalised or not accepted, but no compensation is given for very dry grain. Grain which is overheated when being dried, or in storage, can be spoiled

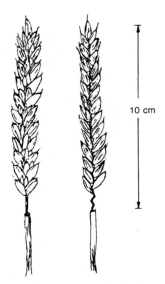

10 cm

Fig. 12.2 Wheat spikelets alternate on opposite sides of the rachis. 1–5 grains develop in each spikelet. A few varieties have long awns.

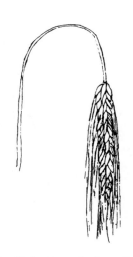

Fig. 12.3 6-row barley, all three flowers on each spikelet are fertile. Awns are attached to the grains.

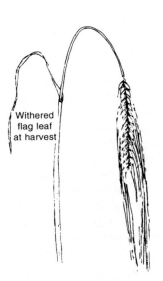

Withered flag leaf at harvest

Fig. 12.4 2-row barley heads hang down when ripe: each grain has a long awn; the small infertile flowers are found on each side of the grains.

Fig. 12.5 Grain easily seen in the spikelets of rye.

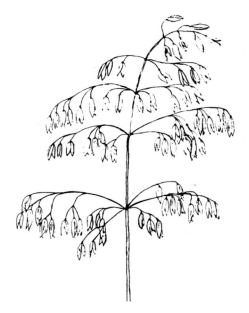

Fig. 12.6   Oats.

for seed, malting or milling. Overheating can also damage the germination ability as well as the protein quality.

2   *Sample appearance and purity.* Good quality grain is clean and attractive in appearance, free from mould growth, pests and bad odours. Sample purity, freedom from contamination with other cereals or weeds will affect the potential value of grain, as will the amount of shrivelled or broken grain. Careful setting of the combine will help to minimise broken grains and produce a cleaner sample. Depending on quality and quantity of grain being sold, some farmers have a drying system with grain-cleaning facilities.

3   *Specific weight (bushel weight).* Specific weight is a measure of grain density. A high specific weight is preferred for all markets, particularly export. Wheat has the highest specific weight and oats the lowest. Specific weight can be affected by husbandry (e.g. time of sowing, disease and pest control) as well as by weather conditions during grain fill and harvest.

4   *Mycotoxins.* Before harvest some fungi present on grain can produce mycotoxins, e.g. *Fusarium spp.* Toxicity of the mycotoxins varies with fungal species. At the moment there are no EU limits though a 3–5 parts per billion for one of the mycotoxins – ochratoxin – may be introduced.

Other standard tests mainly for milling and/or malting include:

1   *Hagberg Falling Number.* This test in wheat measures the amount of breakdown of the grain's starchy endosperm by the enzyme alpha-amylase. A high value indicates low enzyme activity which is required for bread

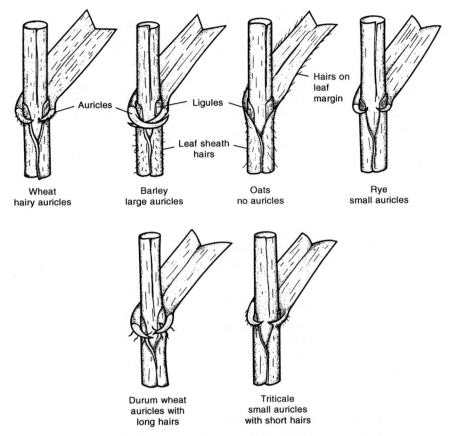

Fig. 12.7 A method of recognising cereals in the leafy (vegetative) stage.

making. If grain has started germinating or sprouting, the Hagberg Falling Number will be low.

2   *Protein content.* A high protein content is required for bread making compared with a low percentage of grain nitrogen in malting samples. Nitrogen content is analysed using the Dumas test or a near infra-red analyser. Results are given at 100% dry matter.

3   *Protein quality.* The quality of the protein in the grain affects the baking characteristics (when the protein in wheat flour is mixed with water it produces gluten). Protein quality can be assessed by several tests, including baking, gluten washing, SDS sedimentation, Zeleny index or the Chopin alveograph test. Tests undertaken will depend on the target market.

4   *Dough machinability.* This is a test which measures the stickiness of dough. It is required for wheat being sold into Intervention.

5   *Germination.* Grain for seed or malting is tested for speed and percentage of germination. Germination should be over 97% for malting.

Table 12.2   EU cereals – main uses 1997/98

Million tonne	Wheat	Barley	Total cereals
Total crop size	87.3	52.5	177
Imports	1.4	0.2	7.8
Exports	10.3	4.7	20.6
Seed	1.3	1.0	6.0
Animal feed	21.4	14.1	101.1
Industrial use	4.4	7.1	15.3
Human consumption	31.9	0.3	41.5
Processing	0.5	0.2	na
Self-sufficiency	145%	232%	106%

Source: Homegrown Cereals Authority (HGCA).

Husbandry of the crop, variety and time of harvesting can affect several of the above quality specifications. Once harvested and stored, little can be done to change or improve grain quality. Gravity separators are available which can help to raise specific weights and the Hagberg Falling Number.

Table 12.3   Minimum quality standards for grain sold into Intervention 2000/2001

Standard	Barley	Wheat
Maximum moisture content (%)	15	15
Minimum specific weight (kg/hl)	62	73
Maximum total impurities %	12	12
of which		
broken grains %	5	5
sprouted grains %	6	4
grain impurities (including shrivelled etc.) %	12	7
of which		
overheated grains %	3	0.5
other cereals and those pest damaged	5	
miscellaneous impurities	3	3
of which		
ergot		0.05
noxious seeds	0.01	0.01
Minimum Hagberg FN	–	220
Zeleny index	–	22
Dough machinability test	–	pass*
Minimum protein content 100% DM basis	–	11.5

*Not applicable if Zeleny > 30
Note: price adjustments are made for moisture contents below 15% mc, protein contents below 11.5% and specific weights below 76kg/hl in wheat and below 64kg/hl in barley
Source: HGCA.

Table 12.3 shows the quality standard for selling cereals into Intervention. Most merchants now only buy grain that comes from farms that have been assured by a registered Assurance scheme. These schemes include strict conditions on crop production and grain storage.

12.3   Cereal growth and yield

A decimal growth stage key is now commonly used to describe the growth and development of the cereal plant (Table 12.4). There are ten main areas of growth (subdivided into secondary stages) from sowing through to the vegetative leaf

Table 12.4   The Decimal Code for growth stages of small grain cereals

Code 0 Germination. Subdivided—00 dry seed to 09 when first leaf reaches tip of coleoptile.

Code 1 Seedling growth
10 1st leaf through coleoptile
11 1st leaf unfolded
12 2 leaves unfolded
13 3 leaves unfolded
14 4 leaves unfolded
15 5 leaves unfolded
16 6 leaves unfolded
17 7 leaves unfolded
18 8 leaves unfolded
19 9 or more leaves unfolded

Code 2 Tillering
20 main shoot only
21 main shoot and 1 tiller
22 main shoot and 2 tillers
23 main shoot and 3 tillers
24 main shoot and 4 tillers
25 main shoot and 5 tillers
26 main shoot and 6 tillers
27 main shoot and 7 tillers
28 main shoot and 8 tillers
29 main shoot and 9 or more tillers

Code 3 Stem elongation
30 ear at 1 cm (pseudostem erect) – visible only if the growing point is dissected
31 1st node detectable (seen or felt after removing outer leaf sheaths)
32 2nd node detectable (seen or felt after removing outer leaf sheaths)
33 3rd node detectable (seen or felt after removing outer leaf sheaths)
34 4th node detectable (seen or felt after removing outer leaf sheaths)
35 5th node detectable (seen or felt after removing outer leaf sheaths)
37 flag leaf just visible
39 flag leaf ligule/collar just visible

(For example, a plant having five leaves unfolded (15), a main shoot and three tillers (23) and two nodes detectable (32) would be coded as: 15, 23, 32.)

Code 4 40–49 booting stages in development of ear in leaf sheath to when awns just visible – 49

Code 5 50–59 stages in the emergence of the inflorescence or ear. Ear fully emerged – 59

Code 6 60–69 stages of anthesis (flowering); flowering complete – 69

Code 7 Milk development stages in grain; late milk – 77

Code 8 Dough development stages in grain; hard dough – 87

Code 9 Ripening stages in grain; grain hard and not dented by thumb nail – 92 and tillering stages, stem elongation, ear emergence, flowering to grain filling and ripening. By dissecting out the growing point or apical meristem, using a microscope, the change from vegetative to ear formation can be seen (Fig. 12.8 and 12.9). These changes in internal development do not always coincide with

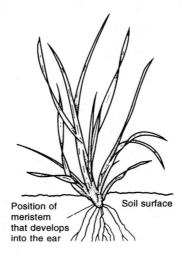

Fig. 12.8   Leafy winter wheat plant with four tillers.

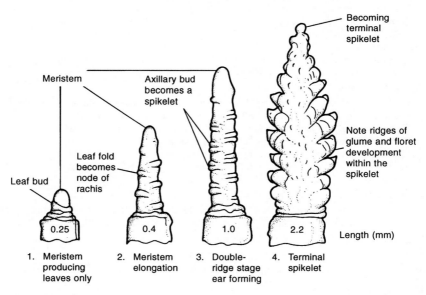

Fig. 12.9   Simplified diagram of an ear forming from the tiller apex of winter wheat (not to scale). (Visible under a microscope or possibly with a good pocket lens.) The apical meristem changes from leaf production to ear formation following cold treatment (vernalisation) and increasing day length (photoperiod). Normally these stages occur in February, March and April and their exact occurrence can be used for efficient timing of pesticides and nitrogen application.

the same growth stage of the cereal plant. Speed of development will depend on type of cereal, variety, temperature, day length and husbandry factors such as time of sowing.

It is very important to be able to identify growth stages correctly as there are only certain stages when pesticides should be applied if they are needed. The main stem should always be looked at when checking the growth stage. A major difficulty is deciding when the plants have changed from the tillering to the stem extension stage. The best method is to use a knife and slice the main stem in half. Growth stage 30 is when the ear is at 1 cm (Fig. 12.10 and 12.11). Distinguishing the nodes is another problem. They should only be counted when there are more than 2 cm between each. They, too, are best studied by splitting the stem with a knife; it can be very misleading just to feel the outside of the stem.

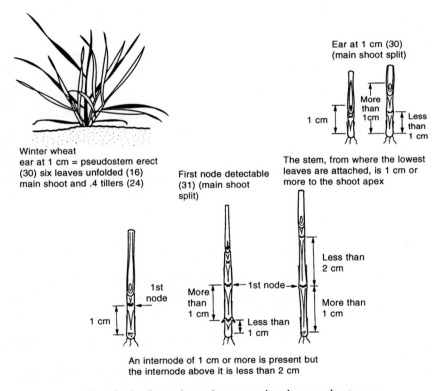

Fig. 12.10 Stem elongation stages in wheat: early stages.

The actual yield of a cereal crop is determined by the contributions made by the three components of yield:

- number of ears per ha,
- number of grains per ear,
- weight (size) of the grains.

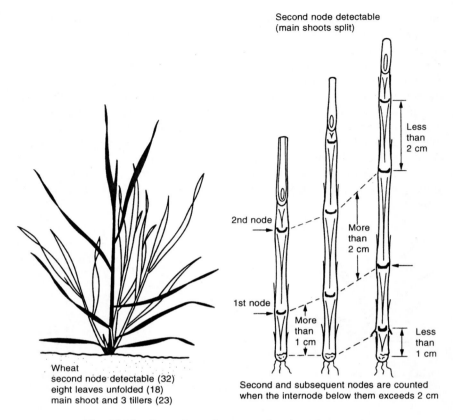

Second node detectable
(main shoots split)

Less
than
2 cm

2nd node

More
than
2 cm

1st node

More
than
1 cm

Less
than
1 cm

Wheat
second node detectable (32)
eight leaves unfolded (18)
main shoot and 3 tillers (23)

Second and subsequent nodes are counted
when the internode below them exceeds 2 cm

Fig. 12.11   Stem elongation stages in wheat: later stages.

These components are inter-related. By increasing the number of ears (e.g. by denser plant populations or more tillering), the number of grains per ear may be reduced and also the size of the grains. Opinions differ on the ideal number or size of each component and it will vary according to the type and variety of cereal, as well as soil and climatic conditions, time of sowing and seed rate, and the occurrence of weeds, diseases and pests.

Winter wheat and 6-row winter barley have the least number of ears/m^2 (500–600/m^2), whereas 2-row winter barley has the highest number of ears (900–1100 ears/m^2). The weight of grain is normally heaviest in wheat and lowest in oats and rye.

12.3.1   Seed rates
Accurate spacing and uniform depth of planting of seed in well-prepared seedbeds are very important if optimum yields are to be obtained. This usually involves doing seed counts/kg and carefully setting the drill. Narrow (10–12 cm)

rows are preferable. Broadcasting seed can be successful provided the seed is distributed uniformly and properly covered. Stony soils, wet and cloddy conditions give crops a poor start.

Seed rates should be chosen with the object of establishing the desired plant population. They can vary between 125 kg/ha and 250 kg/ha. The factors to be taken into account are:

1 Crop – there are different optimum plant populations for the various cereal crops.
2 Seed size, e.g. increase the rate for large seed (low seed counts/kg) and decrease the rate for small seed (high seed counts/kg).
3 Tillering capacity – some varieties tiller more freely than others and so their seed rate may be reduced. Winter cereals have more time to tiller than spring cereals and so fewer plants need to be established.
4 Seedbed conditions – the seed rate should be increased in cloddy and stony conditions; this could result in inferior seedling growth.
5 Time of sowing – for late autumn sowings and very early spring sowings the seed rate should be increased.
6 The possibility of seedling losses by pests such as wheat bulb fly, frit fly or slugs; the greater the possibility, the higher the seed rate.
7 High seed rates can help reduce weed competition.
8 Price and quality of seed; very expensive seed will often mean a reduced seed rate!

The most desirable plant population to aim for is somewhat debatable; it must be related to the potential yielding capacity of the field. Very low plant populations can ripen unevenly and be prone to weed competition, whereas high populations can lead to smaller ears and low grain weight.

12.3.2 Varieties

There are very many cereal varieties now on the market and new ones are introduced every year. However, there are only a few outstanding varieties of each cereal and these are described in the annual UK *Recommended Lists* funded by the HGCA.

Improvements in varieties over the last 30 years have greatly contributed to the increasing yield of cereals. Choice of cereal variety can affect yield, quality, input requirements such as fungicide and growth regulators, and subsequent returns.

Varieties are noted for their susceptibility to the main diseases. Varieties with a rating of 1, 2 or 3 are very susceptible compared with 8 or 9 for varieties that are resistant and where the disease is unlikely to reduce yields significantly.

12.3.3 Time of drilling

Time of drilling will be affected by many factors including crop, variety, possible pest, disease and weed problems, soil type, machinery availability, weather conditions and previous cropping. In winter cereals, sowing in late September will normally give a higher yield than drilling at the end of October. Drilling starts, in most parts of the country, with winter barley and first wheats from the middle of September (occasionally earlier). Second and third wheats which could be affected by take-all, or fields with serious grass weed problems, should be drilled last. Cereal sowing will normally be delayed following root crops. The latest safe date for drilling winter cereals will depend on the variety and its vernalisation requirement. The HGCA gives recommendations on the latest safe date for sowing each variety. Spring cereals (mainly spring wheat) can be sown during the winter. However, the majority of spring cereals are sown in the spring from January onwards. Time of drilling will depend more on suitable soil conditions than on calendar date.

12.3.4   Seed type and dressing
The most commonly grown category of seed sown is C2 – second generation commercial seed. First generation C1 seed is only sown if it is being grown for seed on contract, or for home-saved seed, or if a very new variety is being tried. Most seed is of Higher Voluntary Standard (HVS) rather than the lower EU Minimum Standard (pages 243 and 247).

Unless growing organic crops, seed (purchased or home-saved) may be treated against fungal diseases such as bunt, leaf stripe or those caused by *Fusarium* (Table 7.1). Depending on the previous crop, pests such as wheat bulb fly and wireworm may be a potential threat (Table 6.1). Insecticides can be included in a seed dressing (a dual-purpose dressing).

12.3.5   Fertilisers
Nitrogen is the main fertiliser required. Rates depend on the type of cereal, market, SNS index and the use of organic manures. In wheat, researchers are looking at canopy management as an aid to achieving optimum yields. In this case nitrogen rates are adjusted according to the size of the crop canopy. Average nitrogen fertiliser use in the UK varies according to the crop, i.e. winter wheat 185 kg N/ha, winter barley 140 kg/ha and other cereals 95 kg/ha. To avoid leaching, all nitrogen is now applied in the spring as split dressings between February and early May, depending on crop and quality requirements.

Phosphorus and potassium index levels are assessed by soil analysis and are now fairly high on the majority of arable farms. On fields with high reserves, only maintenance dressings are required or, indeed, none at all. Applications can be made at any convenient time during the growing season. If soil levels are very low (index 0 or 1), the fertiliser should either be applied to the seedbed or combine drilled. Care must be taken with potassium applications if straw is removed from the field rather than incorporated. Barley straw, especially, contains fairly high levels of potassium and removal could well mean that subsequent applications of potassium will need to be increased (Table 12.5).

Table 12.5 All cereals–phosphate and potash recommendations (kg/ha)

P or K index	0	1	2	3	4 and over
Straw ploughed					
in/incorporated					
Winter wheat,					
winter barley (8t/ha)					
Phosphate (P_2O_5)	110	85	60M	20	Nil
Potash (K_2O)	95	70[a]	45M(2–)	Nil	Nil
			20(2+)		
Spring wheat, spring					
barley, rye, triticale,					
winter and spring oats (6t/ha)					
Phosphate (P_2O_5)	95	70	45M	Nil	Nil
Potash (K_2O)	85	60[a]	35M (2–)	Nil	Nil
			20 (2+)		
Straw removed					
Winter wheat, winter					
barley (8t/ha)					
Phosphate (P_2O_5)	120	95	70M	20	Nil
Potash (K_2O)	145	120[a]	95M (2–)	25	Nil
			70 (2+)		
Spring wheat, spring barley,					
rye, triticale (6t/ha)					
Phosphate (P_2O_5)	105	80	55M	Nil	Nil
Potash (K_2O)	130	105[a]	80M (2–)	20	Nil
			55M (2+)		
Winter and spring oats					
(6t/ha)					
Phosphate (P_2O_5)	105	80	55M	Nil	Nil
Potash (K_2O)	155	130[a]	105M (2–)	35	Nil
			80 (2+)		

[a]Reduce the potash recommendations by 25kg/ha on genuine sand-textured soils where the soil has 100mg/l or higher.

Some clay soils can release 50kg/ha of potash annually. On these soils potassium applications should be reduced accordingly.

M indicates a maintenance dressing intended to prevent depletion of soil reserves rather than to give a yield response.

The recommendations for P and K at Index 0 and 1 are for crop requirements as well as for building up soil reserves. If the policy is not to improve soil levels these rates can be reduced by 50kg/ha at Index 0 and 25kg/ha at Index1.

Ref: RB209 – Crown copyright

With the reduction in amount of atmospheric sulphur being deposited, some 30–40% of cereal crops are now at risk from sulphur deficiency. In the spring, 10–20kg S/ha should be applied on these soils. Of the trace elements manganese is the most common deficiency and 15–20% of cereals usually requires treatment with manganese each year. In Scotland up to 30% of cereals needs treating with copper, although in England and Wales the area treated is nearer 5%.

12.3.6 Crop protection

There can be a yield response in cereals from controlling weeds, pests and diseases. The response will depend on the problem and infestation level. On average, a winter cereal crop will be treated with two herbicides, at least two fungicides, an insecticide and a growth regulator during the growing season (Table 12.6). Pesticide inputs are normally much higher in winter cereals than

Table 12.6 Winter cereals – chemical calendar

Month	Crop growth stage*	Herbicides growth regulators	Insecticides/ moll uscicides[†]	Fungicides
September	Crop sowing	Couch and weed control including volunteers, post-harvest pre-drilling	Seed dressing Control of slugs	Seed dressing
October	Crop emergence Leaf emergence	Grass weed control – blackgrass, wild oats, etc. and broad-leaved weeds	Control of aphids (BYDV)	
November December January	Tillering			
February			Control of wheat bulb fly	
March	Stem extension	Broad-leaved weed control		Pre T1 if foliar diseases such as yellow rust or mildew are present
April		Growth regulators and control of wild oats and cleavers		Eyespot and foliar diseases – T1
May	Flag leaf emerging			Foliar diseases – T2
June	Ear emergence			Foliar and ear diseases – T3
July	Flowering Ripening		Orange blossom midge (wheat) Aphids (wheat)	
August	Harvest	Couch control pre-harvest		

* Will be affected by sowing date, crop (wheat or barley) and climate.
[†]Tend to treat when necessary or high risk situations, not on routine basis.

in spring cereals. Prophylactic/routine treatments are commonly used. Rates are often reduced depending on the risk of damage. Accuracy of application is best achieved by following the same wheelings each time. These 'tramlines' are normally introduced at drilling. The loss of land caused by tramlines is small and is more than compensated for by the ease and accuracy of spraying – provided the lines are initially properly positioned. Tramlines also allow easier crop inspection

12.3.7 Diseases

There are many diseases that can affect cereals from those which are seed-borne (e.g. loose smut), to stem-based problems (e.g. eyespot), to foliar diseases (e.g. mildew and rust). The majority of diseases are caused by fungi, although there are a number of important virus diseases such as barley yellow dwarf virus (BYDV) and the mosaic viruses. Some diseases are specific to particular cereals; others can attack most cereals. Routine fungicide programmes are normally used in most crops. Varietal resistance, weather conditions, disease incidence and husbandry will affect the fungicide programme used. The most cost-effective timing for fungicides will vary with the cereal crop. In wheat, for instance, it is essential to keep the flag leaf and ear clean, whereas in winter barley, treatment at early stem extension (GS 31–32) gives the greatest economic return. Care must be taken with fungicide choice as there are diseases which have now developed resistance to some commonly used fungicides (e.g. eyespot resistance to MBCs, mildew to the triazoles and strobilurins, *Rhynchosporium* to the MBCs and some triazoles.). Table 12.7 summarises the main groups of fungicides used in cereals. To avoid resistance, fungicide mixtures should be used with different modes of action. Any one fungicide should not be applied more than twice in a season.

Decision support systems are being developed to aid decision making in relation to disease control in cereals.

12.3.8 Weeds

Weed problems in cereals vary, depending on rotations (both past and present), soil types, cultivation systems and area of the country. Continuous autumn sowing of cereals has encouraged autumn-germinating weeds such as blackgrass and cleavers, especially on the heavier soils. Mixed rotations with autumn- and spring-sown crops have a more varied weed flora. Non-ploughing techniques and early drilling of autumn cereals have favoured the bromes. Where resistant weeds are present a more integrated approach must be taken to control them.

The most important grass weed problems in winter cereals are wild oats, common couch, blackgrass (not Scotland or south-west England), meadow-grasses (particularly in grass cereal rotations) and the bromes. Common broad

Table 12.7 Examples of fungicide/groups for disease control in wheat and barley

Chemical group	Examples of active ingredient	Eyespot – Wheat and barley	Mildew – Wheat and barley	Septoria spp – Wheat	Yellow rust – Wheat and barley	Brown rust – Wheat and barley	Rhyncho-sporium – Barley	Net blotch – Barley
Triazoles	epoxiconazole	(*)	(*)	*	*	*	*	*
Strobilurins	azoxystrobin		(*) (w) *(b)	*	* (w) (*)(b)	*	*	*
Quinolene	quinoxyfen		*					
Spiroketalamine	spiroxamine		*		*	*	(*)	
Morpholine	fenpropimorph		*		*	*	*	
MBC	carbendazim		(*)	(*)			*	
Chloronitrile	chlorothalonil			*			(*)	
Imidazole	prochloraz	*	(*)	*			*	*
Anilino-pyrimidine	cyprodinil	*	*				*	*

Note: Not all active ingredients in the same chemical group have exactly the same disease control profile. Always check the recommendations. The effectiveness of some of these chemicals may vary in the future depending on development of fungicide resistance.
* = Will give control (from good to excellent depending on chemical)
(*) = partial control
w = wheat
b = barley

leaved weeds in winter cereals include chickweed, cleavers, mayweed, speedwell and field pansy. In spring cereals, the polygonums (e.g. knotgrass and redshank) and common hemp nettle (in some areas) are more important. Broad leaved weeds are cheaply and easily controlled in the cereal crop, but this is not always the case with grass weeds. There are an increasing number of cases of herbicide resistance in blackgrass, wild oat and ryegrass populations.

12.3.9 Pests

Cereals can be attacked by several pests (Table 6.1), some of which cause little damage (e.g. leaf miners); others can cause total crop loss if not controlled. The main pests in cereals include aphids, wireworm, slugs, leatherjackets, wheat bulb fly, frit fly, blossom midge, birds and rabbits. Many pests are specific to some, but not all, cereals; wheat bulb fly does not affect winter oats or spring cereals sown after the middle of March. The previous crop can also have a significant effect on the cereal. Pests following grass include wireworm, leatherjacket and frit fly. Some pest problems only occur in certain areas of the country or on particular soil types, e.g. wheat bulb fly is an eastern counties problem, and slugs are associated with heavier soils. Treatment for pests is not routine every year. Research organisations are trying to improve pest forecasting systems. Cultural control methods such as time of drilling, seedbed conditions and varietal resistance should be used before resorting to chemical control.

12.3.10 Plant growth regulators

Plant growth regulators are mainly used in winter cereals to reduce plant height and increase straw strength and so reduce lodging and brackling; leaning is not a problem.

Lodging occurs when the crop is between being flat and up to being at 45 degrees to the ground. Leaning occurs when the crop is standing upright to that leaning no more than a 45 degrees angle from the vertical. Brackling occurs when the straw in barley breaks just below the ear.

There are two types of lodging. Stem lodging is due to the straw buckling and is affected by stem diameter, straw wall thickness and strength. Root lodging is due to the plant actually being uprooted. The amount of soil moisture, soil texture and rooting depth and spread will affect the amount of root lodging.

Lodging is also affected by a number of husbandry factors:

1 Varieties. The HGCA UK Recommended Lists show that there is quite a difference in standing power between crops and varieties. Length of straw is also important. Short-strawed varieties are not as affected by damage caused by the force of high winds.

2 Nitrogen use. Excess nitrogen applications, excess soil nitrogen mineralisation

and/or too early applications of nitrogen can produce lush growth, dense canopies and weak straw. These crops are prone to stem lodging.

3   Stem-based diseases. Fungal diseases such as eyespot and those caused by *Fusarium spp.* can reduce stem strength and increase the likelihood of stem lodging.

4   Weeds. Weed competition can weaken straw strength.

5   Time of sowing and seed rate. Early drilling of winter wheat and late drilling of winter barley and winter rye produce taller-strawed crops that are more prone to lodging. High seed rates produce crops with a narrow spread of roots that are more liable to root lodging.

6   Soil type. Cereal crops grown on shallow, droughty soils can be less affected by lodging than on deeper, more fertile soils. On these soil types care needs to be taken with PGRs as crop damage has been observed.

7   Weather. The amount, timing and intensity of wind and rain, situation of field and exposure affect the amount lodging.

Lodging at early ear emergence causes the greatest yield loss (up to 50% reduction has been recorded). On average, yield losses from lodging in wheat are 2.5 t/ha. Other problems induced by lodging can include the production of secondary tillers and uneven ripening, poorer grain quality, especially if the ears start sprouting, increased weed competition, delayed harvest, increased combining time and drying requirements. By looking at the plant density and canopy size farmers can assess the risk of lodging. Use of growth regulators, rolling in the spring and nitrogen timing can help reduce the lodging risk. Growth regulators tend to be used routinely on fertile soils where high yielding quality cereal crops are being grown, and where there is a history of lodging.

With the increasing numbers of different pesticides and formulations, many farmers rely on independent advisers/consultants and distributor representatives to help with their spray programmes. Many organisations run cereal/arable crop trial centres where it is possible to see new chemicals and husbandry techniques in practice. Organisations such as Arable Research Centres Ltd are farmer funded. Information from these centres is normally only available to members. All farmers selling grain pay a small levy to the Homegrown Cereals Authority (HGCA) to fund cereal research such as the recommended lists.

12.4   Harvesting

Threshing is the separation of the grains from the ears and straw. In wheat and rye the chaff is easily removed from the grain. In barley, only the awns are removed from the grain; the husk remains firmly attached to the kernel. In oats, each grain kernel is surrounded by a husk that is fairly easily removed by a rolling proces – as in the production of oatmeal; the chaff enclosing the grains in each spikelet threshes off. Varieties of naked or huskless oats have now been bred; they thresh free from their husks. The majority of cereal crops are harvested using a combine. Only a few crops are traditionally harvested with a binder where the straw is required for thatching, and some crops are foraged for whole crop silage.

Cereal harvesting can be carried out more efficiently if it is spread out by growing early and late varieties of barley and wheat, starting with winter barley in July, followed by winter wheat and spring barley in August/September. Spring wheat is the last cereal crop to harvest. Good weed control, to minimise green material at harvest and avoidance of lodging, is also very important; combining is easier and less or no drying of grain is required. The combine capacity should be adequate to harvest the crops as they ripen. Delay can result in poor-quality grain (worse if it sprouts in wet weather) and shedding losses. It is preferable to have to dry early-harvested crops than to salvage damaged crops later. Grain losses of 40–80 kg/ha are reasonable (the latter in a difficult season). Slow combining to reduce losses to 10–20 kg can be false economy, more can be lost by shedding due to a delay in harvesting. Timing of harvesting will be affected by area and crops grown as well as quality requirements, combine and drying capacity, labour, use of desiccants and weather conditions.

Grain monitors, if properly used, can be helpful in checking losses. The average rate of working (tonnes/hour) of a combine is about half the maker's rating, which is usually based on harvesting a heavy wheat crop in ideal conditions. Standard combines vary in drum width, straw walker area (the main limiting factor) and header width. Wide headers are desirable for farms where the crops are normally light and the fields reasonably level. Pick-up reels are very useful when harvesting laid crops. The standard combine can work up and down slopes reasonably well, provided the speed is adjusted as necessary. However, combines do not work satisfactorily going across slopes as the grain moves to the low side of the sieves, and little separation takes place because most of the wind escapes on the top side. Some combines are designed to keep the threshing and sieving mechanisms level on sloping ground. The hillside type has a self-levelling mechanism, which can adjust for going across slopes as well as up and down.

The axial-flow combine differs from the conventional machines by having the threshing and separation of grain from straw carried out simultaneously as the material spirals round a large rotor from front to back of the machine. The straw walkers in conventional combines are not the most satisfactory way of separating grain and straw; on some machines these are replaced by a series of drums. This is done in order to increase the output of the combine without increasing the physical size, which is important for access and road transport reasons. These combines, known as 'rotary' combines require more power and tend to damage the straw more than conventional combines.

The conventional header can be replaced by a stripper header which removes the ears only, and the straw is left standing. The header will be more expensive, but the work rate can be very high and laid crops do not present serious problems.

Whatever type of machine is used, the manufacturer's recommendations should be followed for the various settings, e.g. drum speed and clearance and fan speeds. On modern machines adjustments can be made easily.

Combines can be fitted with straw spreaders or choppers, which are helpful in dealing with the problem of straw disposal, although the mounted chopper can reduce combining speed as it requires more power.

Cleaning combines to remove weeds, e.g. wild oats, blackgrass, barren brome, is very important before moving to a clean field. It is essential that the combine is clean before moving into a seed crop.

These days most grain is bulk-handled from the combine into and out of store. In some years, when all the cereals are combined at below 15% mc, no drying will be necessary, only possibly cooling.

12.5   Grain-drying methods

12.5.1   The options available
Below are listed the most important methods of drying grain:

1   Continuous-flow drier. The principle is that hot air removes the excess moisture and ambient air and then cools the grain to 10–15 °C. However, in very warm weather this may not be possible and night air may have to be used to cool the heap after drying. These driers are rated as 'x' tonnes/hour taking out 5% moisture in wheat.
    Wet grain (over 22% mc) may require two passes through the drier, taking out about 5% moisture each time.
2   Batch driers. The drying principle is similar to that of the continuous-flow drier, but the grain is held in batches in special containers during the drying process (extraction rate 6% per hour in small types; 6% per day in silos).
3   Ventilated silos or bins. Cold or slightly heated air is blown through the grain in the silo. This can be a slow process, especially in damp weather (extraction rate 1/3–1% per day).
4   Bulk storage/on floor drying. A large volume of cold or slightly heated air is blown through the grain to remove excess moisture.
    The air may enter the grain in several ways, e.g.:
    (a) From ducts about a metre apart on, or in, the floor;
    (b) From a single duct in the centre of a large heap,
    (c) Through a perforated floor which may also be used to blow the grain to an outlet conveyor when emptying.
    Floor drying is very popular because it can be carried out in a general-purpose building. It is a cheap method and requires very little labour when filling. The rate of drying is 1/3–1% per day. A common mistake with floor drying is to use heat when drying very wet grain. This over-dries the lower layers around the ducts, and the moisture settles out in the cooler

upper layers where it forms a crust which impedes the air flow. The heat should be saved for later stages of drying down to about 14% mc. Problems can arise where rubbish is allowed to form in 'cones' as it is loaded into the store. This could prevent air flowing freely past the grain. The wet and dirty grain should be spread as evenly as possible. Wet grain is more spherical than drier grain; the air spaces are usually larger in the wet grain heap, and so air flows more freely through it than through dry grain. This is a useful self-adjusting phenomenon. Grain stirrers can be installed which overcome the problems of uneven drying. They move automatically through the heap of grain on a framework and thoroughly mix the grain as it dries. The relative humidity (RH) of the air being blown through grain determines the final moisture content of the grain. A 1 °C temperature rise from heaters reduces the RH by 4.5%.

5   Mobile grain driers. These have become popular in recent years. Most are of the recirculating batch type, but portable versions of well-established static batch or continuous flow machines are also in use.

6   Membership of co-operatives. The other option which farmers have is to dry and store grain with a co-operative group. This store will provide drying and sometimes grain wetting facilities (in a dry year) as well as gravity separators for improving grain quality.

12.5.2   Precautions to be taken in grain drying

Whilst using high temperature driers, it is important not to overheat the grain as this may result in loss of quality premiums. Seed, malting and milling samples need to be dried at relatively low temperatures as the seed must remain viable. Feed samples can be dried at higher temperatures and therefore the output of the dryer will be increased. Maximum drying temperatures will depend on the type of dryer used and even the individual make of machine. It is important to follow manufacturers' recommendations.

Grain can be stored in bulk for up to one month at 16–17% mc, but for longer term storage, it should be dried down to 14% mc. Only fully ripe grain in a dry period is likely to be harvested in the UK at 14% moisture. In a wet season, the moisture content may be over 20% and the grain may have to be dried in two or three stages if a continuous-flow drier is used.

Damp grain will heat and may become useless. This heating is mainly due to the growth of moulds, mites and respiration of the grain. Moulds, beetles and weevils may damage grain that is stored at a high temperature, e.g. grain not cooled properly after drying or grain from the combine on a very hot day. Ideally, the grain should be cooled to 15 °C within 2 weeks to prevent sawtooth beetles breeding. This is difficult, if not impossible, in hot weather. For long-term storage the grain should be cooled to below 5 °C to prevent mites and grain weevil breeding.

Before storing grain, it is essential to clean and dry the store thoroughly.

When clean and waterproof, the building should be fumigated with a suitable insecticide to kill any remaining pests. Special formulations of the insecticide pirimiphos-methyl are commonly used for this purpose.

The temperature of stored grain should be checked regularly. Any rise in temperature can indicate an insect infestation. Rodents and birds must also be kept out of the store. When grain is sold, any chemical treatments must be declared. The Pesticide Notification Scheme is now mandatory. Any kind of grain stored for human consumption must be kept under conditions that satisfy the Food Safety Act 1990. This means that the farmer must exercise 'due diligence' in keeping stores rodent- and bird-proof; these are aspects covered in the various Assurance Schemes. Other commodities, such as fertiliser, must not be stored in the same building.

12.6   Moist grain storage

Most grain is stored dry. There are, however, alternative methods of preserving grain that can be simpler and cheaper than more conventional systems. An established method is the storage of damp grain, straight from the combine, in sealed silos. Fungi, grain respiration and insects use up the oxygen in the air spaces and give out carbon dioxide, and the activity ceases when the oxygen is used up. The grain dies but the feeding value does not deteriorate whilst it remains in the silo. This method is best for damp grain of 18–24% mc, but grain up to 30% or more may also be stored in this way, although it is more likely to cause trouble when removing it from the silo, e.g. 'bridging' above an auger. The damp grain is taken out of the silo as required for feeding. This method cannot be used for seed corn, malting barley, or wheat for flour milling.

Harvesting grain when moist and then ensiling is a technique that is gaining popularity on some mixed farms. The cereal is combined three to four weeks earlier than normal at 65% dry matter. The grain is then crushed in a machine called a *crimper* and ensiled. A preservative is normally used to reduce clamp spoilage. Yield of crimped grain is usually higher than grain harvested at the normal time.

A further method of storing damp grain safely and economically is by sterilising it with a slightly volatile acid such as propionic acid. The acid is sprayed on to the grain from a special applicator as it passes into the auger conveying it to the storage heap; 5–9 litres per tonne of acid is required. Grain stored in this way is not suitable for milling for human consumption or for seed, but it is very satisfactory for animal feeding and, after rolling or crushing, it remains in a fresh condition for a long time because the acid continues to have a pre-servative effect.

12.7 Cereal straw

Recoverable straw produced on farms in the United Kingdom is in the range of 2.5–5 t/ha. Virtually all the barley and oat straw is baled for livestock bedding and feed. The feed value for wheat straw is lower than for barley and oats. The use of wheat straw will depend on the area where it is grown and the proximity of livestock enterprises. Each year about 40% of wheat straw is not baled, but is incorporated instead. Other uses include straw for farm and household fuel, mushroom compost, covering overwintered carrots, potato and sugar beet storage. It is also used in horticulture, as well as for thatching and insulation board.

When incorporating straw into the soil it should be chopped, preferably using a combine-mounted chopper. Straw can be incorporated by ploughing to at least 15 cm or, by a non-ploughing method, e.g. heavy discs, to a 10 cm depth. Non-ploughing techniques will tend to encourage annual grass weeds, especially the bromes and blackgrass as well as volunteer cereals. Light cultivations pre-ploughing can encourage weed seeds to germinate, but are not essential for successful straw incorporation.

Applying extra fertiliser or additives has little effect on straw breakdown. Where straw had been incorporated for several years, there appears to be no problem with its decomposition. In practice, there should be a small increase in soil organic matter.

12.8 Wheat

Nearly half of all cereals grown in the EU is wheat. The EU is now a major exporter of wheat as it is 145% self-sufficient. France, Germany and the UK are the main producers. The main markets (Table 12.2) are for human consumption (mainly milling) and animal feed. Of the remainder, some is used for seed, breakfast foods and distilling whisky. Little is sold into Intervention from the UK as there are great difficulties meeting the quality standards (Table 12.3).

All milling wheat should satisfy the following general standards:

1   Be free of pest infestation, discoloured grains, objectionable smells, ergot and other injurious materials.
2   Not overheated during drying or storage.
3   Moisture content of 15% or less.
4   Maximum impurities less than 2% by weight.
5   Pesticide residues within limits prescribed by legislation. A 'passport' describing pesticide application and vehicle cleanliness must accompany all loads taken to the mill.
6   Specific weight – usually at least 76 kg/hl required.

12.8.1   Some important qualities of wheat

In the production and use of flour, quality requirements can be divided into two groups, milling and baking.

12.8.1.1   Milling quality

This refers to the ease of separation of white flour from the germ and bran (pericarp and outer layers of the seed). It is a varietal characteristic and can be improved by breeding. In the milling process, the grain passes between fluted rollers which expose the endosperm and scrape off the bran before sieving. The endosperm is ground into flour by smooth rollers and in this process the good quality endosperms (hard wheats) break along the cell walls into smooth-faced particles which slide over each other without difficulty and so can be easily sieved. However, the poor quality endosperms (soft wheats) produce broken cells with jagged edges that cling together in clumps. This makes sieving slow and difficult; some of the cell contents are also lost. 'Milling value' is a measure of the yield, grade and colour of flour obtained from sound wheat. Varieties favoured for bread making are normally hard wheats. Soft wheats are used for biscuit making.

12.8.1.2   Baking quality

Wheat is the only cereal that produces flour which is suitable for bread making because the dough produced has elastic properties. This is due to the gluten (hydrated insoluble protein) present. The amount and quality of the gluten is a varietal characteristic, but it can also be affected by soil fertility and climatic conditions.

About 20% of the protein is in the wheat germ and is not important in the baking process. However, the other 80% (in the endosperm) is very important. It can be measured in various ways. For example, protein quantity is assessed by the Dumas test and by the near infra-red reflectance (NIR) technique which is a very rapid dry method. In addition, this measures moisture content as well as hardness. Protein quality is best assessed by a mini-baking test, but this takes at least two hours and so is mainly used for testing varieties for bread, biscuit or cake making. The Zeleny test is used for Intervention buying. A white flour is mixed with lactic acid and isopropyl alcohol and the resulting sedimentation is measured – the greater the amount of sediment the better the quality. The SDS method is similar but faster. Another commonly used method is to wash the gluten out of the ground grain and assess it for colour, elasticity and toughness (strength); this takes about 20 minutes. The protein in overheated (more than 60 °C) wheat grain is spoiled (denatured) for bread making.

12.8.1.3   Bread-making quality

To provide large, soft, finely textured loaves, the baker requires flour and dough with a large amount of good quality gluten. This is produced by strong-textured wheats. A good dough will produce a loaf about twice its volume. The small amount of alpha-amylase enzyme present in sound wheat is desirable for changing some starch to sugars for feeding the yeast in bread making. However, wet and germinated wheat contains excessive amounts of alpha-amylase and so excessive amounts of sugars and dextrines are produced. This results in loaves with a very sticky texture and dark brown crusts. Alpha-amylase activity can be measured by the Hagberg FN (falling number) test. The number is the time in seconds for a plunger to fall through a slurry of ground grain and water, plus 60 seconds for heating time before the plunger is dropped. A high FN indicates a low (and desirable) alpha-amylase level. A dough machinability test may also be required for Intervention buying; sticky doughs are likely to be rejected.

If the amino acid cystine is not present in the flour, as may happen if the crop suffered from sulphur deficiency, then the quality of the bread would be poor, even though the grain passed all the usual tests. The amount of water absorption by flour is also important in bread making. It is a varietal characteristic and depends on the protein content and amount of starch granules damaged by milling. A high water uptake by damaged granules of hard wheats means more loaves, which keep fresher longer, from each sack of flour.

12.8.1.4   Biscuit-making quality
A very elastic type of gluten is not required for biscuits, which should remain about the same size as the dough. Weak-textured wheat varieties are suitable for biscuit-making flour.

12.8.1.5   Milling for other uses
As well as flour for bread making and biscuits, there is a large demand for use in cakes, confectionery, soups and household flour. Market specifications will depend on the end product. The Flour Milling and Baking Research Association advises on the suitability of varieties and samples for various baking processes.

The poorer quality grain and the by-products from white flour production, i.e. bran (skin of grain) and various inseparable mixtures of bran and flour (e.g. wheatings), are fed to pigs, poultry and other stock.

Table 12.8 shows typical wheat quality requirements.

12.8.2   Wheat husbandry
Soils and climate
Wheat is a deep-rooted plant that grows well on heavy soils and in the drier eastern and southern parts of this country. Winter wheat will withstand most frosty conditions, but it can be killed out by waterlogged soils. The pH should be higher than 6.0.

Place in rotation
When the soil fertility is good, wheat is the best cereal to grow as its yields

Table 12.8   Typical wheat quality requirements

Market	Maximum moisture content %	Minimum specific weight (kg/hl)	Maximum impurities (% by wt)	Protein content % (100% DM)	Minimum Hagberg FN (sec)
Bread making	15*	76	2	13	250
Biscuit making	15*	–	2	–	–
Export milling	13.5–15	76	2	10.75	min 220
Export feed	13.5–15	72–76	2	–	–
Feed	15	72	2	–	–

*If grain is to be stored for more than four months, maximum moisture content should be 14 %. For Intervention Standards see Table 12.2.

are higher and its returns generally better than those of the other cereals. It is commonly taken for one or two years after grassland, potatoes, sugar beet, beans, peas, linseed or oilseed rape. It has also been grown continuously on many farms as eyespot and grass weeds can now be controlled. After four or five years, the so-called 'take-all barrier' is passed and yields remain fairly constant. However, many farmers now grow as many first wheats as possible (rather than continuous) because of their higher yields.

Seedbeds
A fairly rough autumn seedbed is adequate for winter wheat and helps prevent 'soil-capping' in a mild, wet winter. When soil-acting herbicides are used a fine seedbed is required. In a difficult autumn, winter wheat may be successfully planted in a wet sticky seedbed and usually it still produces a satisfactory crop. Spring wheat should only be planted in a good seedbed.

Time of sowing
Winter wheat can be sown from September to March. Early sowings from the middle of September are usually preferable, but they may not be better than late sowings in some favourable autumns. Occasionally, first wheats are sown at the beginning of September at very low seed rates. These crops normally require higher pesticide inputs and yields can be disappointing. Second wheats should be drilled in October to reduce the risk of take-all. Slow-developing varieties such as *Claire* should be chosen for early drilling, and faster-developing varieties for later sowing (e.g. *Charger*). However, sowing late may be impossible in some years. The HGCA list can be checked for 'latest safe sowing dates'.

Spring wheat can be sown from late autumn to April the following year. An increasing number of farmers are now planting spring wheat after late harvested root crops.

Method of sowing
Ideally, the crop should be sown at a uniform depth of 2.5 cm, depending on the moisture level in the soil.

Methods used:

1   Combine drill with fertiliser in 15–18cm rows.
2   Grain-only drill in 10–18cm rows (10cm preferable).
3   Broadcasting using the fertiliser spinner. This can be very satisfactory (especially if drilling is impossible) provided a good covering of the seed is achieved.

Varieties
(These are detailed in the HGCA UK *Recommended Lists* for cereals.) Winter wheat normally yields very much better than spring wheat, but the latter is usually of very good milling and baking quality. About 98% of wheat drilled is of winter varieties. Most winter varieties have to pass through periods of cold weather (vernalisation) and increasing day length (photoperiod) before the ears will develop normally.

Varieties differ in their susceptibility to disease, e.g. yellow rust and mildew, and it is advisable to grow at least three varieties selected to reduce the risk of spread of these diseases (see Diversification Scheme in the HGCA UK *Recommended Lists*). It is now well established that varieties differ considerably in their management requirements in such matters as the best time to sow; fertiliser timing and treatments for yield and quality; disease susceptibility and control; risk of herbicide damage; straw stiffness and response to growth regulators; risk of sprouting in the ear; ease of combining and saleability.

Examples of recommended varieties:

		Winter	Spring
1	Good quality milling and bread making	*Malacca* *Hereward* *Shamrock*	*Paragon*
2	Fair bread making	*Option* *Charger*	*Ashby*
3	Biscuit quality and others	*Claire, Deben*	

New varieties are continually being developed for higher yields, quality, disease resistance, standing ability. An increasing number of hybrid varieties are becoming available.

Seed rates
The usual range is 100–250kg/ha for winter wheat, and 170–220kg/ha for spring wheat. A number of factors will affect the number of seed sown and hence the seed rate:

1   Seedbed conditions (higher rates in very dry, cloddy and stony soils).
2   Weather and time of sowing. The rate could be 30–50% higher in November than in September. The lowest seed rates are used for the early September drillings.
3   Plant population. A target of 250–350plants/m^2 in the autumn may be reduced

to 150–300 plants/m^2 in early spring. In very good conditions fewer than 100 plants/m^2 can give very satisfactory yields, with fewer lodging and disease problems than in thick crops (e.g. 400 plants/m^2). Yield/ha is determined by the number of plants/ha × number of ears/plant × number of grains/ear × average grain weight (this last can be influenced by variety, growing conditions, disease control and possibly by growth regulators). Low plant populations tend to produce more ears/plant (average about 2) and possibly more grains/ear (about 30–50, with a few to over 90) and a heavier grain weight. These low populations are not very competitive against weeds.

The following is a simple way to calculate seed rate, knowing the thousand-grain weight (g) – TGW.

$$\frac{\text{Seed rate}}{\text{(kg/ha)}} = \frac{\text{no. plants/m}^2 \text{ required} \times \text{TGW}}{\% \text{ establishment expected}} \qquad [12.1]$$

Example: $\dfrac{350 \times 45}{70} = 225$ kg/ha.

Fertilisers

To return those nutrients removed from the soil, 7.8 kg of P_2O_5 and 5.6 kg K_2O should be applied for each tonne of grain/ha expected yield (e.g. at 8 t/ha, 62 kg/ha of phosphate and 45 kg/ha of potash are needed) if the straw is incorporated. If the straw is removed, the potassium rate should be increased to 8.6 kg of P_2O_5 and 11.8 kg of K_2O for each tonne of grain harvested (e.g. at 8 t/ha, 69 kg/ha of phosphate and 94 kg/ha potash is needed). Off-takes are higher in spring wheat when the straw is removed.

The nitrogen rates depend on SNS index (pages 56–7) and the possible use of organic manures. Nitrogen recommendations vary little with changes in grain and fertiliser prices. The nitrogen recommendations for winter wheat in kg/ha are:

Soil type	SNS Index 1	SNS Index 2
Light sandy soils	130	100
Shallow soils over chalk	240	200
Deep silty soils	180	150
Other mineral soils	220	180

Spring wheat, sown in the spring, requires similar nitrogen rates on sandy soils but 150 kg on other mineral soils at SNS Index 1. Grain nitrogen content is a good indication that optimum rates have been used. Nitrogen rates should be increased/reduced by 30 kg/ha for every 0.1% variation from the optimum of 2.0% grain nitrogen. Another system currently being studied is canopy management in connection with optimum nitrogen applications. The target crop canopy at ear emergence is about 6 units of green material per area of ground (GAI 6). As it is known that there are 30 kg/ha of nitrogen for every unit of green area, by finding out the available soil nitrogen level and fertiliser recovery, optimum nitrogen rates can be calculated.

It is important that adequate nitrogen is available when the crop growth rate is rapid at the beginning of stem extension – GS 30–32. If more than 100 kg/ha is to be applied, it will normally be split either two or three times. The first application of 40 kg should be applied during tillering, no earlier than March. Too early an application of nitrogen will encourage tillering and is also more liable to leach. If the crop is late drilled, backward, thin or suffering from pest attack, then the first application should be in February to encourage tillering. The main dressing of nitrogen should be applied at GS 31, in April and no later than early May. Timing for spring wheat, drilled in the autumn, should be as for winter wheat. Spring-sown spring wheat should have the nitrogen split between the seedbed and the three-leaf stage. The earlier the drilling, the more nitrogen should be applied post-emergence and vice versa.

To increase protein by up to 1%, for milling wheat, extra nitrogen (30–50 kg/ha) is usually applied with the main dressing or later. A liquid application of 20% urea is often applied by the milky-ripe stage. Late foliar applications of nitrogen usually have no effect on yield.

Spring grazing of winter wheat is now rarely practised, if there is a need for it and the crop is forward but before GS 31 and the soil is dry, it can provide useful grazing for sheep or cattle in late March. Grazing should cease if the plants are being pulled out of the ground. It should be grazed only once and as uniformly as possible in blocks and then top-dressed with an extra 50 kg/ha nitrogen (unless a reason for grazing was to reduce the risk of lodging). Yields are likely to be reduced by this grazing. Grazed crops have shorter straw at harvest, and sometimes less disease.

Seed dressings

A range of fungicide seed dressings are available to control seed-borne diseases such as bunt, fusarium, septoria and loose smut (see the *UK Pesticide Guide*). There are now some fungicide seed dressings that can reduce the impact of Take-All, e.g. silthiofam and fluquinconazole.

After grass where wireworm can be a serious problem, an insecticidal dressing should be used including tefluthrin or imidacloprid. This last is very effective against aphids and seed treatment may be preferable to an autumn aphicide. Where there is a high risk of wheat bulb fly tefluthrin-treated seed should be used.

Pests

In the autumn, slugs can be a serious problem in trashy, cloddy seedbeds, especially when the crop is late drilled. The slugs can hollow out the seed and shred the young seedling leaves. Methiocarb should be applied if there is a high risk (Table 6.1). Early-drilled wheat can be affected by aphid-transmitted barley yellow dwarf virus (BYDV). The optimum time for control is late October/early

November when a suitable aphicide or an imidacloprid seed dressing can be used. In the new year, in susceptible fields, wheat bulb fly may require controlling either in January at egg hatch or later, when the first 'dead heart' symptoms are seen – limited chemicals are available. Aphids and blossom midges can cause direct damage to wheat, at grain fill. If threshold numbers are present treatment is usually advisable (Table 6.1).

Diseases

Yield and quality of wheat can be very affected by stem-based and foliar diseases. Of the stem base/root diseases take-all and eyespot are the most common. Take-all disease can be a serious problem when growing wheat continuously until the 'take-all barrier' has been passed. On some soil types/rotations second wheats can be affected. On susceptible fields tolerant varieties should be grown, late drilled and/or a seed dressing should be used, e.g. silthiofam or fluquinconazole.

The main foliar diseases in wheat are *Septoria tritici* – leaf spot, yellow rust, brown rust, *Septoria nodorum* – leaf spot and glume blotch, and powdery mildew. The timing of a fungicide spray that gives the largest economic response is at T2 (GS 39). This treatment protects the flag leaf and eradicates any disease on leaf 2. In practice there is usually a yield response from the TI (GS 31/32) timing, which protects leaf 3 and is the best time to control eyespot. The T3 timing (GS 59) is important to control any disease still on the flag leaf and protects the ear. Care must be taken with chemical choice, as some fungicide programmes can encourage ear diseases that cause mycotoxins to be present on the harvested grain.

Fungicide programmes usually give a yield response even on resistant varieties in a low disease year. Rates and types of fungicides should be varied depending on the risk and presence of disease (Table 12.7). Years with low disease pressure and varieties with good disease resistance will only require very low rates of fungicides. Most fungicide programmes contain a maximum of two strobilurin treatments mixed with a triazole. The strobilurins give a yield increase over and above that expected from disease control alone. The strobilurins tend to increase the duration of green crop area. Other fungicides will be included for control of specific diseases such as eyespot and mildew.

The recognition of other cereal pests and diseases and their control are shown in Tables 6.1 and 7.1.

Weeds

In winter cereals, autumn-germinating grass weeds such as blackgrass and meadow-grass should be controlled in the autumn with a residual herbicide such as isoproturon (IPU). Care must be taken with chemical choice if resistant weeds are present. If broad-leaved weeds are controlled in the autumn with chemicals such as diflufenican (DFF) they may only require a reduced dose of herbicide in the spring. Spring treatments, except for wild oats and cleavers, are normally applied by the first node detectable stage GS 31 (Chapter 5).

Plant growth regulators
It has been estimated in wheat that there is a 1% yield loss for every 2% of wheat crop lodged at flowering. Plant growth regulators containing chlormequat have been used by cereal growers for over 40 years. Growth regulators shorten the internodes and strengthen the stem walls. The reduction in internode length varies (7–20 cm) between varieties and with chemicals. To prevent lodging in winter wheat many of the available PGRs are applied at, and/or before, stem extension so that the lower internodes are shortened and strengthened. Recently introduced growth regulators, such as imazaquin and trinexapac-ethyl, offer more flexibility in timing. Most PGRs in wheat are applied early, though occasionally products containing 2-chloroethyl phosphonic acid can be useful for applying after GS 32 to reduce the length of the upper nodes.

Harvesting
Winter wheat ripens before spring wheat. The crop is harvested in August and September.

Yield
Below is listed the yield of winter wheat in t/ha.

	Average	Very good
Grain	8	10
Straw	3.5	5

12.9 Durum wheat

12.9.1 Qualities of durum wheat
This is a different species of wheat (*Triticum turgidum* var *durum*). In Europe it is mainly grown in the more southerly countries. Most of the durum wheat grown is for the internal market; only a small amount is exported. If properly grown, durum wheat has a high protein and hard, amber-coloured vitreous endosperm. On milling it breaks into fairly uniform large fragments called semolina, which is made into a range of high quality pasta products by extruding a stiff dough (semolina mixed with warm water) through dies of various shapes. This high quality market (up to a 50% premium) requires grain which is a light-amber colour and meets minimum standards, i.e. 12.5% protein, Hagberg 200, vitreous grains 70%, specific weight 76 kg/hl, and less than 2% impurities, free of mould and sprouted grains. Some durum wheat is used for breakfast cereals. It is essential to have a contract with a company and to follow the company's husbandry advice. Durum wheat receives 'non-traditional' support in addition to arable aid. In the UK only 5000 ha is eligible for aid.

12.9.2   Durum wheat husbandry

Soils and climate

Durum wheat will grow on a wide range of arable soils, but is probably best suited to medium quality land. It tolerates dry conditions better than ordinary wheat.

It is a Mediterranean crop and the new varieties which have been developed do reasonably well in a good season in the southern and eastern wheat-growing areas in England. The problem is achieving the quality and it yields poorly compared with conventional wheat.

Seedbeds and sowing

Seedbed requirements and methods of sowing are similar to those for wheat crops.

Time of sowing

As durum wheat does not require vernalisation, it can be sown in either October or March. Autumn-sown crops are susceptible to frost damage, so most of it is spring-sown.

Varieties

Varieties all have awns on the ear so are bearded. Current varieties include *Exeldur, Lloyd* and *Tetradur*.

Seed rates

180–200 kg/ha. The target should be 300 plants/m^2.

Fertilisers

Recommendations as for milling wheat.

Pests

Pests as for wheat.

Diseases

Durum wheat is very susceptible to eyespot and ergot.

Weed control

There are only a limited number of products that are approved for use in durum wheat.

Harvesting

An autumn-sown crop is usually combined in August. If possible, it should be given priority (about 20–22% mc for combining) to obtain good vitreous grains and a high Hagberg number. If it is left too late, the endosperm is white and there is a risk of sprouting. Grain should be stored at 15% mc.

Spring-sown Durum wheat is normally harvested in September.

Yield
Durum is not as high yielding as ordinary wheat, averaging between 15–20% less. The range is from 3 to 6 t/ha.

12.10 Barley

12.10.1 Uses of barley
After wheat, barley is the next most important cereal crop in Europe. About a third of the world trade in barley is from the EU; Germany, Spain and France are the main producers followed by the UK. Of the barley grown, 60% is now winter sown. The grain is used mainly for:

1 Feeding to pigs (ground), cows and intensively fed beef (rolled). Some feed barley is exported.
2 Malting. The best quality grain is sold for malting purposes both within the EU and on the world market. This grain should be plump and sound with a high germination percentage (about 97%) but a low nitrogen percentage – less than 1.85% nitrogen. Some varieties are more suitable than others for malting (HGCA UK *Recommended Lists*). The final use for the malt will affect the grain nitrogen requirements. A premium, which varies from year to year, is paid for good malting barley (Table 12.9).

Table 12.9 Typical barley quality requirements

Market	Maximum moisture content (%)	Minimum specific weight (kg/hl)	Maximum admixture (%)	Maximum nitrogen range (%)	Minimum germination (%)
Malting	15	–	2	1.55–1.75	97+
Export malting	13.5–15	64	–	<1.85	95
Export feed	13.5–15	64	2	–	–
Feed	15	60–62	2	–	–

For Intervention Standards see Table 12.3.

In the malting process the grain is soaked, then sprouted on a floor to produce an enzyme (diastase, maltase), dried in a kiln and the rootlets (culms) removed to leave the malt grains. In the brewing process the malt is crushed and then soaked in warm water in mash tuns. The sugars then dissolve in the liquor before being drained off. The remainder of the malt (brewer's grains) is a valuable cattle food. The sugary liquor (wort) is boiled with hops to give it a bitter flavour and keeping quality, and to destroy the enzymes. The strained wort is then fermented with yeast which converts the sugars to alcohol and so produces beer. To make whisky, hops are not added; the fermented wort is distilled to produce a more concentrated alcoholic liquid called malt whisky.

Some malt is used in other products, e.g. malt vinegar and breakfast cereals. Most barley straw is baled for livestock bedding and feed.

12.10.2   Barley husbandry

Soils and climate

Barley can be grown on arable land throughout the UK, provided the pH of the soil is about 6.5; it is more affected by a low pH than is wheat. Barley grows better than other cereals on thin chalk and limestone soils. However, it will grow on a wide range of soils provided they are well drained. On organic and very fertile soils, it may lodge, especially in a wet season, and the grain is unlikely to be of malting quality.

The crop is not exacting as regards climate, but little winter barley is grown in the colder parts of the UK.

Place in rotation

Barley, unlike wheat, is usually grown when the fertility is not very high. On many farms, provided that the soil conditions are right, it has been grown continuously on the same fields producing reasonable yields. However, this was before the mosaic viruses became a serious problem. Reducing the number of winter barley crops in the rotation is now important to slow down the build up of the disease. Spring barley is unaffected. As an early harvested crop, winter barley is commonly grown before oilseed rape.

Seedbeds

Generally a finer seedbed than wheat is required. Shallow sowing at 3–5 cm is important.

Time of sowing

Winter barley can be sown from mid-September to early November. It is important to have the plants well-established before the winter. Too forward crops can be susceptible to frost damage as well as foliar disease and BYDV. Spring barley can be sown from January to early April. A good seedbed is more important than early drilling. Yields are usually reduced from drilling after the end of March. The sowing of barley follows the same lines as for wheat.

Example varieties

(The varieties listed are described in the HGCA UK *Recommended Cereal Lists.*)

	Winter	Spring
Malting quality (two-row)	*Pearl, Regina, Leonie*	*Optic, Decanter, Cellar*
Feeding quality (two-row) (six-row)	*Jewel, Heligan, Angela, Siberia*	*Static, Riviera*
BaYMV resistant	*Jewel, Angela, Antonia Siberia, Carat*	–
BYDV resistant	–	*Spire*

Modern two-row varieties of winter barley yield better than spring barley–especially in areas which suffer from drought in the summer. However, many spring varieties have very good malting quality compared with the winter varieties. To obtain best results with winter barley, the crop must be sown early enough to develop a good root system and become well-tillered before winter.

Seed rates
Winter barley 150–180 kg/ha; spring barley 125–180 kg/ha. The target is for an established 250–300 plants/m^2, which should result from sowing 300–400 seeds/m^2.

Pests
The main pest problem which requires control, particularly in early-drilled winter barley, is to reduce barley yellow dwarf virus transmitted by aphids (Table 7.1). In very susceptible areas the seed should be treated with imadicloprid or an aphicide should be applied at the end of October to the emerged crop. Early-drilled crops may require two applications.

Diseases
The barley crop can be affected by a large number of seed-borne diseases such as leaf stripe, *Fusarium*, smuts and snow rot. Foliar diseases can also be prevalent in both winter and spring barley, such as *Rhynchosporium*, mildew, brown rust and net blotch. A serious problem where a large amount of barley has been grown are the mosaic viruses (barley yellow mosaic virus (BaYMV) and barley mild mosaic virus (BaMMV)). Control is only by rotation and use of resistant varieties.

Many of the seed treatments are similar to those applied to wheat although usually other chemicals are included such as imazalil to control leaf stripe. In barley leaves 2 and 3 contribute more to yield than in wheat. It is the T1 (GS 31–32) fungicide timing in barley that gives the greatest yield response. Normal practice is at least a 2-spray programme at T1 and T2. Occasionally a pre-T1 treatment is required if there is early infection with powdery mildew or brown rust. The T3 treatment gives the lowest response in barley, it is only necessary if the T2 treatment was inadequate – see Table 12.7 for range of products and diseases controlled. For effective control and to reduce the risk of resistance fungicides with different modes of action should be used. Strobilurins are now one of the main components of fungicide mixtures. Low rates of fungicides can be used when growing disease resistant varieties when the amount and risk of disease is low.

Weeds

Autumn weed control is desirable for both annual broad-leaved and grass weeds, using suitable herbicides. Long runs of winter cereals tend to encourage grass weeds such as blackgrass and barren brome, and control can be expensive (Table 5.2). There is an opportunity to control difficult grass weeds before drilling spring barley.

The very early ripening time of winter barley makes it easier to control perennial weeds using pre-harvest glyphosate. This is especially the case with onion couch which senesces earlier than common couch (Chapter 5).

Fertiliser requirements

Nitrogen fertiliser requirements, for barley are detailed below.

	Feed	Malting
Winter barley	40–160 kg/ha depending on SNS Index. Apply 40 kg/ha in February and the rest in April.	40–130 kg depending on SNS Index. Apply by end of March.
Spring barley	up to 150 kg applied by early May.	up to 120 kg applied by end of March.

Early application of nitrogen to winter barley often encourages foliar diseases. The nitrogen should be applied to the seedbed with late-sown spring crops (other than on sands), whereas with the early-sown crop it should be split to avoid leaching.

The phosphate and potash requirements are similar to those for wheat (Table 12.5).

Plant growth regulators

Winter barley particularly can be susceptible to lodging. Stiff-strawed varieties should be chosen. To control lodging in winter barley, chlormequat is not as effective in barley as in wheat; products containing 2-chloro-ethyl phosphonic acid or trinexapac-ethyl can be used. The later applications will control necking and brackling.

Results in spring barley are variable. Highest risk of lodging in spring barley is on thick crops in fertile soil and conducive weather conditions after flowering.

Harvesting
Winter barley is ready for combining from mid-July to early August. Spring barley combining normally starts between mid-August and early September.

In a crop containing late tillers, harvesting should start when most of the crop is ready. Malting crops are usually left to become as dead ripe as possible before harvesting. Harvesting of feed barley is often started before the ideal stage. This is especially the case if a large area has to be harvested and the weather is uncertain. The ears of some varieties and over-ripe crops can break off very easily, resulting in serious losses.

Yield
Below are listed the yields of winter barley in t/ha.

	Average	Good	Excellent
Grain	6.3	7.5	9
Straw	2.75	4	5

12.11   Oats

12.11.1   Uses of oats

Oats are now a minor cereal crop in Europe. The EU is the second largest producer worldwide following the former Soviet and Baltic states. The main countries producing oats in the EU are Sweden, Germany and Finland. The decline of the horse and the higher yields of wheat, as well as no Intervention support for oats, have been the main factors affecting the declining oat area. In Europe only 10% of oats grown are for human consumption, although in the UK this figure is nearer 50%.

The best quality oats may be sold for making oatmeal which is used for bread-making, oatcakes, porridge and breakfast foods. Oats for human consumption should have a high specific weight, 50kg/hl, a moisture content of 14% or less and less than 10% screenings through a 2mm sieve. There are no standards for oil and protein content. Only a very small amount of oats are exported onto the world market.

Oats are particularly good for horses and are also valuable for cattle and sheep. However, they are not very suitable for pigs because of their high husk (fibre) content. There is interest in the 'naked' oats species which has a very high feeding value (better than maize) because the husk threshes off the grain. The normal oat grain contains about 20% by weight of husk (the lemma and palea) but it is easily separated from the valuable groat in naked varieties. New varieties of naked oats have been developed, but they are about 25% lower-yielding than conventional oats.

The major part of oat straw is baled and used for bedding/feed. Quality of the straw is usually better than for wheat.

12.11.2   Oats husbandry
Soils and climate
Oats will grow on most types of soil; they can withstand moderately acid conditions where wheat and barley would fail. If too much lime is present, manganese deficiency (grey-speck) may reduce yields. Oats do best in the cooler and wetter northern and western parts of the country, but even in these areas have been replaced by barley on many farms.

Place in rotation
Oats can be taken at almost any stage in a rotation of crops. They are a useful take-all break.

Seedbed and methods of sowing
These are similar to wheat and barley.

Time of sowing
Winter oats, late September–October. Spring oats, February–March.

Varieties
(These are detailed in the HGCA *Recommended Lists.*) Winter varieties are not so frost-hardy as winter wheat or barley. They usually yield better, especially in the drier districts, and are less likely to be damaged by frit fly, than spring oats. Winter varieties are mainly grown in England and spring varieties further north where frost can be serious. Examples of winter varieties: *Kingfisher, Jana* and *Millennium.* Only *Gerald* is resistant to stem nematodes. Examples of Spring varieties: *Winston* and *Firth.*

Seed rates
190–250 kg/ha for both winter and spring oats. The target plant population is 300 plants/m^2 for spring oats and 250–300 plants/m^2 for winter oats.

Fertilisers
Between 100 and 130 kg/ha nitrogen is usually considered the optimum amount for winter oats on most mineral soils at SNS Index 2/1. This should be applied at stem extension stage. A maximum of 120 kg/ha is required at SNS Index 1 for spring oats. Timing as for spring barley. (See Table 12.5 for phosphate and potash requirements.)

Spring grazing
Very occasionally winter crops are grazed in the spring. This may be desirable if there is a risk of lodging, because grazing results in a shorter straw.

Pests

The main pest problem is that of aphids transmitting BYDV. Oats are the most susceptible crop. An imidacloprid seed dressing should be used or treat with an aphicide in late October. The cereal cyst nematode is encouraged by close cropping with oats (Table 6.1). Nowadays, very little damage is found. Stem nematodes can be a problem in rotations where several susceptible crops are grown. This is controlled by widening rotation and growing resistant varieties.

Diseases

Mildew and crown rust are the major diseases of oats. Oat mosaic virus is not commonly seen; the current winter varieties are resistant. Winter oats may require a fungicide at GS 31 (Table 7.1) against mildew and at GS 39 against crown rust and mildew.

Weeds

Grass weeds are very difficult to control in the oat crop. Few chemicals are recommended. It is preferable to grow oats where there is no blackgrass or wild oat problem. Oats are, however, a very competitive crop against weeds (Table 4 in Appendix 9).

Growth regulators

Oats can be very susceptible to lodging. Choose a stiff-strawed variety and/or apply a PGR in high risk situations. Chlormequat formulations are useful in preventing lodging if applied at GS 32 as is trinexapac-ethyl at GS 30–31.

Harvesting

Winter oats are usually harvested just before winter wheat. Likewise, spring oats (depending on the district) are normally harvested before spring-sown spring wheat. Oats, like barley, are liable to shedding when ripe. The grain is not very dense and requires 40% more storage space than wheat.

Yield

Below are listed the yields of winter oats in t/ha.

	Average	Good
Grain	6.25	8
Straw	3.5	5

12.12 Rye

12.12.1 Uses of rye

Only 3% of the cereals grown in the EU is rye. The EU is 116% self sufficient, the surplus being exported. Germany produces about a half of the EU rye crop. Rye is grown on a small scale in the UK for grain or very early grazing. The grain is used mainly for making rye crispbread and there has been an increased demand for rye in soft grainbread. It is not widely used in the UK for feeding to livestock. Rye should only be grown on contract for milling. The long, tough straw is very good for thatching and bedding, but not for feeding. Usually about half the straw is baled and the rest is incorporated into the soil.

12.12.2   Rye husbandry
Soils and climate
Rye will grow on poor, light, acid soils and in dry districts where other cereals may fail. It is mainly grown in such conditions for grain because, on good soils, although the output may be higher, it does not yield or sell so well as other cereals. It is extremely frost-hardy and withstands much colder conditions than the other cereals.

Place in rotation
Rye can replace other cereals in a rotation, especially where the fertility is low. It can be grown continuously on poor soils with occasional breaks. It is more resistant to take-all and so it can be grown as a third or fourth cereal.

Seedbeds and methods of sowing
The seedbed requirements and methods of sowing are similar to those for wheat.

Time of sowing
For grain production, the crop should preferably be sown in September; there are at present no spring rye varieties grown in the UK.

Seed rate
150–190 kg/ha. 30% lower seed rates are used for the hybrid ryes.

Varieties
Rye, unlike the other cereals, is cross-fertilised and so varieties are difficult to maintain true to type and new seed should be bought in each year. Most of the varieties now grown are hybrids. Seed is expensive but yields can be up to 15% higher than conventional varieties.

   Examples of hybrid rye varieties include *Ursus* and *Esprit*; *Hacada* is a conventional variety.

Fertilisers
Depending on the SNS Index nitrogen requirements are in the range of 40–120 kg/ha. This should be applied as a top-dressing in April. For phosphate and potash requirements see Table 12.5.

Spring grazing
The grain crop can provide useful grazing in late February/March, depending on soil conditions, and before stem extension. Grain yield will be lower.

Crop protection
Rye is a very competitive crop with good disease resistance so inputs tend to be lower than for other cereals. A limited range of grass and broad-leaved herbicides is available (Table 3 in Appendix 9). Because rye is tall and often weak-strawed, a growth regulator is required, e.g. chlormequat (GS 30–31) or trinexapac-ethyl at GS 30–32 or 2-chloroethyl phosphonic acid at late stem extension. Hybrid rye varieties have shorter, stiffer straw.

Rye is less susceptible to take-all than are wheat or barley. It can, however, be affected by eyespot, mildew and brown rust. Hybrid varieties tend to be more susceptible to brown rust than conventional varieties. It is usually worthwhile to apply a fungicide at T1. As rye is open-flowering, it is more susceptible than other cereals to ergot. No very effective fungicide control is currently available, but crop rotation and deep ploughing will reduce the risk.

Harvesting
The crop is normally harvested before winter wheat. It is combined when the grain is hard and dry and the straw is turning from a greyish to white colour. The grain sprouts very readily in a wet harvest season, so it is usually harvested at 20% moisture content.

Yield
Below are listed yields for rye in t/ha.

	Average	Good
Grain	5.6	7
Straw	4	5

12.13 Triticale

12.13.1 Characteristics of triticale
This is produced by crossing tetraploid durum wheat with diploid rye and treating seedlings of the sterile Fl plants so that their chromosome number is doubled and they become reasonably fertile. It is bearded and intermediate between wheat and rye in most of its characteristics. Only the winter varieties are being grown in this country, although other varieties are grown in other countries. There is no EU support. Triticale is in direct competition with other feed grains.

12.13.2   Triticale husbandry

Soil and climate
Triticale is probably best suited to marginal and lighter soils for use as a feed grain of similar value as wheat. It is very winter hardy and tolerant of drought.

Seedbeds and methods of sowing
The seedbed requirements and methods of sowing are similar to those for wheat.

Seed rate
270–330 seeds/m^2 (125–160 kg/ha); minimum plants after winter 150/m^2.

Time of sowing
Mid-September to October. Winter triticale has a requirement for vernalisation so it should not be sown after the middle of February.

Fertilisers
Nitrogen requirements are in the range of 40–120 kg/ha top-dressed at the end of March/early April depending on the SNS indices. For phosphorus and potassium requirements see Table 12.5.

Varieties
Examples include *Fidelio, Trinidad, Vision Partout* and *Ego.*

Crop protection
Some varieties of triticale can be quite weak-strawed, particularly on less droughty soil. The growth regulators such as chlormequat, 2-chloro-phosphonic acid and trinexapac-ethyl will reduce lodging.

Weed control
There are more restrictions on the use of some grass weed herbicides in triticale than with the wheat crop. Several broad-leaved weed herbicides are recommended (Table 3 in Appendix 9).

Diseases
Triticale is susceptible to ergot and eyespot, and fairly resistant to many cereal foliar diseases such as brown rust. A fungicide at T1 for control of eyespot and, where necessary, foliar diseases is recommended. *Rhynchosporium* is common in triticale but usually is not a serious problem. Take-all does not affect triticale to any extent.

Harvesting
The crop is harvested in August, as for wheat. Drying and storage advice is
also the same as for wheat.

Yield
The average yield is 6 t/ha.

12.14   Maize for grain

Maize is a tall annual grass plant with a strong, solid stem carrying large narrow
leaves. The male flowers are produced on a tassel at the top of the plant, and
the female some distance away on one or more spikes in the axils of the leaves.
This separation simplifies the production of hybrid seed. After wind pollination
of the filament-like styles (silks), the grain develops in rows on the female
spike (cob) to produce the maize ear in its surrounding husk leaves.

Climate limits the production of grain maize in the UK. It is a sub-tropical
plant (C4) and in this country it is confined to a very small area in the south-
east where the yields are often low (4–6 t/ha) and not very reliable. The situation
is likely to remain this way until, or if, more cold-tolerant varieties are produced.
Grain maize grows well in many southern countries in Europe especially in
France and Italy. Yields are very high averaging between 8 and 9 t/ha of grain.
With this yield, grain maize is the third most important cereal in the EU after
wheat and barley. Most of the maize crop is used for animal feed with a smaller
amount milled for industrial use (starch) and for human consumption.

There is a limited market for 'corn-on-the-cob' or sweet corn as a vegetable.
This latter is a special type of maize in which some of the sugar produced is
not converted into starch and is harvested when the grain is in the milky stage.

12.15   Further reading

Alford D V, *Pest and Disease Management Handbook*, Blackwell Science, 2000.
Fertiliser Recommendations for Agricultural and Horticultural Crops (RB209), HMSO,
    2000.
Gooding M J and Davies W P, *Wheat Production and Utilisation*, CAB, 2000.
Henry R J and Kettlewell P S, *Cereal Grain Quality*, Chapman & Hall, 2000.
Homegrown Cereals Authority, *The Grain Storage Guide*, HGCA, 1999.
Homegrown Cereals Authority, *Introductory Guide to Malting Barley*, HGCA, 2001.
Homegrown Cereals Authority, *Topic Sheets and Project Reports*.
Homegrown Cereals Authority, *The Wheat Disease Management Guide*, HGCA, 2000.
Homegrown Cereals Authority, *The Wheat Growth Guide*, HGCA, 1998.
Kent N L and Evers A D, *Kent's Technology of Cereals*, Woodhead Publishing Ltd,
    1994.

McLean K A, *Drying and Storing Combinable Crops,* Farming Press, 1989.
Murray T D, Parry D W and Catlin N D, *Diseases of Small Grain Crops,* Manson Publishing, 1998.
Welsh W, *The Oat Crop,* Chapman & Hall, 1995.
Whitehead R, *The U.K. Pesticide Guide,* updated annually, CABI publishing.
Wibberley E J, *Cereal Husbandry,* Farming Press, 1989.

13

Root crops

Cash root crops are an important part of the cropping systems of a significant number of farmers throughout the main arable areas of the UK. They form an important element of the rotation, acting as a break from combinable crops and, if grown well, provide higher gross margins than cereals, oilseeds and legumes. Both sugar beet and potatoes require capital investment in specialist machinery, or the ability to use contractors, and most potato enterprises need investment in buildings suitable for long term storage of the produce.

13.1 Potatoes

The production of potatoes in the United Kingdom was, until 1997, strictly controlled by a quota system administered by the Potato Marketing Board (PMB). A heavy excess levy and possibly a fine were imposed if the area grown by a farmer exceeded the quota. Quotas could be purchased or leased from other growers. The Board also controlled the marketing of the crop annually by setting and supervising quality standards and riddle sizes. Additionally, it organised a support buying scheme to help maintain reasonable prices in difficult seasons, as well as promoting the product, carrying out market research and funding research and development.

On 1 July 1997 a new organisation, the British Potato Council (BPC), was formed to take over some of the activities of the old PMB. The BPC's mission is to 'stimulate, promote and develop the British Potato Industry'. It does this by funding research and development, helping to transfer technology into the industry, collecting and disseminating market information and promoting British Potatoes to consumers in the UK and overseas.

It is a levy-funded body with levies coming from growers with over 1 ha of potatoes (c£36.50/ha) and purchasers each time potatoes are moved (c 15 pence/tonne). There are at present approximately 6000 levy-paying growers with 147000 ha registered. The levy is compulsory and the BPC is robust in its attempts to collect outstanding debts.

The most important change to the industry is the cessation of quotas and Intervention buying. Both of these factors had a stabilising effect on potato prices, although the weather played a part in seasonal price fluctuations even with quotas in place. The last few years have seen a virtually free market place with prices high one year and low the next as growers increase and decrease acreages. The trend is for the bigger and more efficient growers to expand at the expense of the smaller producers, who cannot make use of 'economies of scale'.

The main areas of development for the BPC and the industry are to increase the environmentally sustainable production of ware and seed potatoes and to increase potato quality to meet end-user requirements. Projects currently funded by the BPC include the improvement of input precision and the reduction of production costs, the improvement of understanding of the factors affecting texture, flavour and physical qualities and the identification of novel products, novel characteristics and adding value.

Potatoes are tubers in which starch is stored and, depending mainly on the variety, they vary in shape (e.g. round, oval, kidney, irregular); skin colour (mainly cream and/or red); flesh colour (mainly cream or lemon) and depth of eyes (where sprouts develop).

In this country, potatoes are used mainly for human consumption, but in a glut year some subsidised lots may be used for stock feeding. In several EC countries, some of the crop is used for producing starch, glue, alcohol, etc. The consumption of potatoes in the United Kingdom, at 112 kg/head, is fairly constant from year to year and so the prices obtained may vary considerably according to the supply. An increasing proportion of the crop is processed as frozen chips and crisps or other potato snack products.

Good quality ware potato tubers:

1   Are not shrivelled.
2   Are not damaged, frosted or diseased.
3   Are free of greening (an indication that a toxic substance, solanine, is present); secondary-growth irregularities such as knobs, cracks and glassiness; sprout growth which spoils quality and increases preparation costs;
4   Are of good shape and size (40–80 mm).
5   Have clean skins with shallow eyes for easy peeling.
6   Have not been damaged by pests such as wireworms, slugs and cutworms.
7   Have flesh which does not blacken before or after cooking.
8   Do not break down when being boiled or fried.

Potatoes must have special qualities to be suitable for processing:

1   For crisps, chips and dehydration the tubers must have a high dry matter (starch) content and a low reducing sugar content. Fry colour in the potato snack market is all-important and too much sugar produces dark brown crisps and chips which are not acceptable to the UK consumer.
2   For canning, small (20–40 mm), waxy-fleshed, low dry matter tubers are required which do not break down in the cans.

In the United Kingdom, varieties are classified as first and second earlies and maincrops. The difference in times of maturity between varieties is mainly because the earlies are 'long day' potatoes, i.e. they can produce a crop during the long days of early and mid-summer, whereas the late varieties are neutral or 'short day' types and will not produce full crops unless allowed to grow on until the shorter days in early autumn. Early and some second early potatoes can be considered as a 'high risk/high profit' crop. Yields are lower than maincrops, but prices are often very high, especially very early in the season. Hitting the market at the right time before the price begins to fall is the secret of profitable early potato production. It is no longer the case that old maincrop potatoes coming out of store in the spring are poor quality, soft and 'tired' tubers. Better technology and a better understanding of storage requirements has meant that consumers can buy firm, clearskinned potatoes throughout the year. This means that the novelty of the new crop soon wears off with prices of first earlies falling rapidly by mid-May. There is also strong competition at this time of the year from imported potatoes from southern Europe and the Middle East. Technology and better climates mean that countries such as Cyprus, Egypt and Israel can produce new potatoes and export into the UK virtually all year round, so seasonality has all but disappeared.

Another important change in the UK market is the shift in customer preference. The big supermarkets have pushed 'babies and bakers' with specific varieties being required for each type, currently *Charlotte* and *Nicola* for babies and *Marfona* for bakers. Overseas sources offer better availability of pre-packed, washed potatoes for the supermarket shelf. The traditional 'dirty' early scraper, which needed washing and scraping in the kitchen, has been replaced by small, clear and set skinned baby potatoes which can be dropped into the pan straight from the pack, and which suit our more hectic lifestyle because of the greater convenience.

Many of the maincrop growers enter into contracts with processors or supermarkets. Some growers have entered into agreements with local packhouses, while others have set up their own on-farm pack lines to provide supermarkets with pre-packed, labelled and bar-coded produce. Maincrop growers have less opportunity to 'play the market'. Older maincrop varieties such as *Russet Burbank* are back in favour and grown on contract for the French fries trade.

13.1.1   Varieties
The following are the more important (or promising) varieties:

First earlies: *Arran Comet (d)*, *Maris Bard*, *Home Guard*, *Concorde (r)*, *Premiere (r)*, *Rocket (r)*.

Ware Second earlies: *Wilja*, *Marfona*, *Estima (d)*, *Charlotte*, *Nicola (r)*, *Maris Peer*, *Nadine (r)*, *Penta (r)*.

Ware Maincrop: *Ailsa*, *Cara (r)* (late maturing), *Desiree (d)*, *King Edward*, *Kingston (r)*, *Maris Piper (r)* (versatile variety suitable for ware and chipping), *Maxine (r)*, *Picasso (r)*, *Navan (r)*, *Pentland Squire (d)* (ware, crisps and chips), *Santé (r)* and *Stemster (r)*.

French Fries Maincrop: *Russet Burbank*, *Pentland Dell*, *Victoria (r)*.

Crisps Maincrop: *Lady Rosetta (r)*, *Saturna (r)*, *Hermes*, *Rooster*.

Note: (d) is a deep-rooting (drought-resistant) variety. (r) is resistant to pathotype Ro 1 Golden potato cyst nematode (eelworm). A few varieties show resistance to pathotype Pa2/3, White PCN.

It should be noted that all healthy plants of the same variety are alike because they are reproduced vegetatively.

13.1.2   Seed rates
The decision on the most profitable seed rates for potatoes is complicated by factors such as cost of the seed and expected price for the crop; variety and size of the seed must also be considered carefully. The figures given in Table 13.1 are for normal-sized seed (30–60 mm), but if healthy small seed (20–30 mm) is used the rate can be reduced by at least 25%. In some countries it is common practice to cut larger seeds into pieces by hand or machine before planting. The cut pieces are more likely to rot than whole seed, but they may be treated with thiabendazole, a fungicide.

Table 13.1   Yield and seed rate for potatoes

	Yield of tubers (t/ha)	Seed rate (t/ha)	Time of planting	Time of harvesting
Earlies	10–30	4–5	Feb–Mar	late May/July
Maincrop	35–65	3–4	April	Aug/Oct
Seed	25–40	4–5	April	Sept/Oct
Canning	10–20	5–7	Mar–June	June/Oct

Only basic or certified seed can be bought or sold. Growers can plant uncertified home-grown seed, but this is risky if the crop is from poor stock and is not protected by suitable insecticides against aphids, which spread leaf roll and mosaic virus diseases. Some varieties such as *Cara* and *Pentland Javelin* have a high resistance to virus diseases and so are better suited to growing-on as home-grown seed. A reliable tuber-testing service should be used to check any doubtful stocks of seed before planting.

UK seed crops of potatoes are grown in Northern Ireland, Scotland or at a high altitude in England. Disease and aphid levels are lower in these areas.

13.1.3 Sprouting (chitting) tubers before planting

Fewer than half of the potato crops in the United Kingdom are grown from sprouted seed. It is a necessity with earlies to obtain high early market prices and sprouting should be started in the winter before planting so that single sprouts develop on each tuber (an apical dominance effect). It is desirable for maincrops because, on average, it increases yield by 3–5 t/ha. It also allows more flexibility of planting time, and reduces the risk of yield losses if blight occurs early. However, many maincrop growers have given up the technique of chitting and physiological ageing because of mechanised planting regimes. Chitted potatoes are more difficult to plant and the chits often get knocked off.

Seed crops also benefit from sprouting. Rogues may be removed before planting; the crop bulks earlier and so the haulms can be destroyed early to check the spread of virus diseases by aphids. To obtain high yields from maincrops and seed crops, several sprouts should be encouraged to grow on each tuber (multi-sprouting).

The physiological age of seed tubers can have a marked effect on how most varieties develop. Physiologically old seed is produced from early planted seed crops which are lifted early in warm conditions and, when in store, have heat units added to them by keeping the temperature at levels above 4 °C. Such seed produces earlier crops with fewer tubers, but the crop is more susceptible to drought. Physiologically young seed, produced and stored in cooler conditions, is better for maincrops which grow on to high yields.

The rate of growth of the sprouts is controlled by temperature and is usually quickest at about 16 °C, but varieties differ considerably. The strength of the sprout is controlled by light. Short, sturdy sprouts are formed in well-lit buildings such as glasshouses or barns fitted with warm-white fluorescent tubes. Storage buildings must be frost-proof and well ventilated to avoid high humidity and condensation. Chitting trays, each holding about 15 kg (60–80 per tonne), are still the most popular containers for sprouting, especially for earlies. Higher-capacity white plastic trays, or wire crates and pallet boxes holding at least 500 kg, can speed up the loading of high output maincrop planters.

Well-developed sprouts (2–3 cm) give maximum yield advantage. However, they are easily damaged by mechanical planting and so mini-chitted seed, with sprouts less than 5 mm, is used with automatic machines on at least one-third of sprouted maincrops. Mini-chitting is achieved by temperature manipulation. If well-sprouted tubers are taken from a warm store and planted into cold ground, some varieties are likely to emerge very late due to 'coiled sprout' development, or 'little potato' may develop when no shoots appear and the old tuber is converted into a few new tubers.

13.1.4 Soils and climate

Earlies do best on light, well drained soils in areas free of late frosts, but some frost protection is possible with irrigation. Covering with polythene sheeting can hasten crop establishment. This is removed when the crops grow bigger.

Maincrops are best suited to deep, fertile, loam soils because high yields are very important, especially if prices are low. Potatoes will grow in acid soils (under pH 5.5); common scab can cause problems on high pH soils in a dry season. Irrigation, if it is possible, should be considered in order to suppress the disease.

Place in rotation. Potatoes can be taken at various stages, but usually after cereals. To control cyst nematodes, maincrops should not be grown more often than two years in eight, seed crops one year in five to seven years, but earlies may be grown continuously if lifted before the nematodes develop viable cysts (i.e. before 21 June).

13.1.5 Seedbed preparation

Except on light soils, early ploughing is necessary to allow for frost action. In spring deep cultivations, harrowing, discing and/or power-driven rotary cultivators are used, as required, to produce a fine deep tilth without losing too much moisture. The land is then ridged up using only the weathered soil available. Bringing up unweathered soil will result in smearing of the ridge sides.

Some soils are very stony, which is likely to cause harvesting problems. The stones can be collected mechanically and removed, or mechanically separated into windrows between the ridges. If sufficient tilth has not been formed over winter by frost action, it may be necessary to form mini-ridges to start with and, after planting, build these up to cover the potatoes as the soil dries out. Building up the ridges in several stages also allows the soil to warm up faster and, in addition, this method can hasten the growth of earlies.

13.1.6 Planting

Nearly all potato crops in the United Kingdom are planted mechanically with 2–7 row machines which are either fully or semi-automatic. Hand planting is still practised on a few farms growing earlies. The ridges must be well-formed to protect the new tubers from blight spores and light, which would cause greening. Some farmers who grow on lighter soils are now planting potatoes in beds. There are usually three rows in beds 150–180 cm wide, i.e. the width of two ridges. This is preferable for irrigation (there is less run-off), and it produces more tubers in the 40 mm size and less greening. Modern machinery can deal with the extra throughput of soil (it may, in fact, reduce damage), but it is only feasible in fairly dry conditions on light soils.

Spacing of tubers (sets). Row width for earlies and seed potatoes is 60–70 cm; for maincrop, the commonest width is 90 cm. There is less yield loss and fewer problems in producing good ridges, keeping them free of clods, and fewer 'greened' and blighted tubers with the wider rows. The spacing between the sets will depend on the seed rate and the size of the seed. Table 13.2 gives an indication of optimum seed rate and set numbers per hectare. This assumes that the seed size is 35–55 mm, the rows are 90 cm apart and the seed price about twice that likely to be obtained for the ware. Some land needs to be de-stoned before planting. The de-stoner passes the ridge or bed soil over a moving web and the stones and clods can be windrowed alongside the row. This makes harvesting easier, causes less damage to the tubers and prevents stones in the soil from restricting growth.

Table 13.2 Optimum populations and seed rates

Sets/50 kg	Tuber spacing 90 cm row	Sets/ha	t/ha
630	31 cm	36 000	2.86
650	25 cm	45 000	3.46
600	32 cm	35 000	2.96
700	30 cm	35 500	2.70
800	31 cm	36 000	2.25
750	33 cm	34 000	2.26
750 (*Cara*)	17 cm	65 000	4.30

On highly fertile soils, especially if irrigated, and where higher yields (over 70 t/ha) are expected, some of the tubers may be too large (over 80 mm). In these conditions, the seed rate should be increased by 10% to give more tubers of desirable size. Some varieties such as *Cara* produce fewer tubers than average and so they are often too large, hence the higher seed rates. Other varieties such as *King Edward* and *Maris Piper* produce a large number of tubers and they may not grow big enough (under 40 mm) and so a lower seed rate is acceptable.

13.1.7 Manures and fertilisers

Potato crops benefit from organic manures such as farmyard manure and slurry. Manure can be liberally applied when available at approximately 30–50 t/ha of farmyard manure or slurry equivalent. Normally this is ploughed-in during the autumn. Fertilisers are either broadcast during the spring cultivations or, more efficiently, placed in the ridges by the planter. Too much nitrogen applied to the seedbed usually delays tuber development which can reduce yields in some seasons but is not usually a factor taken into consideration. Too much nitrogen can also delay haulm senescence and skin set. This can lead to more harvest damage and to lower dry matters. The tuber size is increased by nitrogen and potash, whilst phosphate increases the number of tubers (Table 13.3). Dry matter can be decreased by high levels of potash.

Table 13.3   Effect of fertilisers on growth of potatoes

	N kg/ha (assumed SNS index 0–1)	P₂O₅ kg/ha (assumed index 2)	K₂O kg/ha (assumed index 2)
Earlies	120–150	180	145–170
Second earlies and maincrop	180–270	180	275–300
Seed (burnt off early)	180	180	145–170

(i) Recommendations for phosphate and potash are for incorporation into seedbed before planting, although where potash requirements are 300 kg, half should be applied in the autumn and half in the spring. If fertiliser is placed adjacent to the seed a small reduction in the rate of P and K can be considered. Potash recommendations will depend on whether the soil is at the top end or the bottom end of index 2, hence the two figures quoted.
(ii) 25 t/ha FYM could supply 35 kg N, 50 kg P_2O_5 and 100 kg K_2O.

Nitrogen rates for potatoes have recently been reviewed and made more accurate. Growers must take into account the Soil Nitrogen Supply, the length of growing season and the longevity of the haulm. The last factor is a varietal one and varieties fall into one of four categories:

1   Short haulm longevity (determinate varieties).
2   Medium haulm longevity (partially determinate varieties).
3   Long haulm longevity (indeterminate varieties).
4   Very long haulm longevity.

In general, more nitrogen is needed for the determinate varieties and, the longer the growing season, the more nitrogen is needed. This gives nitrogen rates ranging from 0 kg/ha up to 270 kg/ha depending on the above factors. Recommendations are that, unless the soil is prone to leaching, e.g. very light with irrigation, all nitrogen should be applied to the seedbed. No reductions should be made for placement of fertiliser, nor should more or less be used when using beds as opposed to individual ridges.

13.1.8   Weed control

Weeds, by tradition, have been controlled in good weather by cultivations, including ridging after planting. However, the frequent passage of rubber-tyred tractors tends to produce clods in the ridges and this hinders mechanical harvesting. Cultivations also damage the potato roots and stolons, and increase moisture loss from the soil. Consequently, chemical weed control, mainly by contact and soil-acting residual herbicides, has become normal practice on the majority of farms (Table 4 in Appendix 9).

13.1.9   Disease control

13.1.9.1   Blight

This is the worst fungal disease which attacks the potato crop. It can seriously reduce yield by killing the foliage early; during periods of heavy rain the spores of the fungus can be washed into the soil and onto the tubers, so causing them to rot in the ground or during storage. The disease spreads rapidly in warm, moist conditions which, when they occur, are known as 'Blight Periods'. These are recorded by the Meteorological Office, and information can be obtained by farmers by telephone or through the Internet. Growers can then respond by an application of fungicides; a wide range is available. Most fungicides have a protectant and systemic action and are effective if sprayed regularly with the leaves being well covered. A common programme is to use phenylamide mixtures early in the season with follow-up sprays of newer materials such as cymoxanil, cyazofamid and zoxamide. Spray intervals range from 7 to 14 days depending on material used and pressure of disease. The use of irrigation in the summer usually triggers a Blight Period and a spray programme should be started soon after.

Fentin compounds are suitable for killing spores which fall on the soil before they reach the tubers, but may scorch in hot weather. To reduce the risk of spores spreading from the leaves to the tubers, the haulms should be destroyed when about 70% have been killed by blight; this is especially important if heavy rain is expected (Table 7.1). It is also important to think about preventative measures to stop the spread of blight. Potato dumps and volunteers in neighbouring fields are important sources. Reject and waste potatoes should be deeply buried and volunteers should be killed off in other crops before they become a source of infection.

Other diseases of potatoes are mainly seed-borne and can be controlled by treating seed with fungicides, e.g. imazalil and thiabendazole. They include gangrene, silver scurf, dry rot, skin spot and stem canker (Table 7.1).

13.1.9.2   Aphids

Increased physical damage to the crop and the spread of virus by aphids in some hot summers have necessitated the use of soil-acting granular insecticides such as disulfoton or sprays such as deltamethrin or lambda cyhalothrin, on more than half the potato crops in the country (Table 6.1). However, because of the widespread use of organophosphates there are many resistant populations of the Peach–Potato Aphid (*Myzus persicae*) and growers are having to rely increasingly on sprays containing pyrethroids and pirimicarb. Aphids are now showing resistance to these materials as well, so a careful programme of mixtures of active ingredients (sometimes including nicotine) is advised, especially when treating seed crops.

13.1.10   Irrigation

(See also Chapter 8.) In dry seasons, very profitable returns can be obtained from irrigating the potato crop. Quality may be improved by the tubers being more uniform in size and having less common scab, but the dry matter percentage may be lowered if an excessive amount of water is used late in the crop's life. It is important to aim to stop irrigation four to five weeks before the intended lifting date. If water is applied to potatoes which have almost died due to drought, secondary growth may develop as knobs and cracks; this will obviously spoil the crop.

It is vital that irrigation water is used economically and effectively. Restrictions on licences for water abstraction are likely to increase in the future and the cost of water will rise. Accurate irrigation scheduling is vital and over-watering should be avoided. Many growers are investigating the use of trickle irrigation systems and methods of preventing run-off from conventional irrigators.

Earlies may be protected from frost damage by keeping the soil moist on those warm sunny days which are followed by radiation frosts at night. In more severe cases, they can be protected by spraying with about 2–3 mm of water per hour during the frosty period; the latent heat given out, as icicles form on the crop, prevents damage to the leaves. Irrigation may also be used to assist in breaking down clods when preparing spring seedbeds in a dry time.

13.1.11   Harvesting

Earlies are harvested when the crop is still growing. To make lifting and later cultivations easier, the tops should be destroyed with a flail-type machine. Earlies must be treated gently because the skins are very soft. Maincrops, especially if they are to be stored, should not be harvested until the tuber skins have hardened. This is usually about three weeks after the tops have died, or have been desiccated.

Historically, much of the desiccation on maincrops was carried out using acid but, with more and more contractors loath to use such a potentially dangerous chemical, and the threat of revoked approval for acid, growers looked for alternatives. Diquat and glufosinate-ammonium are alternatives cleared for use on maincrop ware potatoes and are widely used. Newer developments include the manufacture of large propane tractor-mounted burners which can straddle three beds at a time and have a high work rate.

Early crops are often picked by hand, but most maincrops are lifted mechanically. One-row, two-row and bed harvesters of various designs and capabilities are available to lift and load the crop into trailers or into bags. Harvesters can be manned or unmanned. Very high rates of working are possible with some machines in stone-free and clod-free conditions, when it is reasonably dry.

Damage to tubers by harvesting operations can result in serious losses due to bad design and/or faulty operation of the machines. Damage can be extremely serious in very dry soil conditions when the tubers are somewhat dehydrated and more susceptible to damage, and the clods are very hard. Machinery and handling equipment is now designed to limit the number and intensity of impacts that the potato suffers during the harvesting, storing and grading process.

13.1.12 Storage

A high proportion of the maincrop has to be stored. Storage may only be for a short period for some of the crop, but in other cases the potatoes may have to be stored until May or June when the earlies come on to the market again. Obviously, badly damaged, diseased and rain-wet tubers will not store satisfactorily and should be sold at harvest time.

Most crops are now stored indoors, either in on-floor bulk stores or in buildings designed to hold large 1000 or 1500 kg boxes which are stacked on top of each other from floor to ceiling. The pallet boxes are expensive, but ventilation is usually very satisfactory and the tubers retain a good appearance. Care is required when filling bulk stores to ensure an even distribution over the floor and to avoid 'soil cones' which can prevent air circulation. If the heap is more than 2 m deep, some form of duct ventilation will be required to prevent overheating. In the more sophisticated buildings, very accurate temperature control is possible by using refrigeration and recirculation ventilation and most growers of contracted maincrop potatoes would have invested in this type. In most ordinary stores it is usual to put a deep (300 cm) layer of straw or nylon quilt on top of the heap to prevent greening and frost damage, and to collect condensation moisture, so keeping the tubers dry. Good ventilation, and allowing the heap to warm up to 15 °C for 7 to 10 days after filling, helps to dry the tubers and to heal wounds. Some farmers apply fungicides to prevent rotting in store.

For late storage (by whatever method), sprout growth has to be prevented in the spring. This can be done by keeping the heap cool (4–7 °C) by cold air ventilation. This method is not satisfactory for crisping and chipping potatoes, because some starch changes to sugars (some reversal of this process is possible by warming the tubers before sale). Alternatively, sprout suppressants such as CIPC on processing potatoes can be applied when the crop is going into store.

13.1.13 Grading

The stored tubers are riddled (graded or sorted) during the winter or early spring when the best potatoes (ware) are separated from chats (small tubers), diseased, damaged, over-sized and misshapen tubers. They are usually sold off in 25 kg paper bags, in bulk to processors or are packed for supermarkets either on-farm or at centralised packing stations.

Yield – see Table 13.1.

13.2   Sugar beet

The sugar which is extracted from the crop supplies the United Kingdom with approximately 60% of its total sugar requirements. About 150 000 hectares are grown on contract by 8500 growers for British Sugar Plc (a private monopoly) which has seven factories in England, five of which are situated on the eastern side of the country, one in Shropshire and one in Worcestershire; however, the Kidderminster factory is due to close in 2002. A contract price per tonne of clean beet containing a standard percentage of sugar (usually 16%) is determined annually by the EU. Weight of raw beet, sugar percentage, dirt tare, transport allowance, early and late delivery allowances, a crown tare allowance and purity bonuses all determine the price which the grower will receive.

There is a world surplus of sugar and, to restrict output, a quota system operates for the production of sugar in the EU and this affects the price the grower receives. The annual beet contract for individual UK growers is expressed as a contracted tonnage, which is itself based on two amounts:

1   A quota – the top price for, at present, approximately 1.4 million tonnes of sugar.
2   B quota – potentially 25% less received per tonne for, normally, 1000 tonnes of sugar, but in practice the A and B price are the same.

Both quotas are adjusted to a 16% sugar level. Any sugar surplus to the A and B quotas is C quota and is sold (usually at a much lower price) on the world market after carrying forward about 70 000 tonnes to the next year's production. This is an insurance against the following year's crop being a poor one for some reason. Nationally the quota should be met, if not, it could mean a lower quota in subsequent years.

Each grower is allocated a quota level based on a proportional figure for A and B quotas. There are good reasons for meeting the contracted tonnage each year:

1   To realise the potential enterprise gross margin of the crop.
2   To avoid a reduction of quota in subsequent years for the grower. The advice is to endeavour to exceed the quota by up to 10% to compensate for yield fluctuations outside the grower's control and which may occur in any one year. Growers accept the fact that some beet will receive C quota price.

The average sugar content is about 17% but this can also differ from year to year depending on the growing season and, in particular, the amount of sunlight the crop receives. Sugar percentages can often be 20% greater in southern Europe because of the extra solar radiation.

There are three useful by-products from the sugar beet industry:

1   Beet tops – a very succulent food, but which must be fed wilted; they are not always easy to utilise but can be ensiled or fed *in situ* in the field.
2   Beet pulp – the residue of the roots after the sugar has been extracted; an excellent feed for stock. This can be sold wet for inclusion in cattle or

sheep rations or as a bagged dry feed very popular for horses as a source of energy.
3   Factory Waste Lime – used by the factories for purification and sold as a liming product back to the growers. It has a useful nitrogen content as well as containing some trace elements.

13.2.1   Soils and climate

Sugar beet can be grown on most soils, except heavy clays which are usually too wet and sticky; thin chalk soils which do not retain sufficient moisture and on which harvesting is difficult, and very stony soils on which drilling and harvesting can be very difficult. Moderately stony soils can be cultivated with newly developed machines which bury the stones below the rooting zone in a one-pass operation.

Sugar beet is a sun-loving crop which will not grow well when there is too much rain and cloud. Sugar is produced by the conversion of the energy of sunlight into the energy of the plant – sucrose in the case of sugar beet. Consequently sunlight, and the amount of crop leaf area which can absorb it, is an important factor in determining the yield of sugar produced by the crop. The grower should see that as much sunlight as possible is intercepted and utilised by a well-developed crop (irrigation may help) in the long days of May and June. This emphasises the importance of early drilling and establishment of the crop as well as a uniform distribution of the plants. However, drilling too early will expose the young plants to periods of low temperature. This vernalisation will make the plants prone to bolting with the subsequent production of a seed head and a dramatic loss of sugar in the root.

13.2.2   Place in rotation

In order to prevent the rapid build-up of *Rhizomania* (Table 7.1), as a condition of the contract sugar beet may not be sown in a field which has grown any Beta species, e.g. fodder beet, mangels, red beet, in either of the two preceding years. This restriction also helps to prevent the build-up of the beet cyst nematode (Table 6.1) and weed beet, although most *Brassica* crops as host crops can perpetuate the nematode. A three or four year interval is, in fact, preferable between sugar beet and any other closely related crops, as well as *Brassicae*.

Ideally, sugar beet will follow a cereal. This allows an opportunity for stubble cleaning but still allows time for careful October/November ploughing with subsequent essential weathering of the topsoil. Care must be taken to avoid certain cereal herbicides which can remain in the soil in sufficient quantities to affect the germination of the beet crop.

13.2.3   Seedbed preparation

The importance of a good seedbed for sugar beet cannot be over-emphasised. The success or failure of the crop can, to a great extent, depend on the seedbed. It must be deep yet firm, fine and level (page 196).

In most seasons, except on heavy soils, the less work done on the seedbed in the spring the better the conservation of moisture and the better the chance of avoiding compaction. The one-pass cultivation/drilling technique with front-mounted equipment and rear-mounted drill has much to commend it. An alternative method is to use a controlled wheeling system where the widths of machines match up and tractor wheelings are confined to specific tramlines across the field. Good early winter ploughing followed by a correctly timed winter cultivation (where possible) is, as always, a prerequisite on the majority of soils. The exceptions are the light peat and sandy soils; they can be ploughed and furrow pressed in the spring, as weathering is unnecessary.

A well-worked under-structure by deep loosening equipment with no soil inversion is, in some situations, preferable to the plough. The latter should be avoided when blowing (wind erosion) could be a problem. Non-ploughing techniques may also help to prevent weed beet germinating in the beet crop. The weed can more easily be dealt with in other crops in the rotation.

13.2.4   Varieties

The NIAB variety leaflet No. 5–*Sugar Beet*, and the factory field officer will help the grower to decide which varieties to use. There is a selection of varieties and they can be grouped according to the yield of roots, the sugar percentage, measured impurities, grower's income and their resistance to bolting. Apart from perpetuating weed beet, bolting is highly undesirable as the roots become woody with a low sugar content. If there are more than 10% bolters in the crop, there is a risk that some of the bolted beet may finish up as part of the sample taken for determining the sugar percentage at the factory gate. Varieties with a high resistance to bolting should be used for early sowing and in colder districts.

Size of top, height of crown (which could affect harvesting losses), establishment in field trials and resistance to rust and powdery mildew are other factors which can be considered in deciding on the variety or varieties to grow. Varieties are also available which show tolerance to *Rhizomania* (see below). On the continent, where *Rhizomania* is endemic in many countries, varieties show resistance rather than tolerance. These varieties have been in trials in the UK but have given poor yields. The tolerant varieties show good yields with one variety, *Concept*, showing a grower's income better than the best fully recommended variety on the current list (*Roberta*). All varieties are purchased from British Sugar under a non-profit making supply scheme.

13.2.5 Manures and fertilisers

The following recommendations can be made:

1   Farmyard manure – 25 tonnes of well-rotted manure per hectare should, ideally, be applied in the autumn. Only about 30% of the sugar beet grown in the country actually receives FYM.
2   Lime – a soil pH of between 6.5 and 7.0 is necessary. Alkaline soils can suppress some plant foods and, on peat soils, which may have a low content of phosphorus, magnesium and manganese, a pH 6.0 to 6.5 is acceptable.
3   Salt – 500 kg/ha should be broadcast a few weeks before sowing (except on poorly structured soils). This can give approximately 500 kg extra sugar per hectare. A subsequent reduction in the amount of potassium can be made if salt is applied.
4   Magnesium – magnesium deficiency has become more evident in recent years, particularly on light, sandy soils. If necessary 500 kg/ha Kieserite (magnesium sulphate) should be applied. Magnesium limestone can also be used if available. This will help to maintain magnesium levels, as well as remedying any lime deficiency.
5   Boron – may also be necessary, particularly (but not always) on well-limed sandy soils where it could be in short supply (Table 7.1). A boronated fertiliser is often recommended in this situation.
6   Manganese – a deficiency of this trace element is found where the pH on peaty soils is above 6 and on sandy soils above 6.5 and also on old ploughed-up grassland. Manganous oxide incorporated with the pelleted seed material is a useful insurance against manganese deficiency, but it may also be necessary to spray manganese sulphate as soon as deficiency symptoms show on the leaves (Table 7.1). A severe deficiency can reduce sugar yield by up to 30%.

Other plant foods required can be summarised as in Table 13.4, assuming phosphate and potash indices at 2 to 3. Kainit can be used at 500–600 kg/ha instead of potash, salt and magnesium (except in severe cases of magnesium deficiency where Kieserite is still necessary). It should be applied some weeks before sowing the seed. The phosphate and potash can be broadcast and worked into the seedbed at any time over the preceding winter months.

Table 13.4 Other plant foods

	N (SNS index 0) kg/ha	P_2O_5 (index 2) kg/ha	K_2O (index 2) kg/ha
With FYM and salt	100	50	75
Without FYM but including salt	120	60	75–100

Except after a very wet autumn and winter which may leave nitrogen reserves on the low side, nitrogen in excess of 125 kg/ha is unnecessary. Excessive nitrogen produces more leaf area with subsequent shading of lower leaves and the possibility of reduced sugar yields. Ammonium nitrate (not urea) must be applied in the spring with half the application broadcast straight after drilling, and the balance after the beet has emerged but before the beet seedling forms a rosette of leaves, which may trap nitrogen prills or granules leading to scorch.

Excess nitrogen, potassium and sodium also leads to higher levels of impurities in the sugar. This makes the sugar more difficult and expensive to extract in the factory. This is now one of the measured factors determining price so it is important that growers tailor their fertiliser applications to the crop requirements and take into account the nutrient contribution from any organic fertiliser applied in the form of slurry or manure.

13.2.6   Seeds and sowing

Genetical monogerm seed has now replaced the hitherto used 'multigerm natural seed' – a cluster of seeds fused together which usually produced more than one plant when it germinated and meant that the crop required thinning by hand (labour intensive and expensive). Monogerm seed contains a single embryo from which only one plant will grow.

All monogerm seed is pelleted, i.e. it is coated with clay to produce pellets of a uniform size and density – 3.50–4.75 mm. Unpelleted monogerm seed is rather lens-shaped and, as such, cannot easily be handled in the precision drill. The pelleting of seed allows the use of seed dressings. All seed (apart from that sold for organic production) is now sold treated with an insecticide against aphids and more and more seed is being sold as 'primed', such as 'Advantage' seed where part of the germination process has been completed before pelleting. Seed is also colour-coded according to the seed dressing.

13.2.6.1   Seed rate
This depends on seed spacing and row width. The recommendation is that pelleted seed should be sown at 1.1 units/ha (each unit has 100000 seeds) to give a final optimum plant population of about 80000 per hectare allowing for field losses. Higher seed rates are being investigated by UK researchers and growers following the example of French and German farmers who have increased plant populations and achieved significantly higher yields with the newer varieties available.

13.2.6.2   Time of sowing
Mid-March to mid-April. Sugar beet seed will not germinate at soil temperatures below 3 °C and only very slowly between 3° and 5 °C. In most years 5 °C is not reached until about 20 March and sowings before that date will normally result in poor establishment and many bolters. It is best completed by early April, after which the yield penalty gradually increases until, in May, about 0.6% is lost for each day's delay in sowing (Fig. 13.1).

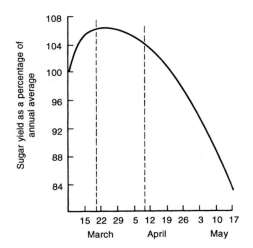

Fig. 13.1 Effect of drilling date on sugar yield; the average of 12 years of trials at Brooms Barn, West Suffolk. Maximum yield, with minimal bolting, is most unlikely to be achieved by drilling between 20 March and 10 April.

13.2.6.3 Drilling-to-a-stand

All of the national sugar beet crop is now drilled-to-a-stand, i.e. the individual seeds are placed separately in the position required for the plant, using the precision drill. This attempts to give each plant an equal growing area so that the crop produces uniformly sized roots (making harvesting more efficient and effective) and minimises shading of one plant by another. The number of plants which make up a crop of sugar beet is one of the most important factors in determining not only the yield, but also the sugar content of the roots. The main aim of drilling-to-a-stand is to have a reasonably spaced, yet adequate, plant population. Ideally, this should average 75–80 000 plants/ha; when it is less than 65 000 there can be a serious reduction in final yield.

The seed spacing should be decided on the basis of the germination percentage of the seed (obtainable from the factory which supplies the seed), the possible losses before and after plant emergence (soil conditions, spray damage, inter-row cultivation, wireworm and other pests) and the row width. As far as is consistent with any inter-row work and the harvesting of the crop, the row width should be as narrow as possible, but at present it is unlikely to be less than 46 cm because of the restrictions imposed by harvester design. If, as is more usually the case, a 50 cm row width is being used, the seed is spaced between 12 and 17 cm apart depending on the desired plant population.

Shallow-sowing will aid germination; 18–25 mm is ideal with a well-prepared seedbed. A press wheel fitted immediately behind the seed coulter effectively compresses the soil over the seed and in dry weather this will help germination.

13.2.6.4 The bed system

The development of the bed system for growing beet has logically followed the now common practice of overall spraying of the crop for weed control and the use of multi-row harvesters. One of the most important advantages of the bed system is that all tractor wheelings are avoided where the beet is actually grown. The wheelings are concentrated on, at the most, 15% of the land surface (i.e. the area between the beds – suggested width of 60cm) and, if necessary, any compaction there can be dealt with by soil-loosening tines following the tractor wheels.

A reduction in compaction should lead to better establishment and a consequent quicker crop cover and earlier interception of essential sunlight. To compensate for the larger beet growing in the rows adjacent to the 60cm gap (less competition), the plants should be closer-spaced than those in the inner rows. In this way a more uniform-sized crop should be obtained. Harvesting of the beds (at least four rows at a time) could be a problem in some seasons, and the heavier soils in the wetter West Midlands, for example, could be a limiting factor for the success of the system. With the right conditions, beet grown in beds will almost always do better than conventionally grown crops, both in terms of yield and sugar percentage.

13.2.6.5   The prevention of wind damage

A recent survey by British Sugar showed that 20% (about 40000 hectares) of the national sugar beet crop is at risk from wind damage. However, only about 16% is actually protected. Unless preventive measures are taken, it is not unusual for crops in vulnerable situations to be completely destroyed or damaged beyond an economic recovery. This necessitates a late re-drilling with a consequent serious loss of yield, quite apart from the extra cost involved. It is on the lighter and/or organic soils that the beet is at greatest risk from the wind, from sowing until the 6-leaf stage.

Straw cover, as a wind barrier, can be provided to reduce the wind speed at ground level by creating a straw planted 'hedge' about 110–120cm apart between or across the beet rows, using a straw planting machine. Another type of barrier is a spring cereal, normally barley, drilled at about 60kg/ha some three weeks before the beet is sown (the latter being drilled between the cereal rows) or, using an ordinary corn drill, the cereal can be sown at right angles to the intended direction of the beet rows. Before there is any competition, and when the beet has established itself sufficiently to withstand any damage, the cereal is sprayed out using, for example, cycloxydim.

Instead of a straw cover the beet can be drilled direct into ploughed and furrow-pressed land. Compaction is avoided and the press helps to stabilise the top few centimetres of the soil. Direct drilling into the undisturbed soil, usually stubble, using precision drills modified for the purpose, is sometimes used as a technique to prevent blowing.

Chemicals such as polyvinyl acetate, which bind the surface particles together, are commercially available. They are, however, expensive (more than £100/ha) and are not widely used. A barley cover crop is much cheaper.

13.2.7 Weed control

(See also Chapter 5.) Band spraying with herbicides at the time of drilling is no longer the usual method of controlling weeds within the row. When it was introduced it was cheaper than conventional overall spraying and also it had the advantage of leaving weeds for a time between the rows (to be removed later). These provided alternative food for pests and on light soils the weeds, acting as a barrier, helped to reduce wind damage. However, band spraying is a slow operation and it has to be followed up by inter-row cultivations. Consequently, most growers now use a low-volume low-dose technique for overall application of the herbicide. Normally, a sequence of sprays is necessary, but application should always be carried out when the weeds are in the cotyledon to first true-leaf stage.

A method of weed control brought across from the continent is the FAR technique. This involves spraying low rates of a mixture of chemicals onto flushes of cotyledon weeds. The components of the mixture are:

F = phenmedipham
A = activator (usually ethofumesate)
R = residual component (e.g. metamitron, chloridazon or lenacil)

Typical costs of this programme are £72–£90/ha.

Where overall application has not been carried out, steerage hoeing (in the early stages of plant establishment) will be necessary, perhaps twice, to assist in weed control. Further hoeing, after the beet has reached the 5–7-leaf stage, should be undertaken, but only as much as is necessary. Weeds have to be kept in check, but any inter-row cultivation does mean a loss of moisture. It is another point in favour of overall herbicide application.

13.2.8 Disease control

(See also Chapter 7.) *Virus yellows.* It may be necessary to spray the crop to kill the aphids which spread virus yellows (Table 7.1), or to apply the insecticide as a seed dressing, e.g. imidacloprid. Viruses which cause the leaves to turn yellow are particularly damaging to sugar beet because they restrict the conversion of solar energy into sucrose. Warnings are issued by British Sugar about the likelihood of infected aphids appearing in crops.

Powdery mildew. A fungicide should be used if the disease becomes prevalent in late July or August (Table 7.1). Some of the new, very active cereal triazole fungicides are approved for use in sugar beet for the control of mildew as well as rusts and *Ramularia* leaf spot. Sulphur can also be used for powdery mildew control.

Rhizomania. (See also Table 7.1.) This notifiable viral disease is spreading through the UK having been imported into the UK in 1987. It is spread by a soil fungus and there is no chemical control. Its effects on beet are severe. Roots are shrivelled, fanged and covered in fine hairs and cankers. Sugar percentage is drastically reduced. It can be spread in soil on machines, vehicles and any root crop grown in the field. A field quarantine system exists and British

Sugar has recently introduced a 'Rhizomania Stewardship Scheme' where growers with infection on their land can have their contracted tonnage grown elsewhere without losing ownership of it.

13.2.9   Irrigation

Although it is a deep-rooted crop, the beet plant will respond to supplementary water in the majority of seasons. This will obviously depend on summer rainfall as well as soil type.

If, by irrigation, the plant can develop a reasonable canopy of leaves early on in its life, it will be able to make more efficient use of sunlight for growth and sugar production. But the crop will also need water during June and July. In hot, dry and bright weather it may require up to 350 mm during these two months. Many soils (notably the sands), particularly in eastern England, are unable to supply this need. With irrigation on the farm and, as part of an overall irrigation plan, a water-balance sheet will have to be kept to determine the soil moisture deficit and when it will be necessary to irrigate the sugar beet crop (pages 192–3). However, irrigation should not be necessary after August, even if there is a water shortage; by then the deep root should easily be able to cope.

13.2.10   Weed beet

Weed beet is a big problem for sugar beet growers. About 50000 hectares of the total sugar beet area are affected. Weed beet is any unwanted beet within and between rows of sown beet and, of course, other crops. Unlike true beet, which is a biennial, it produces seed in one year. It originates from several sources, but mainly from naturally-occurring bolters in the commercial root crop, from contamination of seed by cross-pollination from rogue plants, from beet ground keepers either growing in crops following beet, or from old clamp and loading sites, and from seed shed from weed beet itself.

Once weed beet becomes established it is self-perpetuating (although it can remain dormant for many years) and on average it produces 2000 seeds in the year. However, only about 50% survive, but nevertheless a light infestation (perhaps 1000/ha) can mean one million weed beet/ha the following year.

Control measures. It is very important to prevent any weed beet seed returning to the soil and so, to start with, inter-row work should be carried out to clear any seedlings which are between the rows (at present there are no selective herbicides to control weed beet in sugar beet). Following this, as far as possible, all bolters should be removed from the beet field, no later than the middle of July, i.e. before the seed has set. Rogueing, i.e. by hand pulling, is the most economical method if there are fewer than 1000 weed beet/ha. If rogueing is not possible, the bolters can be cut down mechanically. This is done usually twice, the first time 7–10cm above crop leaf height and then, 14–20days later (about the end of July), to prevent lateral shoots and late-flowering plants from producing seed, cutting is repeated–this time at crop leaf height.

Instead of mechanical cutting, the 'weed wiper' fitted to the front or rear of the tractor can be used. This consists of a boom containing a herbicide (at present glyphosate is recommended) which permeates through a nylon rope 'wick' attached to the boom. As it goes through the crop the wick wipes the taller (than the beet plants) weed beet and bolters and so deposits the herbicide. Usually more than two treatments are necessary, starting at the beginning of July and then repeating treatment at 10–14 day intervals until the middle of August.

In addition to preventing the propagation of seed from bolters or weed beet which have already contaminated the crop, it is important that weed beet in crops other than sugar beet should also be controlled. Weed beet found growing on old clamp and loading sites, headlands and waste land must also be destroyed.

The effect of cultivations on weed beet is being examined to try and find out the ideal sequence of cultivations necessary to reduce the weed beet population more rapidly. Direct drilling, tine cultivation and ploughing are all being studied.

Finally, with a bad infestation in a field, it will be necessary for the grower to widen his rotation to six to seven years, rather than the more usual three to four year interval.

13.2.11 Harvesting

In theory, early to mid-November is the right time to harvest the beet crop. In most years it is still slightly increasing in weight, with the sugar percentage reaching a maximum at that time before beginning to fall off.

However, apart from the fact that, in most years, conditions for harvesting deteriorate as autumn proceeds, if all the beet were delivered at this time there would be tremendous congestion at the factory. Therefore a permit delivery system is used, whereby each grower is given dated loading permits which operate from the end of September until about the end of January. It means that growers must be prepared to have their beet delivered to the factory at intervals throughout the period. It is important that the permits are used and are not lost through failure to have sufficient beet out of the ground for delivery to the factory. In the early part of the harvesting period (up to about mid-October), when the beet is still growing in the field, the aim should be to keep just ahead of the delivery permit, in case bad weather stops lifting. As far as possible, the poorest part of the crop should be harvested first. It will never be able to respond to open weather at the end of the year in the way that a full crop can. More beet should be clamped in November and, depending on soil conditions, all lifting should be completed by mid-December.

To minimise dirt tares, direct delivery from field to factory should be avoided, especially when the beet is harvested in wet conditions. If the beet can be clamped at least a week before delivery, and then reloaded using an elevator with a cleaning mechanism, dirt tare will be considerably reduced; it should not exceed 10%. Very dirty loads may not be accepted by the factory.

Factories are also discouraging farmers delivering their beet in small loads by tractor and trailer.

Topping. As far as possible, the topper should be set to remove the tops to the level of the lowest leaves or buds on the plant (Fig. 13.2). A top tare of between 2% and 5% is acceptable; below 2% probably means that the crop has been over topped with an excessive loss of beet. Adjustment to the topping mechanism will normally be necessary throughout the period of harvesting. It should not be neglected.

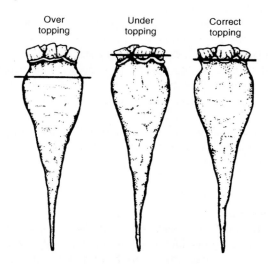

Over topping   Under topping   Correct topping

Fig. 13.2   Correct topping of sugar beet.

There are two main types of topping mechanisms:

1   The feeler-wheel topper. This unit consists of a feeler wheel which moves up and down (floats) according to the height of the beet. This controls the position of the knife in relation to the crown of the beet.
2   Flail topper. This consists of metal flails, which remove most of the leaf and stem, followed by scalper blades which, with the aid of guide bars, trim off the remaining leaves.

Following topping, the leaves are swept off the row prior to its being lifted. 'Top savers' which windrow the tops (away from the line of the harvester), leaving them in a clean condition for subsequent grazing or ensiling, are not widely used. Most tops are ploughed-in.

Harvesters. There are a number of different types of harvesters and harvesting systems. These range from the single-row trailed tanker harvester to the multi-row (2–8 rows) self propelled tanker, as well as the single and/or multi-row side-loading trailed machine. Depending on the distance to the clamp, the tanker harvester can be operated as a one- or two-man system. This can fit in well with the steady supply of beet needed for the permit delivery arrangements without seriously interrupting other work on the farm. A recent survey by British Sugar indicates an average 8% loss of yield by machine harvesting, i.e. about 3.5 t/ha per 45 t yield of crop per hectare. Over topping and breakage of roots below the ground and leaving whole roots on the ground are the main reasons.

13.2.12 Storage

Beet stored on the farm should be done carefully, with the clamp adjacent to a hard road and accessible at both ends so that beet which has been clamped longest can be delivered first. Plenty of air circulation should be allowed. Ideally, use a ridge-type clamp for beet which is clamped early whilst the temperature is still high and there is a risk of overheating in the pile. However, as it gets colder this is less important and large square clamps, or heaps against a wall, can be built.

It is important to prevent the beet being damaged by frost, either in the ground or in store, as this has a very quick and deleterious effect on the sugar percentage. Badly frosted beet will be rejected by the factory as it is useless for sugar extraction. Depending on the district and incidence of frost, the clamp may have to be covered. Plastic sheeting, rather than straw, is now being used. It has the advantage that it can be removed quickly, with a rise in ambient temperature and easily replaced when necessary. It is also important to avoid clamping damaged or diseased beet for long periods. Although the beet root is more robust than the potato, it is still possible to break the skin by careless handling. Sucrose will then begin to leak out with a consequent reduction in sugar percentage.

13.2.13 Yield

- Roots (washed) 50 t/ha.
- Sugar at 17% 8.5 t/ha.
- Tops 25–35 t/ha depending on the variety. A stock farmer may prefer a large-topped variety, although fewer are grown now.

13.2.14 The future

British Sugar and the NFU have recently negotiated the Inter-Professional Agreement (IPA). This is a new form of contract between BS and the grower. Important changes to the old contract include:

- A crown tare allowance – growers are now paid for crowns up to 10% per tonne of clean beet.
- A purity bonus – linked to average level of amino nitrogen for each grower.
- A new sugar content payment scale.
- Earlier factory opening dates.

- Additional contract tonnage to be released.
- Factory reception times increasing and less sampling to be carried out. British Sugar is looking to reduce sampling to 60% of current levels. This means that not every load will be sampled. All samples will be processed through the Wissington factory.
- An Outgoers' Scheme. Growers will be paid for releasing their contracted tonnage.

British Sugar has also gone on-line. Growers can get technical information from the web as well as receiving personal information about their own beet deliveries.

13.3   Further reading

Burton W G, *The Potato*, Longmans, 1966
Harris P M, *The Potato Crop: The Scientific Basis for Improvement*, Chapman & Hall, 1991
Jaggard, R W, *Sugar Beet: A Grower's Guide*, British Sugar, 1995

14

Fresh harvested crops

14.1 Vegetable production on farms

There have been major technological advances and changes in the market for vegetable crops that have affected how vegetables are grown in the UK. Many large farms now specialise in vegetable crop production, growing produce both in the UK and abroad to meet the needs of the supermarkets for 12 months of the year. Crop protocols have been drawn up between growers and retailers to promote good practice, enable crop traceability and give assurance to the consumer. Over 80% of produce is now sold through the multiples and an increasing proportion of that is sold ready prepared. The demand for organic produce has increased greatly and some organic crops can command a premium in both the fresh and processed markets.

It is very important for the farmer to know the requirements of the market, to produce only what is needed when it is needed, in suitable quality and quantity and following crop protocols. It is usually advisable to arrange contracts with buyers, such as processors, prepackers, supermarkets, chain stores or co-operatives in advance of growing the crop. Alternatively, all or part of the produce may be sold direct to the public at the farm gate, through pick-your-own, or door-to-door delivery such as the 'box scheme' used by some organic growers.

There have been many developments in the farm-scale production of vegetables that have resulted in a high level of mechanisation from seed to supermarket shelf. Seed choice and sowing and planting systems combine to result in a uniform crop that comes to maturity evenly and which often can be harvested mechanically at a single pass. Husbandry techniques aim to optimise crop growth while reducing inputs to the benefit of the farmer, the environment and the consumer. Harvesting and post-harvesting methods aim to transfer the crop from field to customer as efficiently as possible and in the best condition. These developments may be summarised as follows:

1   F1 hybrids give uniformity and vigour.
2   Seed is available graded or pelleted to help with drilling, and treated or primed to aid and speed germination.
3   Modules allow efficient transplanting and promote quick establishment in the field.
4   Bed systems make best use of the available land and give good access to the crop for all tractor operations.
5   Plastic mulches and fleece crop covers extend the growing season in the field.
6   Integrated crop and pest management reduces inputs by applying fertiliser and water only as needed for crop growth and targeting pesticides at those weeds, pests and diseases actually present in the crop.
7   Specialised harvesting rigs ensure fast and efficient removal from the field; mobile packhouses and modern grading, washing and packaging facilities ensure the crop is then transferred from field on to distribution quickly while maintaining quality.

Limiting factors on the individual farm may be:

1   Need for specialist machinery.
2   Unsuitable soils, poor drainage or lack of irrigation.
3   Difficulty obtaining suitable casual staff and supervision.
4   Lack of a suitable water supply for washing and effluent disposal.
5   Distance to local wholesale market.

Useful information can be obtained from crop consultants, agronomists and research institutes.

14.2   Harvested fresh peas

Pea crops can be a useful part of a rotation on the farm, peas are harvested fresh for canning, freezing or in the pod or grown on and used for drying, for human or animal consumption. Rotation and growing details can be found in Chapter 16, Combinable Break Crops.

1   *Vining peas.* Vining peas are used for canning fresh ('garden peas'), quick-freezing or artificial drying and are normally grown under contract, or by co-operating groups. The contracting companies usually supply the variety they require and decide on times of sowing (February–May) based on a 'heat unit' system. This is in order to spread the harvesting period to ensure a regular and constant supply to the processing factories.
Some examples of varieties are:

- First earlies: *Avola, Cabree, Enterprise, Misty, Span, Zamira*;
- Second earlies: *Ripon, Jaguar, Tacoma; Minado* for *petit pois*;
- Early maincrop: *Barle, Bikini, Novella, Oasis, Tristar; Waverex, Bastion, Jewel* and *Paso* for *petit pois*;
- Maincrop: *Ambassador, Balmoral, Pinnacle, Polo, Puget.*

2   *Fresh picking (pulling) peas.* These make up approximately 3% of the market.
- For sale in pods, e.g. *Feltham First, Early Onward, Ambassador* and *Onward;*
- For edible pod peas, e.g. *Oregon Sugar Pod.*

Strict EC quality standards apply to the sale of picking peas.

The growing of vining and pulling peas is broadly similar to that of dried peas (see Chapter 16) However, the following points should be noted:

1   Vining and pulling peas are best suited to the main arable areas and where there are local processing plants or markets (most pea growers are concentrated in East Anglia and Humberside).

2   Most vining pea growers are members of co-operatives, often linked to a specific processing factory. The factory must ensure a continuous and reliable supply of produce so that it can operate efficiently. Growers must ensure, therefore, that the crop comes ready for harvesting over several weeks. This is achieved by: (a) growing different varieties with different maturity dates, or (b) staggering the drilling of the peas based on the calculation of accumulated heat units (AHUs). Thus, if the difference between one day and the next in East Anglia at harvest time is 11.5 AHUs, and the vining capacity of the co-operative is 40 hectares per day, each block of 40 hectares must be drilled 11.5 AHUs apart back in the spring. (This could mean delays in drilling of several days if the weather is cold.) This should result in the shared harvesters moving around the pea crops just at the right stage of ripeness. If there is a delay because of weather or breakdown, then the peas can often get too tough for freezing or canning. They can then be allowed to grow on for harvesting dry.

3   Vining peas are ready for harvesting when the crop is just starting to lose its green colour and the peas are still soft. The firmness of the peas is tested daily (near harvest time) with a tenderometer. This gives a guide to the best time to start cutting. The reading for freezing peas is about 100 and, for canning peas (which can be a little firmer), at 120. When ready, the crop must be cut as soon as possible. The pods are picked up and put through a special vining machine that gently separates the peas from the pods. The shelled peas are rushed to the processing plant for freezing, canning or drying. The haulms (vines) left by some of the harvesters may be made into silage or hay, or, more usually, incorporated into the soil to increase organic matter.

4   Pulling peas are harvested by removing the pods when the peas are in a fresh, sweet condition. This requires a large gang of casual workers or very expensive machinery and so the market is declining. However, a number

of farms have set up 'pick-your-own' enterprises that include fresh peas among their crops.

The yield of both vining and pulling peas is in the range of 3–8 t/ha.

14.3   Broad beans

14.3.1   General
Broad beans belong to the same species, *Vica faba*, as field beans and the husbandry, diseases and pests are similar (see Chapter 16). Broad beans quick-frozen or canned are an excellent product and, as with peas, close co-operation between grower and processing plant is required. Pick-your-own broad beans are popular with customers as they are quick and easy to pick.

The crop can be grown throughout the UK on a wide range of soil types; however, high rainfall can lead to excessive vegetative growth and drought at flowering and pod swelling reduces yield. Broad beans can be useful in a crop rotation – the haulm provides organic matter and root nodules fix nitrogen – but a four-year break should be practised between all leguminous crops to help prevent the build up of pests and diseases.

14.3.2   Broad beans husbandry
Types and varieties
Continuity of cropping can be achieved with the use of different varieties, or sequential sowings of one variety based on accumulated heat units.

1   For processing. These types are sown from early February to May as directed by the processors. Drilling normally starts in early April to fit in with pea harvesting. Some broad bean varieties bruise during harvesting and coloured-flower varieties give a pink-brown tinge to the beans and water during the canning process, so white-flowered varieties may be stipulated. In recent years there has been a development of smaller-seeded varieties marketed as 'baby broad beans' and there is also growing interest in green-seeded varieties.
    - standard sized seed, e.g. *Listra, Metissa, Statissa C*;
    - smaller seeded, e.g. *Danko, Diamant, Gold*;
    - green-seeded, e.g. *Jade, Verdy*.
2   Fresh market and pick-your-own. Some types are sown November/December and are harvested late May/June, e.g. *Aquadulce Claudia*. Other types are sown in spring, February to May, for harvest from June to August, e.g. *Acme, Dreadnought, Express*.

Fertilisers
Nutrient status should be determined by soil analysis. Table 14.1 gives a guide

as to the plant nutrients necessary assuming phosphate and potash indices at 2. Nitrogen-fixing *Rhizobia* on the roots provide the crop with nitrogen; farmyard manure will delay, and may suppress, nodulation. The pH should be maintained at 6.0 to 6.5; overliming can induce manganese deficiency.

Table 14.1 Nutrients required for peas and beans

Crop	N kg/ha	P_2O_5 kg/ha	K_2O kg/ha
Broad beans	0	100	100
Green beans	120	100	100
Peas	0	85	90

Sowing
The large seeds are likely to be damaged in some drills. Vacuum drills and some types of belt-fed drills are more suitable. Seed should be sown at least 7.5 cm deep to avoid damage by simazine if used.

Seed rates
For processing types see Table 14.2. Fresh and pick-your-own types are sown about 10–12 cm apart in 45–60 cm rows at a rate of 125–160 kg/ha.

Table 14.2 Seed rates for peas and beans

Crop	Row width cm	Plants/m^2	Seed rate kg/ha
Broad beans	40–50	16–20	125–160
Green beans	30(20–45)	35(30–40)	100–135

Weed control
Only approved herbicides must be used, processors may stipulate which ones. Inter-row cultivations may be practised; thistles and volunteer potatoes can be a problem.

Irrigation
The crop is sensitive to soil moisture deficit and this should be checked; higher yields are obtained from irrigating at early flowering and pod fill.

Pests and diseases
Pea and bean weevil, stem and bulb nematode, black bean aphids, bean seed fly, chocolate spot, foot rot and rust are usually the main problems in broad beans (Tables 6.1 and 7.1).

Harvesting
The fresh and pick-your-own crops are normally picked over two or three times.

Fresh produce is usually sold in 8–10 kg boxes. Average yield is 4 t/ha, although some late crops can yield as much as 8–10 t/ha. Dates of havesting are give in Table 14.3.

Table 14.3   Dates of sowing and harvesting for peas and beans

Crop	Dates of sowing	Dates of harvest	Yield (t/ha)
Broad beans	April/mid-May	Mid-July/end August	2–6
Green beans	Mid-May/July	July/late October	4–9

Processed crops are harvested according to the tenderometer readings (TR), as with fresh harvested peas. For instance, for freezing the TR should read 110–140, for canning the TR should read 130 and over.

14.4   Green beans

14.4.1   General
Green beans (Dwarf French) *Phaseolus vulgaris* are an important farm crop grown chiefly for processing to meet the expanding demand for convenience foods. The use of efficient one-pass mechanical harvesters has enabled the expansion of the crop, with freshly harvested pods taken direct from the farm to the processing plant for canning, freezing or drying. Green beans may also be grown for the fresh market or pick-your-own and are not as perishable as peas or broad beans.

Green beans need to be grown in loams and the lighter types of soil. It is a late-sown, short-season crop and so requires a deep, free-draining, moisture-retentive soil. Avoid soils that cap, and those of a high organic matter content (peat soils), which will produce excessive vegetative growth. A site offering shelter and a southerly aspect should be selected. A four-year crop rotation between all leguminous crops should be practised. Most of the crop is grown in the eastern counties south of the Wash.

14.4.2   Green beans husbandry
Types and varieties
Green beans are usually classified according to pod width. Consumer preference has moved away from sliced beans obtained from larger podded varieties towards whole beans and to the extra fine types grown on the Continent. Some of these fine bean varieties can be grown in the UK. Variety and performance may differ with site and season and small areas of a new variety should be evaluated before planning a large production programme.

There are many varieties on the market and where growing for processors, advice should be sought from them regarding their requirements. Consideration should be given to yield, quality and ease of harvesting, together with the growing environment on each individual farm.

- Extra fine (5.0–6.5 mm pod width). *Safari, Piccolo.*
- Very fine (6.5–8.0 mm pod width), less than 13 cm long (cut to 6 cm). *Masai, Nickel* and *Flevoro* have been grown successfully in the UK.
- Fine (8.0–9.0 mm pod width):
  - short (10 cm pod length). Grown for processing whole, but can be used for slicing or cutting. *Lesso, Jamaica, Mondeo, Yukon;*
  - intermediate (10–13 cm pod length). Varieties in this group are mainly processed as cut beans. *Catch, Marcio, Magnum, Milagrow* and *Paulista.*
- Medium fine intermediate (9.0–10.5 mm pod width). Varieties in this group are processed as cut or sliced, the wider varieties are more suitable for slicing only. *Nomad, Scuba* and *Laguna.*
- Medium fine long (9.0–10.5 mm pod width; long pod length >13 cm). *Matador, Sigma, Tasman, Ursus, Green Arrow.*

Flat-podded beans are being developed by crossing runner and green bean types to give the quality and pods of runner beans and growth habit of green beans. Popular for the fresh market, e.g. *Atlanta.*

Weed control
Efficient weed control is essential to avoid difficulties with machine harvesting and if weeds are not controlled, the crop may be unsaleable. Perennial weeds should be dealt with before planting. Some varieties are susceptible to herbicides and advice should be sought from the Processors and Growers Research Organisation (PGRO).

Fertilisers
Fertiliser requirements should be determined through soil analysis. French beans do not normally develop nitrogen-fixing nodules, as is the case with other legumes. Nitrogen, therefore, has to be applied (see Table 14.1 for index 2). No more than 100 kg/ha of nitrogen is applied to the seedbed, the rest applied once the crop is fully established. Top dressing of 40 kg/ha is irrigated into the whole rooting zone. *Rhizobium* bacteria, which nodulate and fix nitrogen, can be used as an inoculant.

Dates of sowing
These are determined by the contracting processing factory. Green beans are sown between mid-May and July. The crop can be damaged by late frosts.

Seeding
Seed is usually sold on a unit cost basis and this has the advantage of a direct cost comparison between small and large seed varieties. Seed rates are calculated according to factors such as number of seeds/kg, plant population required,

percentage germination, and an estimate of likely seedbed losses. Distance between rows and plant density influence final yields (final plant population for processing 30–40 plants per m^2, fresh market 20–25/m^2) therefore it is essential to achieve optimum plant arrangement. The seed testa is easily damaged and seed must be handled with care. Pneumatic precision or belt feed-drills are used.

Irrigation
This can be very beneficial in a dry season. Severe drought reduces yield and pod quality especially during flower initiation when the pods are developing. Care is required, however, because irrigation may spread some fungal diseases, especially botrytis during flowering, when the wet petals stick to developing pods.

Diseases and pests
Green beans are subject to attack by a number of pests especially bean seed fly, aphids, slugs and cutworms. Halo blight, botrytis and sclerotinia can be a problem in some seasons.

Harvesting
Runs from July to October. The harvested bean pods must be whole, undamaged, separated (not in clusters) and free of stems, leaves, soil and stones. As the crop matures, the pods lengthen rapidly and they then enlarge as the seeds develop. Maturity should therefore be judged according to pod width and not length. The beans should not be allowed to over-mature or quality will be reduced. This is assessed by taking a random sample of 10 plants, taking the most mature seeds from the most mature pods on each plant and measuring the length of 10 seeds. As a guide, for all beans the length of one seed should be no more than the pod width.

- Fine beans (8.0–9.0 mm pod width)   for freezing 10 seed length 80 mm
                                       for canning 10 seed length 90 mm
- Medium fine beans   for freezing 10 seed length 90 mm
  (9.0–10.5 mm pod width)   for canning 10 seed length 110 mm
- Medium fine/large beans   for freezing 10 seed length 100 mm
  (>10.5 mm pod width)   for canning 10 seed length 120 mm

Flat-podded beans must be harvested before the development of seeds becomes obvious externally.

14.5   Carrots

14.5.1   General
Carrot production for human consumption in this country is now a large and

specialised enterprise for most growers. Carrots are grown for many different markets – fresh, frozen and processing – and quality requirements of the buyers can be exacting. Uniformity and freedom from damage and disease are very important. High capital investment in grading, washing and packing plants is involved.

Some farmers grow the crop and pack it on the farm, whereas others grow for packing to be carried out elsewhere. It is normally desirable to have contracts for the various types of production. Carrots are not grown now for stock feed but outgrades and rejects (up to 30%) may be fed to cattle.

14.5.2 Carrots husbandry

Soils and climate

The climate in most arable areas of the country is suitable for carrots; the main limiting factor is soil type, which should not restrict root growth. Sandy soils and peats are the most favoured as they normally give good yields of well-shaped roots that may be harvested at any time and are easily washed. Most crops are grown in the eastern counties.

Types and varieties

There are 5 main types of carrot (based on shape) with over 100 varieties (see NIAB *Vegetable Variety Handbook*). The cylindrical stump rooted types such as *Nantes* are popular for pre-packing and some processing; small finger carrots for freezing are usually *Amsterdam Forcing* types; small *Chantenay* types are used for canning; where large, uniformly-coloured carrots are required for slicing and dicing and dehydration, *Autumn King* and *Berlicum* types are suitable.

Varieties are also grouped by date of harvest: first earlies end of May, second earlies end of June and maincrops from July.

Rotation

A rotation of one crop in five years is preferable to avoid disease. Sugar beet and potatoes host violet root rot.

Sowing: timing

To provide for supplies over a long period, it is necessary to sow at different times. First earlies are sown under polythene in October. Varieties grown in this way must have good resistance to bolting. Second earlies are also sown under polythene from December to February. Maincrop carrots are sown in the open from March to early July, depending on the variety and intended harvest date.

Sowing: seedbed

To produce more than 70% of carrots of a desired size is very difficult, and every effort has to be made to prepare a really good, uniform seedbed. The seedbed also needs to be moist. Carrot seed is difficult to germinate because of seed coat 'inhibitors' and these must be dissipated away from the seed in the soil solution.

Sowing: drilling and plant populations

Graded seed should be precision-drilled at a uniform depth 'on the flat' or in a bed system, the final row spacing depending upon the harvesting machinery being used. The seed rate will depend on the germination percentage, field conditions at sowing, and crop spacing for the particular carrot size required. Where small carrots are required for freezing, there should be about 1100 plants/m²; large carrots for dicing require about 55/m²; ware 150/m²; pre-pack or bunching 160–180 m²; canning and small pre-pack 380–400 m².

Manuring

For a soil, tested to have a phosphate and potash index of 2, the fertiliser recommendations would be 200 kg/ha N, 100 kg/ha of P_2O_5 and 175–125 kg/ha of K_2O. Leaf analysis should be taken for manganese and copper.

Weed control

Any perennial weeds should, if possible, be controlled in the previous year. Pre-drilled and post-emergence herbicides should be obtained from the current *UK Pesticide Guide*; only approved products for the crop must be used. Machine and hand weeding may be needed to control fool's parsley, hemlock and wild carrot, which are difficult to control with herbicides.

Pest and disease control

Carrot fly, cutworms, willow-carrot aphids and nematodes, and the diseases violet root rot and cavity spot are the main problems in the growing of the crop. Some varieties are more susceptible to cavity spot than others.

Irrigation

Irrigation is often required in dry periods, but great care should be taken with water management to avoid root splitting.

Harvesting and storage

The crop is harvested from June onwards, according to market demand and size of roots. The roots are very easily damaged and careful handling is required at all stages. Modern machines are designed to minimise damage. Those machines which lift by the leaves, and then top the roots, work well during the summer and up to about the end of October, when the tops become too weak. From late October onwards, share-elevator type diggers have to be used which require the tops to be flailed-off first.

To maintain freshness, the time from lifting to dispatch to market should be as short as possible. When storage is required, some carrots are stored in buildings, but they lack the fresh appearance associated with newly lifted carrots, therefore much of the crop is left in the ground during the winter and lifted as required. This can be achieved by earthing-up or covering with straw and/or black polythene sheets to protect from frost. These protections may also help to stop regrowth in the spring. Lifting may continue until the crop becomes too woody or otherwise unsaleable, usually by early May.

Yield
Yield can vary from 20 t/ha for early crops to well over 60 t/ha for maincrops, however rejects in grading can be 30% or more.

14.6   Bulb onions

14.6.1   General
Precision drilling, onion sets, improved chemical weed, pest and disease control, and better methods of harvesting and storage have all contributed to making bulb onions an attractive crop to grow.

14.6.2   Bulb onions husbandry
Soils and climate
Bulb onions can be grown on a wide range of mineral and peat soils provided that they are well drained, have a good available-water capacity, a pH of 6.5 or more and can provide a good seedbed. Bulb onions can be grown in many parts of the United Kingdom, but do best in the eastern and south eastern counties where it is often drier at harvesting time, which helps promote skin quality. An even supply of water to the bulbs is needed while they are developing, for when growth is halted by a period of drought and then followed by heavy applications of water, the outer skins may split, reducing quality.

Rotation
Onions, together with crops such as field beans, oats and parsnips, should not normally be grown more often than one year in five in the same field, because of the risk of nematode, onion fly and white rot disease.

Fertiliser
Fertiliser application rates should follow soil analysis. Fertiliser recommendations for soil index 2 are given in Table 14.4.

Table 14.4 Nutrition requirements for bulb onions

Type of bulb onion	N kg/ha*	P_2O_5 kg/ha	K_2O kg/ha
Spring-sown	75	100	150
Autumn-sown	60	100	150

*Base dressing.
Top dressings of up to 100kg/ha may be needed depending on crop growth. Split applications may be more beneficial to maintain plant growth.

Time of sowing and variety choice
A comprehensive list of varieties can be found in the NIAB *Vegetable Variety Handbook.*

Overwintered onions should be sown between the first and third weeks in August. Bulbing starts about mid-May for harvest in July/August. Yields are not higher than for spring crops but there may be a premium for early crops, and they can help in providing a continuity of supply. August-sown varieties: *Buffalo, Keepwell, Shakespeare.*

Spring-drilled crops should be sown as soon as possible after mid-February. They begin to bulb in late July, regardless of growing conditions, and are harvested in August/September. Spring-sown maincrops varieties: *Barito, Spirit, Durco, Marco.*

Sowing
Overwintered and spring-sown onions can be grown successfully on good soils by precision drilling in shallow rows, 1–2 cm deep. The bed system is now used with the seed spaced in various ways, as below.

The recommended seed rate for dressed, graded seed is 4–6 kg/ha. The target is 75 plants/m^2 for the spring-sown crop and 100/m^2 for the August-sown crop. Alternatively, where spring crops are sought on late and difficult soils, multi-seeded transplants, grown in glasshouses, can be used to give an earlier (up to a month) and more reliable crop. The seed is sown in late January/early February in peat blocks or small-celled trays (about 5–7 seeds in each cell). These are transplanted as sturdy plants in late March/early April, usually in 4- or 5-row beds, aiming for 10 units of 6 seedlings/m^2. Transplanted crops are expensive to produce but yields can be up to 50% higher than from drilled crops.

Growing from sets
Sets are produced from high-density crops and stored at 10–15 °C over winter to prevent bolting. At thumbnail size, they can be planted quickly with special machines and are not likely to be affected by soil capping. Sets are planted late September/October for harvesting in June, or spring planted for harvesting in July/August.

Varieties
These are *Alpha, Sturon, Turbo, Orion.*

Weed control
Perennial weeds should, wherever possible, be controlled in the previous year. Bulb onions are not very competitive and grass weeds can be a problem post-emergence. Approved herbicides for the crop can be obtained from the *UK Pesticide Guide.* Only approved chemicals may be used.

Irrigation
Late irrigation should be avoided as it delays harvesting and gives rise to softer bulbs with more bacterial rots.

Diseases and pests
The following can affect bulb onions: leaf blotch, downy mildew, onion white rot, neck rot, onion fly, cutworms, bean seed fly, stem and bulb eelworm, thrips and bulb and stem nematode.

Harvesting
For long-term storage the crop is usually sprayed with meleic hydrazide, applied pre-harvest as a growth depressant at 10% leaf fall. Bulb onions are ready for harvesting when about 80% of the tops have fallen over. This is usually from June to September, depending on the time and method of planting. If left later, yields might increase slightly, but the outer skins are more likely to crack which can lead to disease loss in store.

Before harvesting, the bulk of the tops are cut off with a flail harvester, leaving only 80–120 mm and then the bulbs are lifted, windrowed and left to dry. The bulbs are then elevated into a trailer and taken to the store where they can be stacked to a height of 4 m. Preparation for storage is completed in three separate stages. Throughout the process, temperature and humidity control are crucial for the control of quality. In the initial drying stage, warm air (25–30 °C) is blown through ventilation ducts until the moisture is removed from the leaves and tops of the stems (approximately three days), to bring the onions to 'rustle dry' condition. In the curing stage, in the next 2–4 weeks, air at about 25 °C is blown intermittently through the stack and the relative humidity is reduced to 65%. This completes the drying of the neck tissue and the skins should now be a golden colour. Once the bulbs are fully dried, in the cooling stage, the stack temperature is lowered slowly to prevent sprouting.

When the onions are then required for grading and sale, air of 2.5 °C greater than the stack temperature is blown through the stack to prevent condensation forming on the onions. Refrigerated storage will be needed for late-marketed crops. EC grading regulations apply to this crop. Onions are graded for quality and size before sale.

Yield
Yield varies with conditions and timings from 30 to 50 t/ha.

14.7   Cabbages

14.7.1   General
Cabbages can be grown throughout the UK. Few are grown these days specifically for stock feeding though outgrades of cabbages grown for the domestic market can of course be used this way.

14.7.2   Cabbages husbandry
Soils
Cabbages require a soil pH of 6.5–7.0 but can be grown on a wide range of soils that are well drained, moisture retentive and provide a good firm tilth. Irrigation may be required, especially on sandy soils. A three-year rotation should be practised to avoid the build-up of soil-borne diseases, especially club root.

Nutrition
Fertiliser requirements should follow soil analysis and will vary according to the type of cabbage to be grown. Table 14.5 gives the requirements for soil index 2 for different *Brassicae*. No more than 100 kg of nitrogen in the base dressing should be applied or germination and seedling growth may be affected.

Table 14.5   Nutrition requirements for *Brassica*

Type of *Brassica*	N kg/ha	P_2O_5 kg/ha	K_2O kg/ha
Spring cabbage	75	100	200
Summer/autumn cabbage	270	100	200
Winter/savoys	210/260	100	200
Brussels sprouts	270	100	200

(i)   A phosphate and potash index of 2 is assumed.
(ii) Appropriate adjustments downwards should be made when farmyard manure is applied.
(iii) No more than 100 kg of nitrogen should be applied in the base dressing; the rest should be applied as a top dressing when the crop is fully established.

Irrigation
Irrigation is generally applied to aid crop establishment. If irrigation is sparse, apply up to 50 mm of water 20 days before cutting to improve yield and crop quality.

Plant production
Production is predominantly through transplanting modules. Transplanting gives a better crop production programme and more even crop maturity. If direct drilling, drill two seeds per station at the final spacing for the crop, then remove the weaker plant by hoe after germination. If modules are used, they should be given a high nitrogen feed and a cabbage root fly drench before planting.

Weed control is as for Brussels sprouts.

Varieties
Cabbages are grouped according to leaf colour, structure, density, and time of harvest.

- Spring greens/spring hearts: These are traditionally harvested from February to May although the supermarkets now require the crop all year, sold bagged either as leaf or semi-hearted. Spring cabbage is mainly drilled direct in August, transplanted in September or direct drilled in April to June at its final crop spacing 38 × 13 cm.
  Varieties: *Mayfield, Duncan, Durham Elf, Duchy, Pyramid, Advantage.*
- Early summer cabbage: These are sown in February and planted April/May at a final spacing of 46 × 23 cm for harvesting in June/July. Some varieties are prone to bolting when sown early.
  Varieties: *Compactor, Combinor, Junior.*
- Summer cabbage: These are drilled in March/April or planted May/June, for harvesting July/September, at a spacing of 46 × 23–46 cm.
  Varieties: *Derby Day, Duchy, Stonehead, Castello.*
  Savoy types: *Daphne, Atlanta, Capriccio.*
- Late autumn/winter and Savoy: Generally the round-headed types are grown, sown March to May, or planted April/July and harvested September onwards. Spacing 46 × 30–46 cm.
  Varieties: *Guardian, Metro, Colt, Cutlass, Winchester.*
  Savoy: *Midvoy, Darvoy.*
- Winter white or red storage cabbage: Grown for home consumption (cooked or raw) and for processing. The preferred size for the retail market is 0.5–1.2 kg, for processing a minimum of 2.5 kg, optimum weight around 3 kg. These types can be available 12 months of the year with field or controlled storage. Sown April/May or planted May/June, harvested October onwards. Spacing 46 × 36 cm.
  Winter white varieties: *Regent, Marathon, Kilor, Lion, Piton, January King.*
  Red Cabbage varieties: *Robus, Roxy, Rodima.*
- Winter hybrid green cabbage: Sown mid-March, transplanted late June/July, at a spacing of 60 × 38 cm.
  Varieties: *Roulette, Celtic, Tundra.*

Where winter cabbage is intended solely for stock feeding, a variety such as *Flagship* or *Marabel* should be used. For this purpose moist, heavy soils are preferable for maximum yield, which can be up to 90 t/ha, although harvesting may be more difficult.

Pest and disease problems
Cabbage root fly can be a problem, particularly with summer and winter cabbage, and treatments will be required from April onwards. Aphids, caterpillars, cutworms, slugs, downy mildew, dark leaf spot, ring spot, alternaria, white blister can all be a problem. Consult the current *UK Pesticide Guide* for approved products.

14.8   Brussels sprouts

14.8.1   General
Brussels sprouts can be grown on a commercial scale in most parts of the country, but the bulk of the crop is grown in Bedfordshire and Lincolnshire.

The crop is grown for the fresh market and quick freezing. The hectarage of Brussels sprouts grown over the past 10 years has halved, but it is considered that there will be an increase in demand in the future due to the crop's stated health benefits and with new varieties having improved flavour.

Varieties
Recent developments in breeding Brussels sprouts have aimed to give disease resistance and improved flavour with some new super-sweet varieties on the horizon. The NIAB *Vegetable Variety Handbook* gives details of the current recommended varieties. Where a continuity of supply is required, the examples in Table 14.6 can be used. Most varieties are now F1 hybrids offering a more uniform plant with an even distribution of buttons along the stem length and evenness of maturity (ideal for machine harvesting), compared with traditional open pollinated types.

Table 14.6   Varieties of Brussels sprouts

	Examples of varieties
For the fresh market	
Early–up to mid-October	*Oliver, Maximus, Angus*
Mid-season–mid-October to end December	*Corinth, Romulus, Helemus, Adonis, Genius*
Late–after December	*Stephen, Exodus*
For freezing	
Early to mid-season	*Icarus, Corinth*
Late	*Louis, Ariston*

For the fresh market, a smooth, round, dark green button, free from disease, and of a good flavour is required. A large proportion of the crop is size graded for different markets.

Plant production

Traditionally, Brussels sprouts were grown from bare root transplants but now the greater part of the crop is produced in modules as transplanting is faster, the crop establishes more easily and uniformly and plant growth in the module can be controlled. Some direct drilling is still done in North Lincolnshire and Humberside on land with no irrigation. Before planting, the modules are given a liquid feed of 200 mg/l N: 200 mg/l K_2O and a cabbage root fly drench.

Bare root transplants. The plants are raised from seed in a seedbed with a seed rate of 0.5–2 kg for every hectare to be planted out. The actual rate will depend on percentage germination, the conditions in the field at the time of drilling and the plant population required for the crop. It is important that the seedbed used for transplants is free from disease, particularly club root. The pH should be 6.5–7.0. Seed dressings are recommended against the flea beetle, cabbage stem weevil and soil-borne fungi.

The seed is sown:

1   Under cold frames in February and March for transplanting late April/May, for crops to be harvested in mid-August to November.
    or
2   In an open seedbed drilled March/April for transplanting end May/June, for crops to be harvested in October to March.

The transplants should be handled carefully and graded when pulled. Exposure to any wind and sun must be kept to a minimum, as should the time between lifting and replanting.

Transplanting is normally done by machine. However, to ensure a full and uniform stand, any gaps should, if possible, be 'dibbled in' by hand within 10 days after the main planting.

Direct drilling

The seed is direct drilled in the field at two seeds per station (final spacing for the crop) and the weaker plant removed by hoe after germination. Crops sown from mid-March to mid-April are harvested October to March and are drilled in rows 45 to 90 cm apart, with the two seeds sown 3 cm apart, every 90 cm down the row.

Plant population

This is important as it affects sprout size, stem length, maturity date, uniformity and disease susceptibility. For hand harvesting, sufficient access is necessary for the pickers. A plant density of 2.5–4.5 plants/m², up to 45 000 plants/ha is suitable for machine harvesting.

Soils
The ideal soil is a well-drained medium to heavy loam with a pH of at least 6.5.

Crop rotation
The crop fits well into a cereal rotation on land that is clean and fertile. The earliest crops are cleared by the end of September and so it is possible to follow with winter wheat. Because of club root, it is inadvisable to grow sprouts or other brassica crops more often than one year in three.

Manures and fertilisers
Nutrition requirements should follow soil analysis. In Table 14.4 a phosphate and potash index of 2 has been assumed.

Nitrogen top-dressing will be required to make a total nitrogen application up to 300 kg/ha, depending upon the appearance of the crop. A top-dressing is generally applied within two months of the base-dressing application.

Weed control
A programme of herbicide treatment may be necessary to keep the crop clean of weeds. This could include a pre-planting treatment or the use of residual herbicides, followed by post-planting treatments. Only approved products should be used, see the current *UK Pesticide Guide*.

Pest control
Control measures may be necessary against the cabbage root fly aphid and cabbage caterpillar. Powdery mildew, ringspot and white blister may need to be controlled; some varieties are more resistant than others. Cabbage root fly control will be required from the end of April.

Irrigation
Irrigation may be applied before and after planting to aid establishment. Up to 40 mm of irrigation to early and mid-season sprouts, when lower buttons are 15–18 mm in diameter, will increase yield and quality.

Stopping the plants
By removing the growing point (or terminal bud) from the plant, or by destroying the bud with a sharp tap using a rubber hammer, sprout growth is stimulated and this produces a 5–10% higher yield, especially in earlier-maturing varieties. Some later-maturing varieties and more recent introductions tend to be more cylindrical in set and stopping is not needed.

Harvesting
The major part of the crop is machine harvested, usually beginning in August

and extending until March, according to the variety and season. Occasionally, the earliest maturing varieties are picked over by hand before the whole plant is cut for stripping. Where hand harvested, the crop is usually picked over three or four times during the season.

The EC Standard for Fresh Sprouts applies for produce sold fresh to the consumer.

Yield
17–25 t/ha (average about 20 t/ha).

14.9 Swedes (for the domestic market)

The growing of swedes for human consumption follows the same lines as swedes for livestock feed (pages 350–3). Farmers may grow the crop for the domestic market but in many cases a proportion of the crop will be used for stock feed.

Varieties

- *Ruta Otofte* is suitable both for the domestic market and for stock feed.
- *Joan*, a standard variety grown at a high density for pre-packing and stew packs.
- *Marion*, suitable for dicing and resistant to clubroot and mildew.

Harvesting
Although machine harvesting is increasingly being used for swedes, a sizeable proportion of the crop is still pulled by hand.

14.10 Further reading

Brewster J L, *Onions and Other Vegetable Alliums*, CABI publishing, 1994.
Maynard D N and Hochmuth G J, *Knott's Handbook for Vegetable Growers*, 4th Ed., John Wiley & Sons, 1997.
Processors and Growers Research Organisation, leaflets and publications.
Rubatzky V E, Quiros C F and Simon P W, *Carrots and Related Umbelliferae*, CABI publishing, 1999.

15

Forage crops

15.1 Crops grown for their yield of roots

Fodder beet and, now to a much lesser extent, mangels are grown for feeding to cattle and sheep in the more southerly parts of the country where they are better croppers than swedes and maincrop turnips which are mainly grown in the north. Occasionally fodder beet is grown as a cash crop, for sale, but the area has declined in recent years. With its lower dry matter content, the mangel has now almost died out as a forage crop. The financial difficulties facing the livestock sector, the absence of machinery suitable for root growing on stock farms and the development of contractor operations focused on forage maize has caused a major decline in the use of these crops.

15.1.1 Fodder beet
This has been bred from selections from sugar beet and mangels. At one time it was quite popular as a feed for pigs. Fodder beet should not be grown on heavy and/or poorly-drained soils, nor on stony soils. There could well be establishment and harvesting problems in these conditions.

Varieties
An important consideration, apart from dry matter yield, is the cleanliness of the roots after harvest. Dirty roots may be less palatable and cause digestive upsets. Another important factor is the dry matter percentage. *Kyros* has been a popular variety for many years and gives average yields of clean roots of about 16% dry matter. *Magnum* is an example of a variety with very high yields

and a higher dry matter of about 19%. Sheep and young cattle, if they are to be fed on fodder beet, do better on the lower dry matter varieties. *Feldherr* at about 14% dry matter is suitable for feeding to younger animals, but has a lower yield than the others. Herbicide-tolerant varieties are being tested but have not, so far, been approved for farm use in this country. The NIAB annual *Fodder Beet Variety Leaflet* gives more details of current varieties. Kingshay Farming Trust has also recently undertaken a comparison of varieties.

Many aspects of the growing of fodder beet are similar to those of sugar beet, but there are some important differences and the chief points to remember are described below.

Seed rate
Seed is usually sold in acre packs of about 50 000 seeds (125 000 seeds or 2.5 packs/ha). Graded (3.50–4.75 mm) pelleted seed is used, which should be precision-drilled if possible.

Time of sowing
Early to mid-April is best; sowing earlier will cause bolting (some varieties are more susceptible). Later sowing will reduce yields.

Most crops are now drilled-to-a-stand with monogerm seed which will preclude the need for thinning out. However, a proportion of monogerm fodder beet seed will produce more than one seedling and so the crop may need 'rough singling'. The drill width will depend on the harvesting system; it is normally at 50 cm with a 15–17 cm spacing in the row. The aim is to achieve at least a 65 000/ha plant population, but because of inevitable plant losses through pests and diseases, post-emergence herbicides and/or accidents with inter-row work, if carried out, a target figure of 85 000 plants/ha is not unrealistic.

Manures and fertilisers
As with sugar beet, fodder beet responds to sodium; agricultural salt is often used as a basal fertiliser. About 375 kg/ha (supplying 200 kg/ha Na_2O) should be applied and worked into the soil at least a month before the crop is drilled. The phosphate and potash can also be applied at this time. If salt is not applied potash fertiliser should be increased by 100 kg/ha K_2O. Normally half the nitrogen will go on the seedbed and the balance at the 3–7 leaf stage. Fertiliser recommendations are shown in Table 15.1.

A pH of 6.5–7.0 is necessary. Acid conditions cause stunted and misshapen roots. However, as with sugar beet, overliming can induce boron and manganese deficiencies. A soil test for boron availability is a useful guide.

Crop protection
It is important to keep the crop clean for at least the first six weeks and details of suitable pre- and post-emergence herbicide programmes suitable for fodder beet are given on page 106.

Table 15.1   Fertiliser recommendations (kg/ha) for fodder beet and mangels

SNS*, P or K index	0	1	2	3	4	5
Nitrogen (N)	120	100	60	30	30	0
Phosphate (P$_2$O$_5$)	100	75	50M**	0	0	0
Potash (K$_2$O)	150	125	100M(2−) 75(2+)	40	0	0

* Soil Nitrogen Supply index–see MAFF RB209.
** M–maintenance dressing only.
Agricultural salt is recommended for all soils except fen silts and peats. About 375 kg/ha of salt (200 kg/ha Na$_2$O) should be worked into the seedbed well before drilling. If the salt is not applied potash fertiliser should be increased by 100 kg/ha K$_2$O.
Where FYM and slurry have been used the above figures should be reduced to take account of the available nutrients supplied. Serious pollution can be caused by over application of organic manures.
These recommendations have been based on MAFF RB209, 2000.

Weed beet can be a problem. In fodder beet it tends not to be treated as seriously as in the sugar beet crop. Apart from the fact that the problem is perpetuated if the weed is allowed to grow unchecked (obviously a serious problem if sugar beet is also grown on the farm), it does compete with the fodder beet itself and with the following crops in the field. Hand roguing is the best form of control.

Fodder beet seed should be dressed with thiram against footrot. It is also possible to guard against the wide range of soil-borne insect pests (including flea beetle) which attack both fodder beet and sugar beet by having the seed dressed with imidacloprid. Mangel fly and aphid vectors of beet yellows virus can also be controlled by this treatment. Where fodder beet is being grown after ploughed-out grass, leatherjackets may be a serious problem.

Virus yellows can be very serious in some years, particularly if an early attack by aphids occurs (Table 7.1). In the absence of imidacloprid seed treatment a spray of pirimicarb may be necessary. Powdery mildew, appearing as a white powdery growth on the leaves in early autumn, is not usually considered serious enough to treat with a fungicide.

Harvesting
This usually takes place in October and November, although, depending on the season, the crop can continue to grow through the autumn. However, late harvesting generally increases the risk of more difficult conditions, particularly on heavier soils. Many growers will also harvest earlier at the expense of yield in order to follow with winter cereals (usually wheat) and to achieve a clean root sample.

The crop is lifted either by specialist harvesters or modified (or even out-dated, cheap) sugar beet harvesters. Top savers can be used when necessary or a forage harvester can be used for topping the beet (an inaccurate technique, unless the crop is very evenly grown). Topping must be carried out correctly when it is intended to store the beet for any length of time. The roots should be topped at the base of the leaf petioles. Overtopping soon results in mould growth; leaving too much green material causes high losses through increased respiration in the clamp.

Fodder beet can be fed fresh to stock, but normally it has to be stored. This must be done carefully and for outside storage a temperature of 3–6 °C should be maintained in the clamp. Heating in the clamp should be avoided, although the beet should be adequately protected against frost. A straw covering or plastic sheet can be used, but a strip along the top of the clamp should be left uncovered for ventilation, except in freezing conditions. The storage capacity needed will vary according to the size of the beet but is usually in the range 700–750 kg/m^3.

Yield

- Fresh yield: 80–100 + t/ha.
- Dry matter yield: 11–16 t DM/ha–up to 20 t DM/ha in favourable conditions.
- Dry matter yields of tops may also be 3–4 t DM/ha.

Wholecrop fodder beet for silage

Details of this technique can be found on page 452. Early harvesting in dry conditions is advised and the selection of a variety yielding clean roots is also important since the amount of soil entering the clamp should obviously be minimised.

15.1.2   Mangels (mangolds)

As shown in Table 15.2, there is a significant difference in dry matter yield and energy content when comparing mangels with fodder beet. Very few mangel crops are now grown.

The husbandry of mangels is similar to that of fodder beet. Most of the varieties such as *Wintergold* (a low dry matter traditional variety) are multigerm, with the seed producing more than one plant when sown (*Peramono*, a monogerm variety, is an exception). The seed, although graded and pelleted, will produce a relatively full row which should be thinned. The target population is 65000 plants/ha.

Although the mangel grows with much of its root out of the ground, harvesting, except by hand, is difficult. The root 'bleeds' very easily and so the tops (not the crowns) are either cut or twisted off. The tops are small and are rarely fed. The roots are not trimmed. Ideally, they should be left for a period in the field to sweat out in small heaps covered with leaves to protect them from frost damage. Following this they can be clamped in the same way as fodder beet.

Table 15.2   Typical yields and nutritional values of forage crops compared with grass

Crop	Typical fresh annual yield (t/ha)	DM (%)	Dry matter yield (t/ha)	ME* (MJ/kgDM)	Crude protein (%)
Fodder beet	90	17.0	15.3	12.5	7
Mangels	120	11.0	13.2	12.4	9.2
Yellow turnips	70	8.5	6	12.7	11.2
Forage turnips	40	10	4	12.5	11.5
Swedes	90	11.0	9.9	13.5	9.5
Kale	55	15.0	8.3	11.4	20
Forage rape	45	11.0	5	10.4	19
Fodder radish	50	10	5	9	2.3
Forage maize	40	30.0	12	11.5	9
Wilted grass for silage**	40–60	25.0	10.0–15.0	11	13.7

* ME–Metabolisable energy.
** Grass yields in particular may vary substantially according to growing conditions.
Published figures regarding the yields and compostion of all these crops vary substantially and
    farm experience too can be extremely variable. The figures presented represent the potentials
    in average conditions with near optimum inputs.

15.1.3   Swedes and turnips

Although their popularity has declined in the past 30 years, swedes especially
and maincrop turnips are still grown, particularly in the northern and western
parts of the United Kingdom. In the north of England and Scotland these crops
are also known as 'neeps'. In appearance, the difference between the two crops
is that swedes have smooth, ashy-green leaves which grow out from an extended
stem or 'neck' whilst turnips have hairy, grass-green leaves which arise almost
directly from the root itself.

Both crops are valuable for cattle and sheep and, depending on the variety
grown, they can also be used as table vegetables. There has been a small-scale
revival in their popularity for this purpose. A limited market has also developed
for contract growing turnips for freezing.

The growing of swedes and main crop turnips is similar.

Varieties
Varieties of swedes and main crop turnips are shown in Table 15.3.

Climate and soil
The crops like a cool, moist climate without too much sunshine. Powdery mildew
can be a problem in the warmer and drier parts of south and east England;
varietal resistance and fungicide treatment are not yet the complete answers.
With the exception of heavy clays, most soils are suitable.

Seedbed
A fine, firm and moist seedbed is necessary to get the plants quickly established.

Table 15.3 Varieties of swedes and maincrop turnips

Type	Examples of varieties	Average dry matter (%)	Remarks
Swedes (grouped according to skin colour and type of root)			
Purple Globe	*Angela*	10.2	Could be described as
	*Ruta Otofte**	9.6	semi-tankard. Useful
	Magres	11.9	for grazing.
	Marian	10.3	
	Airlie	10.3	
	Brora	11.2	
*Green Globe***	*Melfort*	12.7	Good winter hardiness.
Bronze Intermediate	*Angus*	10.9	Good winter hardiness.
Turnip			
Yellow-fleshed	*Yellow Top Scotch*	8.6	Slower to mature than
	The Wallace Aberdeen	9.1	other types; hardy and
	Greentop Imperial	8.8	good keepers.
	Green Globe	7.8	

* Suitable for the domestic market.
** Varieties such as Seegold (Green Globe) should be used where clubroot is suspected.

In very wet districts the crops can, with advantage, be sown on a ridge. This traditional practice is mainly carried out in northern areas and in Scotland. It is not widespread now, mainly because of the use of precision seeding and improved chemical weed control.

Manures and fertilisers

The average fertiliser requirement is summarised in Table 15.4. If possible, 25–40 t/ha farmyard manure should be applied in the autumn. This is especially important for improving the water-holding capacity of lighter soils. Slurry can be used instead, in the spring, at about 35 000 l/ha.

The lower amount of nitrogen is recommended in wetter areas, although exactly how much is used will depend on the soil nitrogen supply status. More

Table 15.4 Fertiliser recommendations (kg/ha) for maincrop forage swedes and turnips

SNS*, P or K index	0	1	2	3	4	5
Nitrogen (N)	100	80	60	40	0–40	0
Phosphate (P_2O_5)	100	75	50M**	0	0	0
Potash (K_2O)	200	175	150M(2–) 125(2+)	80	0	0

* Soil Nitrogen Supply index – see MAFF RB209.
**M – maintenance dressing only.
Where FYM and slurry have been used the above figures should be reduced to take account of the available nutrients supplied. Serious pollution can be caused by over application of organic manures.
These recommendations have been based on MAFF RB209 issued in December 2000.

phosphate may be needed on heavier soils, but this will depend on the soil phosphate index; if the fertiliser is placed 5–10 cm below the seed, the phosphate can be reduced. Potash is important, but savings can be made by using organic manures. As with fodder beet, boron deficiency (also known as brown heart or raan in swedes) may occur in some soils and, particularly if the pH is high, a soil sample to determine boron availability is advisable.

Lime is important as swedes and turnips are often grown on potentially acid soils. Soil pH should be above 6.0. Club root (finger and toe disease) can be prevalent under acid conditions. However, overliming is equally serious as it can induce boron deficiency symptoms (Table 7.1).

Time of sowing

Mid-April to end June. Swedes can now be sown earlier than has been the case in the past, thus increasing the yield potential. Powdery mildew, which is more likely with earlier sowing, can be reduced by chemical treatment, e.g. with tebuconazole. In addition, an increasing number of swede varieties show reasonable resistance to mildew.

The yellow-fleshed turnip is normally sown in May and June.

Seed rate

Rates vary between 0.5 and 4.5 kg/ha. The lower amount is used with precision drilling. Pelleted seed is sown 7–15 cm apart in the row at row widths of 17–35 cm. This is particularly applicable where overall chemical weed control is carried out and/or the crop is to be folded. The aim is for a high plant population of up to 100 000/ha.

With inter-row cultivation, or if the crop is to be lifted, the plants are spaced (or rough-singled) at 20–22 cm apart with row widths of 50–75 cm. A plant population of 60 000/ha is acceptable, although, if the roots are rather large and dry matter content is low, they may not keep well. Both swedes and turnips can be broadcast at 3.5–5 kg/ha.

Seed treatments containing thiram for protection against soil-borne fungi should be used. The only approved treatment for flea beetle control is a pre-sowing application of carbofuran granules. Leatherjackets are likely to be a problem in fields recently ploughed from grass.

Weed control

The stale seedbed technique, using a contact herbicide to kill the weeds at the time of drilling, followed, if necessary, by the use of a post-emergence herbicide such as propachlor applied when the crop has three to four leaves, should keep it clean of weeds (Table 4 in Appendix 9). However, inter-row cultivations or harrow combing are carried out by some growers, particularly on organic farms.

Harvesting swedes

Swedes, like turnips, are very often grazed *in situ* in the field, but both crops can deteriorate under wet and frosty conditions. Varieties with globe-shaped roots, rather than tankard-shaped, are less prone to damage when harvested by machine. In most districts they are lifted in late autumn, before they are fully matured. It is advisable to allow the roots to ripen in a clamp to minimise scouring when fed to stock. In mild districts, swedes may be left to mature in the field; for this purpose, high dry matter, more winter-hardy types are recommended. In this case the swedes would only be stored for two to three weeks, with the grower endeavouring to keep harvesting just ahead of any possible bad weather. Swedes are vulnerable to rotting under poor storage conditions. They should be handled carefully and not clamped in wet conditions. Clamps should be no higher than two metres and adequately protected against frost. Because of the low dry matter content and probable soil contamination, the tops are usually wasted. However, they are quite high in protein.

Harvesting turnips
The main crop (yellow-fleshed) is normally harvested in October when the outer leaves begin to decay. They can also be grazed *in situ*, but are often lifted and stored in one main clamp. Machines which merely top, tail and windrow the roots for subsequent collection are now being replaced by complete harvesters. On mainly livestock farms harvesting is often contracted out.

Yield
Swedes, 65–110 t/ha (7–12 t DM/ha).
Turnips (yellow-fleshed) 60–80 t/ha (5–7 t DM/ha).

15.2 Crops grown for grazing

15.2.1 Kale
Kale is grown for feeding to livestock, usually in the autumn and winter months. It can either be grazed in the field or cut (normally with a forage harvester) and carted off for feeding green. In this case a heavier yielding crop is needed. It is mostly grown in the south and south west of England. It would probably be true to say that the majority of kale crops at present are grown for game cover. There has, however, been a small revival of interest in kale for ensiling in recent years.

Recently published figures from the Institute for Grassland and Environmental Research (IGER) showed that the area sown to kale, cabbage (for stock feeding) and rape had declined from nearly 190000 ha in 1960 to about 12000 ha in 1994. This decline reflects changes in the ways in which livestock have been kept and fed during the winter months; for example from being outwintered on hay and strip-grazed kale, to being inwintered on silage.

Climate and soil
Kale is an adaptable crop, although under very dry conditions it may be difficult to establish. For grazing it is essential that it is grown on well-drained fields.

Place in rotation
On livestock farms kale is often direct drilled into an old ley after glyphosate spraying. This may be after an early grazing or a silage cut has been taken. Kale can also follow a catch crop put in after cereals the previous autumn. Club root disease (finger and toe) (Table 7.1) can be a problem. Kale should not be grown in the same field more often than one year in three.

Seedbed
A fine, firm, and clean seedbed is required. Kale is very suitable for direct drilling and this has the important advantages of conserving moisture at sowing time, leaving a much firmer surface for grazing and good annual weed control. After sowing in conventional seedbeds good consolidation is important, for moisture conservation and good seed–soil contact. Rolling is also important immediately after direct drilling has taken place in order to close the slits where the seed has been placed.

Manures and fertilisers
Up to 50 t/ha of farmyard manure can be applied in the autumn. It is especially important when a heavy-yielding crop is desired. Slurry at about 35000 l/ha can also be used in the spring, prior to cultivation. A light dressing of slurry applied to a direct drilled field after drilling can also be beneficial.

Total plant nutrients required are summarised in Table 15.5. This assumes phosphate and potash indices of 2. Less nitrogen is required if kale follows grass. Fertiliser is usually applied during final seedbed preparations. The nitrogen application can be split and part top-dressed when the young plants have up to five leaves. A pH of 6.0–6.5 is recommended.

Table 15.5 Fertiliser recommendations (kg/ha) for kale

SNS*, P or K index	0	1	2	3	4	5
Nitrogen (N)	130	120	110	90	60	0–40
Phosphate (P_2O_5)	100	75	50M**	0	0	0
Potash (K_2O)	250	225	200M(2–) 175(2+)	130	0	0

*Soil Nitrogen Supply index – see MAFF RB209.
**M – maintenance dressing only.
Where FYM and slurry have been used the above figures should be reduced to take account of the available nutrients supplied. Serious pollution can be caused by over application of organic manures.
These recommendations have been based on MAFF RB209 issued in December 2000.

Varieties
There is no current recommended list of kale varieties available from NIAB. *Maris Kestrel* and *Hereford* give good yields of dry matter and are highly digestible. *Camaro* is one of the highest yielding varieties and *Keeper* (as its name suggests) is favoured for game cover. Some farmers have used blends of *Keeper, Pinfold* and *Hereford* for ensiling.

Time of sowing
Kale can be sown from the end of March until mid-July. The heaviest crops are normally associated with early sowing.

Seed rate
Precision-drilled: 0.5–2.0 kg/ha, the seed spaced at 2.5–10 cm apart in 35–75 cm rows. Drilled: 3.0–5.0 kg/ha in 35–75 cm rows. Broadcast: 5.0–7.5 kg/ha. For ensiling, seed rates of from 7.5 kg/ha if drilled to 9 kg/ha broadcast are recommended.

Weed and pest control
The herbicides propachlor and trifluralin are approved for use on kale crops. Steerage hoeing or harrow combing of drilled crops is a possibility on organic farms. Alpha-cypermethrin spray is an approved post-emergence treatment against flea beetle. Leatherjacket damage is likely when the crop is spring-sown after grass.

Utilising the crop
Strip grazing behind an electric fence reduces the need to handle the crop. Light or well-drained soils are essential, otherwise both stock and soil suffer. Wastage can be high at 15–30% and up to 70%. Starting to graze kale in the early autumn and mowing before grazing can both improve utilisation. If strip-grazing is not possible a forage harvester may be used to chop the crop coarsely and blow it into a trailer for feeding in the yard.

For ensiling, the crop should be utilised by mowing and baling in August/September when dry conditions suitable for wilting usually obtain. Full details of the ensiling technique are given on page 452. If the field is not required for reseeding or winter cereals, a useful option is to allow the stubble to regrow. This can provide a useful extra grazing for young cattle or sheep later in the autumn.

Kale contains an imbalance of trace elements such as iodine and manganese and a chemical called S-methyl cysteine sulphoxide (SMCO); very high intakes should be avoided as they can lead to haemolytic anaemia and fertility problems.

Yield
35–75 t/ha at about 15% dry matter (5–11 t DM/ha).

15.2.2   Forage catch cropping

This is the practice of taking a quick-growing crop between two main crops. The term is usually applied to crops of quick growing stubble turnips, or rape, or mixtures of the two, sown for autumn or early winter grazing after an early harvested cereal (usually winter barley).

Forage catch crops are usually direct drilled since this helps to conserve soil moisture. It is also a valuable technique for weed control since the grass weeds and cereal volunteers, which almost invariably germinate alongside the forage crops, can be effectively controlled either by a glyphosate spray pre-drilling, or a post-emergence application of an effective approved graminicide. Apart from weed control and the probable need to control slugs, the only other important treatment for these crops is nitrogen fertiliser so they have the additional advantage of being very cheap to grow.

15.2.2.1   Quick growing white turnips (stubble turnips)

The quick growing white turnips originated in The Netherlands and are also known as Dutch turnips. The name stubble turnips has arisen from the practice of growing them as a catch crop on the stubble following an early-harvested cereal crop. However, in the north of England and in Scotland it is seldom possible to obtain a reasonable yield by stubble growing. In colder regions they are sown earlier and, indeed, in England, these turnips can be sown in April to provide a useful midsummer feed in July. There is nothing against establishing a second crop of turnips after the first crop from a spring sowing has been utilised. However, repeated sowing in the same field would inevitably lead to a problem with clubroot disease.

Sowing should be completed ideally by mid-August. Later sowing produces much reduced yields and may not be worthwhile. Direct drilling is now the norm; it helps to conserve moisture at what is usually a dry time of the year. Broadcasting seed into a cereal crop before harvesting using a pneumatic fertiliser spreader or autocasting seed when the cereal is combined are time-saving establishment techniques which have been tried in recent years. Success depends very much on the effectiveness with which the straw is chopped and spread as well as the incidence of rainfall. In dry conditions these techniques are unlikely to be successful.

Direct drilling and grazing a crop of forage turnips is a valuable way in which to obtain additional dry matter production from a poor grass ley prior to reseeding in July or August. It is also a useful technique in that it enables the old sward to be completely eliminated prior to establishing the new seeds.

Seed rate
2–5 kg/ha drilled; 6–9 kg/ha broadcast. A common practice is to mix with forage rape for autumn sheep feeding.

Varieties
There is no NIAB recommended list for these turnips. Popular varieties are the tetraploids *Vollenda* and *Taronda*. *Barkant* gives some of the highest dry matter yields. *Appin* and *Tyfon* are varieties with low root yields, mainly leaves. These varieties, however, are unsuitable for spring sowing as they are likely to bolt. There are some varietal differences in susceptibilities to clubroot, powdery mildew and *Alternaria* diseases.

Fertiliser
The main need is for nitrogen; turnips will respond to between 75 and 100 kg N/ha. This should be broadcast at the time of sowing or top-dressed shortly after the crop has emerged. Phosphate and potash are not usually needed although this will depend on soil indices.

Weed control
Spraying with paraquat or glyphosate prior to direct drilling into an old grass ley is normal. A useful technique is to spray glyphosate prior to the final utilisation of grass. This will not harm the stock and ensures a complete kill of the old sward. The use of an approved graminicide such as fluazifop-P-butyl can be helpful for controlling grass weeds and cereal volunteers when sowing after a cereal crop.

Utilisation
The crop is ready for grazing about three months after drilling. Whereas most crops are utilised by sheep, cattle, including dairy cows, can also benefit considerably. Some dairy farmers regularly sow turnips for summer and early-autumn grazing to supplement grass growth and both milk yield and quality can benefit.

Yield
30–40 t of fresh yield/ha is the norm for autumn sowings at about 10 % dry matter. Much higher, but quite variable yields are possible when sowing in favourable conditions in spring or in July.

15.2.2.2 *Forage rape*
This is a quick growing palatable crop which is ready for grazing 12 weeks after sowing. It generally lacks winter hardiness and should be used before the end of the year.

Varieties
Relatively little plant breeding has been undertaken in forage rapes in recent years and so *Emerald* which was first introduced in the 1970s is still in general

use today. A common failing of forage rape varieties is susceptibility to club root (finger and toe) disease. *Sparta* and *Dinas* both have moderate resistance to this disease.

Seed and sowing
The seed is either drilled at 4.5 kg/ha or broadcast at 10 kg/ha from the end of April until mid-August, usually following, in this last case, an early harvested grain crop. Direct drilling is a suitable method of establishment. Rape is often sown as a mixture with quick growing turnips for feeding to sheep.

Fertiliser
Up to 100 kg N/ha can be used. Phosphate and potash are not always necessary, but this will depend on soil indices.

Yield
30–50 t/ha at 11–12% dry matter.

15.2.3   Rape kale and Hungry gap kale
These are not true kales but are related to, and similar to, rape. Both are particularly susceptible to clubroot and mildew. They are grown on a small scale only in the southern part of the United Kingdom, and can be used for a limited period after Christmas. They can produce useful regrowth for grazing in March and April–the so-called 'Hungry gap' before spring grass is available.
   Seeding and growing requirements are similar to those for normal forage rape.

15.2.4   Fodder radish
This crop is grown for forage, green manuring (page 76) and game cover. By producing quick ground cover it can also be used as a protection against wheat bulb fly. It is very quick growing (some varieties grow a metre high) and is suited to most soil conditions. *Nerys* and *Slobolt* are two available varieties. Fodder radish is normally sown in July for forage at 8 kg/ha drilled, or 13 kg/ha broadcast. For green manuring it may be sown earlier, with a heavier seed rate of 17 kg/ha. Up to 75 kg N/ha can be applied with the seed. Phosphate and potash should not be necessary.
   Fodder radish should be grazed off before flowering and whilst it is still palatable, being normally utilised within 8–12 weeks of sowing. This crop is very susceptible to frost but resistant to clubroot and mildew. It can also provide a good rotational break on farms where sugar beet is grown and there is limited evidence that fodder radish can reduce soil populations of sugar beet cyst nematode.

15.2.5 Winter cereals as grazing crops

Rye and triticale are often sown specifically as grazing crops. Early sowing in August or September can result in a worthwhile crop for grazing as early as November. The seed rate for rye should be about 200–250 kg/ha and for triticale, 180 kg/ha.

Grazing in March or April following an early (February) application of about 50 kg N/ha is the normal time for utilising these crops. On many stock farms this period often coincides with a shortage of winter feed and occurs prior to the availability of spring grass for grazing. When an early grazing takes place it may be possible also to graze a regrowth three to four weeks later or even to take this regrowth for a silage cut. Neither crop is particularly palatable to stock and it is common practice also to sow about 10 kg/ha of an Italian ryegrass variety to improve this.

Average dry matter yields of around 3.5 t/ha are possible from rye and triticale. After the last grazing the fields would be suitable for cropping with kale or forage maize, following an application of FYM or slurry.

Well-grown crops of winter wheat are also occasionally grazed and the practice is widespread in other countries. Care should be taken to ensure that poaching does not occur and that grazing does not take place after stem extension (Zadoks growth stage 31 see page 267) or yields may be seriously affected.

15.3 Crops grown for ensiling

15.3.1 Forage maize

Apart from grass, forage maize has become the most important forage crop grown in the UK. In an emergency, maize can be cut early for feeding green to stock, but most of the crop is now used only for ensiling. Some support from the Common Agricultural Policy (CAP) in terms of area payments has increased interest in maize growing and at the turn of the millennium the area sown was just over 100 000 ha, most of it in the Midlands and southern England.

Maize was introduced into this country at the beginning of the twentieth century and early trials on its suitability as a crop for ensiling were carried out at Wye College in Kent. Although high yields were shown to be possible, two main factors were responsible for the relatively slow uptake of the crop. The first was that the varieties available at that time were mainly of American origin and did not mature early enough for British conditions. In particular, the development and yield of cobs (the most valuable part of the plant, rich in cereal starch) was not good. The second reason was the absence of suitable and reliable mechanical systems for harvesting, ensiling and feeding out.

The rapid development of the crop from the early 1960s has taken place in response to two main stimuli. One has been the introduction of high-yielding, early maturing hybrid varieties, capable of developing good yields of cobs and high starch percentages in many parts of the UK. The other has been the advent of high capacity, contractor-based growing and harvesting machinery systems. The increasing use of contract harvesting systems, not only for maize but also for grass silage, has also developed concurrently with the decline in the labour force on many livestock farms.

A further reason for the rise in the maize area has been that it satisfies the need for a high energy forage for winter feeding. Good maize silage has an ME value of 11.2–11.3 MJ/kgDM and a D value of between 70–75%. Cereal starch normally accounts for between 25 and 30% of the dry matter. This augers well for the production of high yields of high protein milk, and maize silage fits in well with the move to complete diet feeders which are the basis of such systems. Tower silos have been tried, but now largely abandoned, by UK farmers, in favour of long narrow bunker clamps, which better suit current handling systems, and which minimise the likelihood of secondary fermentation at feed out.

Climate, site and soil types

Maize is a C4 plant of sub-tropical origin. This means that is has a different way of photosynthesising compared with most other crops grown in the UK. In particular, it responds well to high temperatures and requires a minimum soil temperature of 10 °C before active growth commences. Crops grown at low altitude in the coastal areas of southern England and Wales will, therefore, have much higher yield potentials, because they experience higher levels of temperature during the growing season.

Maize is best grown on fields with a southerly aspect, and does not normally do well in cold conditions at high altitude. Fields with steep slopes should also be avoided because of the risk of soil erosion. Maize is best suited to light to medium textured soils with good drainage. Growing maize on poorly drained clay soils is not advisable. Good crops can be grown, but in a wet autumn substantial harvesting difficulties may be encountered.

Forage maize should form part of a rotation of crops. On some farms this may not be possible and it may have to be grown for several years in the same fields. Where maize provides a very high (up to 100%) proportion of the winter bulk feed ration there may be no alternative to repeated cropping. However, such practices are risky and may lead to problems with trash-borne diseases such as maize eyespot (*Kabatiella zeae*) which can have devastating effects on yields (Table 7.1).

Cultivations

Winter ploughing following soil loosening in a reasonably dry autumn is the best preparation. However, where maize fields are to receive large quantities of organic manures, ploughing may be delayed until the spring, to facilitate application. In dry areas it is preferable to roll immediately after ploughing, to conserve moisture. Power harrowing to produce a medium tilth then takes place immediately before the maize is precision-drilled. Rolling after drilling should then complete the operation.

Choice of variety

Apart from the choice of field and seedbed preparation, as described above, the choice of a suitable variety is the main factor influencing the success of growing maize. A great many varieties (all hybrids) are available to UK farmers. NIAB produce an annual descriptive list with data drawn from a range of about ten sites in England and Wales. Varieties are categorised according to a range of maturity classes from 3 to 10, with 10 being the earliest maturing (usually the lowest yielding) group, suitable for growing in the most northerly or marginal conditions. At the other end of the scale, varieties in maturity class 3 are much later maturing and suitable only for growing in the most southerly and favoured situations.

The correct choice of variety for a site depends on calculations of accumulated temperature which the crop is likely to experience. Farmers can categorise their growing conditions by referring to published climatic data expressed as 'Ontario heat units' (OHUs). ('Maize heat units' in some publications also describe temperature accumulation, but in a slightly different way.) The Meteorological Office has produced a map that depicts the areas of southern England and Wales which achieve various levels of Ontario heat units in nine years out of ten (see Appendix 10). It should be remembered that altitude and aspect, in particular, will also affect temperature on a local basis.

The most favourable areas for maize are those which achieve more than 2500 OHUs. These areas have the highest yield potential and would be suitable for the varieties in maturity classes 4 and 5. Earlier harvesting (but with a probable yield penalty) would be possible by choosing varieties from maturity classes 6 or 7 in these favourable areas. Maturity class 3 varieties are only suitable for exceptionally favoured areas in the extreme south of England.

In less favourable areas, with 2300–2500 OHUs, varieties from maturity classes 6–10 should be the main choices. Using varieties in the earliest maturity classes (9 and 10) will enable very early harvesting, or successful maize growing in the least favourable sites, but these varieties are normally the lowest yielding of all.

Apart from comparing dry matter yields, the NIAB information also gives details of likely levels of ME and the percentage starch content. Not all of the varieties available in the UK are currently listed by NIAB. For information about the suitability of non-listed varieties for growth in specific farm situations, information should be sought from the seeds company concerned or from the UK Maize Growers Association Ltd.

Seed rate and sowing date

Almost all maize is precision sown with fungicide dressed seed. Optimum seed rates are between 120000 and 140000 seeds/ha with seeds being sown in rows about 75 cm apart. Trials have been carried out to compare the conventional row spacing with closer drilling or drilling in double rows (alternately close drilled and conventional row widths) from which some slight advantages are apparent. Some farmers are adapting cereal drills to sow maize seed in 37.5 cm rows and saving the contractor costs of precision drilling. Maize seed is best sown at between 5 and 10 cm deep, depending on soil moisture content.

Sowing date is important and the majority of maize crops are drilled in the latter half of April in southern England. A more precise guide is soil temperature and the soil at 10 cm should have reached 8–10 °C before drilling commences. When drilling is delayed beyond the first week of May serious reductions in yield are inevitable.

Sowing maize under plastic film

There has been some interest in the use of biodegradable plastic film for establishing maize, particularly in marginal areas. Increased dry matter yields and accelerated maturity have been achieved; however, the technique has not become widespread mainly for reasons of cost-effectiveness.

Fertilisers

pH is an important consideration for maize growers since the crop is often grown on light or sandy soils which have a predisposition to acidity. If lime is required it is best applied in the autumn prior to sowing or on the ploughed furrow prior to final seedbed preparation in the spring. The field should be limed to a pH of 6.5.

The response of maize to added fertilisers is variable because the crop is often grown in conditions of very high fertility. Theoretically, an average (say 12 t DM/ha) crop, would remove about 150 kg N/ha, 55 kg/ha P_2O_5 and 170 kg/ha K_2O. After deducting the theoretical values of nitrogen, phosphorus and potassium delivered by the large quantities of FYM and slurry applied to maize fields, it is probable that the average crop requires no additional fertiliser apart from about 50–60 kg N/ha. However, it is very common practice for maize to receive at least 125 kg/ha of monoammonium phosphate (MAP) worked into the seedbed or placed close to the seed at drilling. This provides readily available nitrogen and phosphorus to enhance the vigour of seedling growth in the early stages of crop development. However, there is evidence to suggest that, on fertile sites especially, where large quantities of organic manures are habitually applied, and in spite of a strong apparent visual response in the early stages, there is seldom a worthwhile final yield response. Extra nitrogen, where required, may be worked into the seedbed or top-dressed in the early stages of growth. The current recommendations for maize are summarised in Table 15.6.

Crop protection

Seedling fungal diseases are normally protected against by a fungicidal seed dressing. Until recently, bendiocarb seed dressing served both as a guard against attack by frit fly and wireworm larvae and as a bird repellent. Since

Table 15.6 Fertiliser recommendations (kg/ha) for forage maize

SNS*, P, or K index	0	1	2	3	4	5
Nitrogen (N)	120	80	40	20	0	0
Phosphate (P$_2$O$_5$)	110	85	60M**	20	0	0
Potash (K$_2$O)	230	205	180M(2–) 155(2+)	110	0	0

* Soil Nitrogen Supply index – see MAFF RB209.
** M – maintenance dressing only.
Where FYM and slurry have been used the above figures should be reduced to take account of the available nutrients supplied. Serious pollution can be caused by over application of organic manures. This is particularly important on fields which have grown maize for several years.
Where maize is being grown without organic fertilisers some response may be gained from placing most of the phosphate and about 15 kg/ha of nitrogen below the seed at drilling.
These recommendations have been based on MAFF RB209 issued in December 2000.

approval for this material is soon to be withdrawn maize growers will have to resort to other approved soil-applied insecticides to control these important pests.

Weed control using the residual herbicide atrazine has been the standard for many years. However, weed strains resistant to this herbicide (notably black nightshade and fat hen) have developed and other materials, such as pyridate, bromoxynil and bromoxynil/prosulfuron, applied post-emergence have become popular. Atrazine is a persistent chemical in the soil and its residues may damage subsequent crops; ploughing after maize is always advised to minimise this problem. Atrazine use has been banned in some other European countries. The use of a stale seedbed, coupled with steerage hoeing or harrow combing, are viable alternatives for those who wish to grow maize on organic farms.

So far, disease has not been a major factor limiting maize yield in the UK. Smut (*Ustilago maydis*) may occur in maize, especially in very hot dry conditions. Maize eyespot (*Kabatiella zeae*) has caused severe yield loss in some instances in the south-west. However, the disease is incompletely understood and no chemical control has been approved so far (Table 7.1).

Undersowing forage maize
The residual effect of atrazine herbicide gradually wears off after the initial application, especially in warm conditions. Italian ryegrass, sown about six weeks after maize drilling, may be affected to some extent but will still produce a worthwhile crop. ADAS trials indicate a 5% reduction in ryegrass establishment following half rate atrazine compared with a 23% reduction following full rate. In any case, the establishment of Italian ryegrass in maize stubble can provide valuable additional grazing, as well as reducing, to some extent, the risk of soil erosion and nutrient leaching. In an alternative technique involving the withholding of atrazine and the use only of the non-residual bromoxynil for post-emergence weed control, it has also been possible to establish lucerne undersown in forage maize.

Harvesting

Harvesting usually takes place in the early autumn (weather permitting) and the best time for most crops is in early October. The ideal dry matter level is about 30 % which can readily be assessed by the thumbnail test (i.e. when a thumbnail can just be pressed into the grain on the cobs). At this stage the yields of dry matter and starch are likely to be optimum. Earlier harvesting may lead to a very acid fermentation because of high sugar levels and, almost certainly, to a loss of dry matter yield. Later harvesting of very dry material may result in hard uncracked grains and difficulty in achieving the very fine chop required for good fermentation and efficient utilisation.

Details of the ensiling process for making maize silage are given on page 451.

Yields

30–50 t/ha fresh yield (9–15 t DM/ha).

Cob-only maize harvesting options

The harvesting of maize cobs with a forage harvester equipped with a maize picker-header, for chopping and ensiling, gives rise to a very high energy ensiled product which has become known as ground ear maize (GEM). Corn cob mix (CCM) comprises ensiled moist maize grain after harvesting with a combine harvester. In 2001 a project was started to evaluate the potential for dry grain maize in the UK. In an unpromising season plot yields of up to 9 t/ha at 15 % moisture content were achieved from several sites in southern England.

Environmental concerns

A loss of soil structure and a subsequent loss of yield can follow from repeated harvesting of maize in the same fields in wet autumn conditions. Furthermore, nutrient enrichment can follow from the repeated heavy dressings of FYM and slurry usually applied to maize fields. This, coupled with the erosion of soil which inevitably takes place, both before and after harvest, is leading to quite serious pollution problems in some areas involving not only nitrate leaching, but also substantial losses of phosphate.

15.3.2   Other cereals as silage crops

Good silage can be made from virtually any cereal, harvested as a whole crop. The most popular is winter wheat, for which the husbandry is virtually the same as would be the case if it were destined for grain production. Making wholecrop silage provides the farmer with the ultimate in flexibility since the precise tonnage required may be harvested for ensiling and that part of the crop not required left for combining. Wholecrop cereals usually substitute for maize in the northern part of the UK but there are two important benefits which also make it a popular option for farmers in the south. One is that the full cereal arable area payment may be claimed for a crop grown on eligible land. The other is that the crop is harvested much earlier than maize (July or August at the latest) leaving the grower with many more post-harvest cropping opportunities.

Yields
15–40 t/ha fresh yield (dependent on the dry matter % at harvest – see also page 452) 10–16 t DM/ha.

Spring cereals, too, can be taken for wholecrop silage and there is a tradition in some parts of the UK of sowing mixtures of spring cereals with legume crops as 'arable silage mixtures'. Such mixtures may contain spring barley or oats mixed with forage peas or vetches. Currently, mixtures of cereals and peas, so long as they contain a minimum of 10% of cereals, may be eligible for cereal arable area payments. These crops are normally harvested when the cereal grain is 'cheesy' in July and yields of 6–8 t DM/ha are possible.

15.3.3 Forage peas

Specific varieties of peas grown for forage purposes (such as *Magnus*) have been introduced in recent years. However, good yields may also be obtained from conventional grain varieties. Peas for ensiling are normally harvested when the pods are fully formed and the grain has started to fill. The yields are not particularly high (6–8 t DM/ha) but the great virtue of peas is their speed of growth. A crop sown in early May with reasonable average rainfall could be harvested in mid-July at about 8 t DM/ha and 20% crude protein. Earlier-sown crops could be harvested in June.

Peas are an excellent crop for undersowing and the semi-leafless grain varieties in particular, sown at about 50 seeds/m^2, can offer a good cover for establishing a ryegrass/clover ley or even lucerne. In an emergency, peas can be an excellent crop for grazing and, in view of their high tannin content, there is little fear of bloat.

The husbandry of peas grown for forage is very similar to that for grain crops. The main difference is that since the crop is intended for grazing or ensiling there is little point in spending heavily on weed or disease control. In fact a good stale seedbed coupled with drilling in late April can preclude the need for a herbicide altogether. Pea and bean weevil damage to undersown legumes can, on occasions, be a problem.

15.3.4 Other annual forage crops grown for ensiling

Vetches are a traditional legume crop. Winter and spring varieties are available. Vetches are not normally grown alone but, most commonly, mixed with cereals such as oats or barley and used as an arable silage crop. They are a good smothering crop and suitable for inclusion in organic rotations for reducing weed competition and increasing the soil nitrogen status.

Forage lupins and forage soyabeans are both crops under development in the UK with dry matter yield and protein levels apparently similar to forage peas. Lupins, in particular, are very sensitive to soil conditions, especially pH, and care should be taken to avoid high pH sites. Soyabeans are extremely temperature sensitive and failure to nodulate effectively is a problem that may necessitate the use of some nitrogen fertiliser.

Other crops grown for ensiling on occasions include linseed and field beans. Little information is available to date on the yield potential or feeding value of these crops when grown for forage.

15.4   Further reading

Heron S, *Wholecrop Cereals*, Zeneca Bio Products, 1996.

Hocknell J and Carter C, *Spring Sown Forage Crops: Five Year Results*, Kingshay Farming Trust, 1998.

Hocknell J, *Fodder Beet – Growers Guide and Variety Comparison*, Kingshay Farming Trust, 2001.

Jarvis P, *Alternative Forage Crops*, Milk Marketing Board FMS Information Report 44, MMB, 1985.

Kaleage Development Association, The, *Growers Guide*, 1997.

Lane G P F and Wilkinson J M (eds.), *Alternative Forages for Ruminants,* Chalcombe Publishing, 1998.

MAFF, *Fertiliser Recommendations for Agricultural and Horticultural Crops*, RB209 The Stationery Office, 2000.

NIAB (Annual), *Forage Maize Variety Leaflet*, NIAB.

NIAB (Annual), *Fodder Beet Variety Leaflet*, NIAB.

Royal Agricultural Society of England (The), *Fodder Beet, its Full Potential*, RASE, 1985.

Wilkinson J M, Allen D M and Newman G, *Maize: Producing and Feeding Maize Silage*, Chalcombe Publishing, 1998.

16

Combinable break crops

16.1 Introduction

As a group the fortunes of combinable break crops tend to fluctuate, sometimes on an annual basis because of adverse weather and changes in arable area payments or the world supply situation. Generally, they should not be grown more often than 1 year in 5 nor within 4 years of each other. For some of the crops it is essential to ensure market availability before growing the crop and buy-back contracts are sometimes available. Industrial crops with a contract for non-food use may be grown on set-aside and in some parts of Europe bio diesel production from oilseed rape is being encouraged.

The range of crops, often with autumn- and spring-sown varieties, offers an option for most soil types, pH and crop rotations. Seasonal weather may adversely influence crop establishment, harvesting and sample quality. Home-saved seed has been an option for some of the crops but with specific quality requirements for the harvested product and a trend to lower seed rates/plant population, high quality seed with an appropriate treatment is advisable.

On a world-wide basis a large range of crops produce edible oil with soya, palm, rape (canola) and sunflower currently supplying over 75% of the total. However, the market for vegetable oils cannot be separated from the market for oilseed meals which are in great demand for intensive livestock systems and therefore influence the popularity and the price of oilseed crops. Soya beans grown on a large scale in both North and South America are a major influence on world trade and price but their importance has been reduced by the introduction of double low oilseed rapes which may be used as direct substitutes for both soya oil and meal.

For the future, events such as the entry of China into the World Trade

Organisation (WTO) and the accession of the first group of Eastern European countries into the EU are likely to influence the market. The issue of GM crops is yet to be resolved and has the potential to impact on both the quantity and quality of oil produced. In the EU, the major oilseed crop is rape followed by sunflower and a smaller area of soya. The UK is the third largest producer of rape after France and Germany whereas Spain is the largest sunflower producer with Italy and France being the main soya producers.

16.2   Oilseed rape

Oilseed rape is now established as an important and profitable crop in the United Kingdom. The small black seed contains 38–40% oil which is extracted by crushing and used for the manufacture of margarine and cooking fats. Oilseed processors will only accept seed which is low in both erucic acid and glucosinolate for human or animal consumption. The protein-rich residue left after the oil has been removed can be included in animal rations at a higher rate now that the glucosinolate levels are below 25 micromoles per gramme of seed but still limit the amount of meal which may be fed to non-ruminants. These varieties are known as 'double lows' and all recommended varieties are now double lows. Set-aside land can be used to grow oilseed rape for non-food use and these may be high in erucic acid rape (HEAR) varieties. Care must be taken to avoid cross-contamination, e.g. by volunteer plants and an islolation gap of 50 metres is required.

Soils and climate
The crop will grow in a wide range of soil and climatic conditions provided the land is well drained, pH over 6, and the soil and subsoil structure is good.

Rotation
Ideally, rape and other *Brassica* crops should not be grown more than one year in five, in order to avoid a build-up of diseases such as clubroot and also of pests. It is an alternative host for sugar beet nematode and this could affect the place taken by sugar beet in a rotation.

Varieties
These are detailed in the HGCA UK *Recommended Lists for Oilseeds*.
    Winter and Spring Swede rape (*Brassica napus*) are used for most of the UK's rape crop with winter varieties being the most important. There is now some interest, particularly in the north of the country, in spring and winter turnip rape (*Brassica rapa*) varieties and some spring varieties are being grown for their earlier maturity. Varieties are replaced on a regular basis and this has allowed for improvements in disease resistance, plant types and the introduction

of hybrid varieties which are now filling a significant market share. Varieties vary in their yield, oil and glucosinolate content, resistance to disease and lodging and some have regional recommendations.

Examples of recommended varieties:

	Winter	Spring
Conventional	*Escort*	*Heros*
	Madrigal	*Senator*
	Recital	
Hybrids	*Gemini*	*Concept*
	Pronto	
	Royal	*Mistral*

Time of sowing
Winter crops early August to early September, usually following winter barley with a trend towards the earlier time. For spring crops, early March, if possible.

Seedbeds
The seed is very small and so a quality seedbed with fine, moist soil conditions is required. Economic pressure on growers has increased the popularity of the Autocast system (mounted on the combine) and of direct drilling into, or broadcasting on, stubble which can be very successful on soils with good structure. Whichever method is used, seedbed condition and crop establishment must be good and the seed needs to be about 2 cm deep. Pans and large flat stones restrict the growth of the deep taproots.

Seed rates and plant population
Seed size varies considerably between varieties and seed types and should always be taken into account. For conventional winter varieties 80–120 seeds/m^2 should be sown to give 50–70 plants/m^2 in the spring resulting in a seed rate of 5–7 kg/ha. For the hybrids, 60–70 seeds/m^2 should give 40–50 plants/m^2 in the spring; the seed rate in this case is about 3–5 kg/ha. For spring varieties the conventional varieties should be sown at 150 seeds/m^2 and the hybrids at 120 seeds/m^2 to give the required higher plant populations and result in similar seed rates per hectare as the seeds are smaller in size.

For the hybrid varieties, the breeder or seed company will provide appropriate advice.

Fertilisers
To return those nutrients removed from the soil 14 kg of P_2O_5 and 11.0 kg of K_2O should be applied for each tonne of seed/ha expected yield (e.g. at 3.5 t/ha, 49 kg/ha of phosphate and 38.5 kg/ha of potash are needed) if the straw is incorporated. If the straw is removed the rate should be increased to 15.1 kg/ha of P_2O_5 and 17.5 kg/ha of K_2O for each tonne of seed harvested (e.g. at 3.5 t/ha, 52.8 kg/ha of phosphate and 61.2 kg/ha of potash are needed).

The nitrogen rates depend on SNS index (page 56) and the possible use of organic manures. Unlike winter wheat a 30 kg/ha application of seedbed nitrogen may be justified for winter oilseed rape on soils with 0–2 SNS index. Spring nitrogen rates may vary from 0–220 kg/ha according to SNS index. For rates less than 100 kg/ha the whole dressing may be applied in late February or early March. If more than 100 kg/ha are to be applied split dressing is recommended with half in late February or early March and the remainder by late March or early April.

For spring oilseed rape rates vary from 0–120 kg/ha according to SNS index. All the nitrogen can go in the seedbed except on light sand soils with rates greater than 80 kg/ha where 50 kg/ha should be applied in the seedbed and the remainder by early May.

If sown too early or at too high a seed number, or excessive nitrogen is available, winter rape may develop too much leaf and this may result in a yield reduction. Current research advocates a lower seed number than used conventionally and assessing spring fertiliser nitrogen requirement from crop size (Green Area Index, GAI) which indicates amount of nitrogen already in the crop (about 40 kg/ha per unit of GAI) and future soil supply predicted by measuring the Soil Mineral Nitrogen (SMN) in the top 90 cm of soil in February. Most crops need a total nitrogen supply of 150–200 kg/ha for optimum yield. However, in deciding rate of fertiliser nitrogen required to make up any shortfall a recovery of 60% should be assumed.

In some crops, e.g. on light soils with a high pH, boron deficiency may occur (stunted growth, curled leaves with rough mid-ribs and some stem cracking), 20 kg of borax or 10 kg of 'Solubor' per hectare can be applied to the seedbed (if a deficiency has been diagnosed) or sprayed on the leaves in early spring. Oilseed rape has a high sulphur requirement and much of the UK is now at high risk of sulphur deficiency because of the reduction in deposition from the atmosphere. Where a deficiency has been diagnosed 50–75 kg/ha SO_3 (20–30 kg of elemental sulphur) should be applied in the early spring as a sulphate-containing fertiliser.

Seed treatments

Iprodione/thiram protect against *Alternaria* and damping-off diseases and tend to be used as standard. Betacyfluthrin plus imidacloprid is a new insecticidal seed treatment for the control of cabbage stem flea beetle in winter oilseed rape with the added advantage of reducing aphid attack (they may transmit beet western yellows virus) and slug damage and treated seed should be drilled to ensure soil cover.

A follow-up pyrethroid spray may be required in October/November if scarring by the larvae is found on more than 70% of leaf petioles. (Scarring may be seen as brown or brownish-purplish pitting on the upper surface of the leaf petiole.)

Pests

Slugs may be a serious problem of winter oilseed rape particularly in trashy or cloddy seedbeds or when late established. Methiocarb or equivalent should be applied where there is a high risk and method of establishment may influence the timing of this application. The peach-potato aphid may transmit beet western yellows virus into the crop and a pyrethroid spray may be required in October/November.

In the spring, seed weevil and bladder pod midge may require control and in spring rape, pollen beetle may also be a problem (Table 6.1).

Pigeon damage

This can be very serious in some years, especially in areas with a high pigeon population, near woods, where only a small area of rape is grown and where there is little else for the pigeons to eat in the winter.

Although the crop is often eaten in late autumn and early winter, the greatest damage is caused by grazing in late winter and early spring when the new shoots and buds are developing. This leads to uneven flowering and ripening and creates problems for timing of spraying and harvesting, as well as loss of yield and a lower oil percentage. Many sound and visual scaring devices such as bangers, kites, balloons and scarecrows are available and when used they should be moved around from time to time. Shooting is also helpful, particularly when the weather is severe and in the more vulnerable periods. A full crop is always likely to be less damaged than a thin patchy crop.

Rabbit damage

This is an increasing problem during winter where rabbit populations are getting larger. Electric fencing and gassing are possible solutions.

Diseases

Light leaf spot and *Phoma* may be a problem in both autumn and spring time and varietal resistance should be taken into account. A prediction system is available for light leaf spot and a rape pest and disease prediction service is being developed (Password) funded by the HGCA and other organisations. Light leaf spot should be sprayed if bleached spots (no black picnidia) are seen in the autumn/winter. *Phoma* has leaf spots with black picnidia and early attacks, before the end of October, are most damaging and should be treated with e.g. tebuconazole when 10–20% of plants are affected, with a follow-up spray 4–6 weeks later. An attack at later growth stages may be treated with the same group of fungicides which may also give useful control of *Alternaria* and *Sclerotinia* (Table 7.1).

Weeds

An early-drilled vigorous crop may smother many weeds; however, if managing the crop to optimise canopy size then weed control may require extra attention. Treatment should be targetted to difficult weeds like cleavers. In some situations volunteer cereals may be a problem and should be effectively controlled. For broad-leaved weeds a range of herbicides such as propyzamide and benazolin are available either alone or in combination with another active ingredient. For some herbicides, restrictions apply for broadcast crops and weather conditions. For grass weeds, herbicides such as tepraloxydim (Aramo) and fluazifop-P-butyl are available but careful choice is necessary if herbicide resistant grasses are present and restrictions may apply (Table 4 in Appendix 9).

Growth stages

A knowledge of the growth stages of the oilseed rape plant will be helpful in deciding the best time to treat the crop for a specific problem (Table 16.1).

Table 16.1 Growth stages in oilseed rape

	Definition	Code
0	Germination and emergence	
1	Leaf production	
	Both cotyledons unfolded and green	1,0
	First true leaf	1,1
	Second true leaf	1,2
	Third true leaf	1,3
	Fourth true leaf	1,4
	Fifth true leaf	1,5
	About tenth true leaf	1,10
	About fifteenth true leaf	1,15
2	Stem extension	
	No internodes ('rosette')	2,0
	About five internodes	2,5
3	Flower bud development	
	Only leaf buds present	3,0
	Flower buds present but enclosed by leaves	3,1
	Flower buds visible from above ('green bud')	3,3
	Flower buds level with leaves	3,4
	Flower buds raised above leaves	3,5
	First flower stalks extending	3,6
	First flower buds yellow ('yellow bud')	3,7
4	Flowering	
	First flower opened	4,0
	10% all buds opened	4,1
	30% all buds opened	4,3
	50% all buds opened	4,5
5	Pod development	
	30% potential pods	5,3
	50% potential pods	5,5
	70% potential pods	5,7
	All potential pods	5,9
6	Seed development	
	Seeds expanding	6,1
	Most seeds translucent but full size	6,2
	Most seeds green	6,3
	Most seeds green-brown mottled	6,4

Table 16.1   *(Cont'd)*

	Definition	Code
	Most seeds brown	6,5
	Most seeds dark brown	6,6
	Most seeds black but soft	6,7
	Most seeds black and hard	6,8
	All seeds black and hard	6,9
7	Leaf senescence	
8	Stem senescence	
	Most stem green	8,1
	Half stem green	8,5
	Little stem green	8,9
9	Pod senescence	
	Most pods green	9.1
	Half pods green	9,5
	Few pods green	9,9

Note: To estimate later leaf stages judge the number of lost leaves by their scars. Note that stages from bud to seed development should normally apply to the main stem and seed development stages should normally apply to the lowest third of the main inflorescence. Otherwise, branch position on the inflorescence should be stated. Senescence stages apply to the whole plant.

Harvesting

Shedding losses may become significant if harvesting of oilseed rape is delayed. Consequently, whilst some growers combine direct most prefer to either desiccate or swath the crop. The drying-out period for desiccated crops is about 10 days compared to 15–20 days for swathed crops. However, it is important that swathed crops be combined when fit or shedding losses may again be significant. Combining early or late in the day or during dull weather may reduce shatter losses.

Drying and storage

The moisture content of the seed at harvest time will be in the range 8–15% for swathed crops and 10–25% for direct-combined crops. The contract price is usually based on 9% moisture content and so it should be dried to a slightly lower percentage. Damp rape seed must be dried as soon as possible, either by a continuous drier (not above 60°C, and then cooled quickly) or by bulk drying on the floor or in bins. The undried seed should not be piled more than 1.25m deep, and the drying ducts covered with hessian.

Normally, dry cold air is used for drying, but some heat may be required to lower the relative humidity to 70% so that the seed dries to 8% moisture content. When rape seed is being conveyed in trailers or lorries, great care should be taken to block all holes through which it might escape.

Yield

Winter oilseed rape has a yield range of 2.0–4.5 t/ha with 3.25 t/ha being average.

For spring oilseed rape the yield range is 1.0–3.0 t/ha with 2.00 t/ha being average.

16.3   Linseed and flax

Linseed and flax are different varieties of the same plant which has been grown in this country since Roman times. Figure 16.1 shows the main difference between the two varieties. Linseed is short-strawed with capsules (bolls) to give a higher yield of seed than flax which is long-strawed and grown for its fibre.

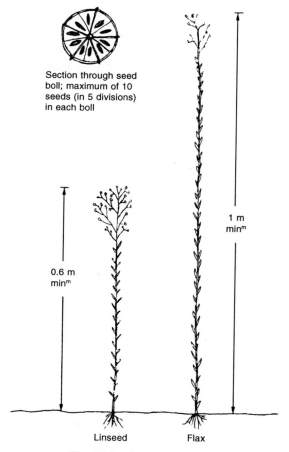

Fig. 16.1   Linseed and flax.

16.3.1   Linseed
The seed contains about 40% oil and the residue is a high protein animal foodstuff.

The drying oil produced from linseed is mainly used for making paints, putty, varnishes, oil cloth, linoleum, printer's ink, etc. Over the last 20 years the introduction of plastics and latex paints and the use of other oils have reduced its importance. However, with a move towards more 'natural' products the use of linseed oil has increased. Better dyeing techniques have meant that linoleum can be made bright and attractive to compete with modern PVC floor coverings. Unlike PVC linoleum is biodegradable and it does not give off poisonous gases when burnt. The majority of varieties are spring sown but developments in plant breeding have resulted in the introduction of autumn-sown varieties and of edible oil (linola) types which are usually grown on buy-back contracts. There is also a small but expanding market for high quality seed to be used for human consumption for which traditional varieties are suitable.

The reduction in arable area payments has reduced the attractiveness of linseed as a crop and the future may depend on the creation of an industrial crops regime including linseed.

Soils and climate
Linseed can be grown in any of the arable areas of the UK but moisture-retentive soils are preferable because of the small root system of the plant. The pH should be about 6.5.

Rotation
Because of the comparatively small area grown, pests and diseases are not generally a serious problem. Nevertheless, linseed should not be grown more often than one year in five.

Varieties
Modern linseed varieties are a great improvement on older varieties, especially with regard to yield, shorter and less fibrous straw, earlier and more uniform ripening, less shattering of seed and disease resistance. Current varieties include *Oscar*, *Biltstar* and *Jupiter* for spring linseed with edible oil varieties such as *Windermere* and *Rydal* whilst *Oliver* and *Fjord* are autumn-sown varieties (HGCA UK *Recommended List for Oilseeds*).

Seed and sowing
The seed should be treated to control damping-off and drilled 1–2 cm deep into a fine, firm and moist seedbed. For autumn-sown crops 800–950 seeds/m^2 should be sown in late September/early October. For spring-sown crops 650–750 seeds/m^2 should be sown in late March/early April to obtain an established population of about 500 plants/m^2. Seed rates will range from 40–65 kg/ha because of variation in seed size between varieties.

Fertilisers
On fertile soils there is little response to nitrogen, phosphate and potash and

at SNS index 1 the crop has a nitrogen requirement of 50–80 kg/ha, phosphate 55 kg/ha and potash 50 kg/ha.

Weed, disease and pest control
A well-grown crop is essential to reduce the likelihood of problems. Crop protection products are available but generally with 'off-label' approval and advice for the best options should be obtained, for example from the Pesticide Safety Directorate (PSD). Flea beetle control may be a problem because gamma-HCH has been withdrawn (Table 6.1).

Flowering
Most varieties have pale blue (a few have white) flowers which appear in early July. The plant flowers early in the morning, but only for a few hours. The flowering period for the crop may last several weeks. It is normally self-pollinated.

Harvesting
Linseed is ready for harvesting when the whole plant is dry, the stems are yellow-brown and the leaves have fallen from the base. The seeds should rattle in the bolls, be plump and brown and, for direct combining, show the first signs of shattering. Linseed straw is very tough and wiry and so the knife must be kept sharp. The crop may be desiccated before combining, especially where there are weeds. The secret of successful desiccation is achieving penetration below the canopy of capsules so that the chemical can dry out the stems. Higher than normal volumes of water should be used and 'crop tilters' are useful as long as they do not damage the crop.

Plant breeders are attempting to produce varieties which have their capsules near the top of the crop. The cutter bar of the combine may then be set high to keep as much of the fibrous stem as possible out of the machine – this will reduce blockages and tangles. The long straw can then be ploughed under after harvest and allowed to rot. Combining should only be carried out when the crop is dry and a stripper header is advantageous.

Yield
Linseed has a yield range of 1.0–2.5 t/ha with an average yield of about 1.5 t/ha.

Drying and storage
The seed must be carefully dried to about 8 % moisture content for safe storage. It is harvested at about 12–16 % moisture. Linseed is a small slippery seed which can easily fall through tiny holes; this must be watched carefully at all stages, i.e. at sowing, combining, in trailers and in the drier and storage buildings.

16.3.2   Flax

This is grown mainly for the fibres in the stem which are present as long bundles around a woody core. Traditionally, flax has been grown for the linen market (long fibre) whereas UK-produced fibre is classified as short fibre and is used for textile blends and for industrial markets. There has been an increased interest in both long and short fibre but as with linseed the reduction in area support has had an adverse effect on the crop area.

The growing of the crop is similar to linseed but in the UK the crop may be harvested early by mowing or later, after desiccation, by combine therefore also obtaining some seed. After cutting it is important for the crop to 'ret' which is a natural process which results in the woody part of the stem being easily removed. This retting in the field is very important and in-field conditioners are available to speed up the process.

16.4   Sunflowers

Sunflower plants grow successfully throughout the UK in gardens and coverts for game birds, but the agricultural crop is restricted mainly to the south eastern part of England, mainly south of a line drawn from the Wash to Exeter. The crop is restricted in area because of its need for warmth to reach maturity and its dislike of wet conditions during seed ripening.

Modern sunflower varieties are potentially high-yielding, semi-dwarf varieties with a requirement from drilling to harvest of 1300 to 1400 accumulated day degrees (over a base of 6 °C). They are produced in a similar way to maize hybrids, by using male sterility and restoring gene technique.

The area sown in the UK is very small, approximately 500 hectares, yielding up to 3.5 t/ha. The decorticated (dehusked) seed contains over 40% of valuable, high quality edible oils which can be used in food manufacture, margarine and cooking oils, but the major part of UK production goes into pet foods because the soft shells of the seed make it suitable for native birds.

A well-drained soil with a pH of over 6.0 is essential and the crop is best sown after cereals or fallow. A rotation of one year in five or six is recommended to reduce the risk of infection from *Sclerotinia sclerotiorum*. Fertilisers used are 50 kg/ha each of nitrogen, phosphate and potash and should be applied to the seedbed. Too much nitrogen can cause lodging, favour disease development, delay maturity and lower seed oil content.

The crop is sown in late April/early May when soil temperature in the top 10 cm is 10 °C. Sowing rate depends on seed size and the aim should be to achieve 9–15 plants/m^2 using either a precision or air drill at a depth of 2.5–5.0 cm in 25–50 cm rows. Herbicides are available for broad-leaved weed control, although inter-row hoeing can be successful.

The crop is ready for harvesting in mid-September/October when seed has reached 30% moisture content. At this stage the back of the head is yellow and the bracts have begun to turn brown. Desiccation is not used. A conventional cereal combine can be used with similar settings as for harvesting beans. Seed is dried on a drying floor at a maximum depth of 1 m using cold air until the moisture content reaches 15%. After cleaning, seed should be dried to 9% using heat.

16.5   Soya beans

These are potentially attractive to replace imports particularly with the concern about using products from genetically modified crops. New varieties from Eastern European breeding programmes are now available and have potential for production in southern England and the seed should be inoculated with *Rhizobium* bacteria. In the EU the crop is supported under the oilseeds area payment scheme.

16.6   Evening primrose

The seed contains about 20% oil which is a very rich source of gamma-linolenic acid (GLA) and this is highly valued for certain pharmaceutical uses. The market for evening primrose oil is very small and is dominated by Chinese and Eastern European production, although the plant is still grown on a small scale in the UK. It is recommended that this crop should only be grown on contract for a reputable end-user.

16.7   Borage

Borage is a naturalised annual plant which has grown wild or in gardens for many centuries. It was developed as a crop plant for its seed, with an oil content of about 30% which is very rich in GLA. It therefore competes with evening primrose oil and again, the crop should only be grown on contract for a reputable end-user.

16.8   Combinable pulses

More than 40 species of grain legumes belonging to the Leguminosae family are grown throughout the world and unlike other crops do not need nitrogen fertiliser because the *Rhizobium* bacteria found in nodules on the roots are able to fix atmospheric nitrogen. They are excellent break crops providing a yield advantage in the following crops which benefit from the residual nitrogen. The nutritional value of the grain legume seeds is of major importance in both human and animal nutrition providing protein, starch and fibre. The grains of these crops are traded as pulses but this term excludes leguminous oilseeds such as soya beans and under EU regulations this crop is supported as an oilseed.

The EU imports about two-thirds of its vegetable protein requirement and the area sown to pulse crops is less than 4% of the arable area. EU support resulted in an increase in area of these crops in the late 1980s to 1.2–1.5 million ha with a decline in the last few years.

The EU divides pulses into two categories:

1   Protein crops – peas, field beans, lupins.
2   Other grain legumes – chickpeas, lentils, vetches.

In the 1990s the EU became the world's major producer of dry peas with France providing 70% of total production. China is the major producer of field beans but Australia and the UK, the major EU producer, are the main exporters. Australia is the major producer and exporter of lupins whilst India is the major producer of pulses, mainly dry beans, chickpeas and lentils but has to import from countries such as Australia, Canada and Turkey to try and meet demand. In the EU Spain is the main producer of other grain legumes.

In the UK, the main pulses grown are the various types of beans and peas with a developing interest in lupins. The grain from these crops is primarily used for animal feed but premium markets are available for human consumption, mainly peas, and for pigeon feeding mainly maple peas and small seeded beans. Favourable area payments have maintained interest in these crops but yield can be variable and harvesting difficult. The current interest in traceabililty of food products, the aversion to GMO's and the interest in organic farming may encourage the production of these crops. There is also now an interest in using these crops, mainly peas, for whole-crop silage.

16.8.1   Field beans

Soils and climate

Winter field beans do well on the heavier soils provided they have a good structure and are well drained. In the north of the country there is an increased risk of frost damage and later harvest. Spring beans are more suitable on medium soils but may then suffer from drought.

Place in rotation

Field beans are a good break from cereals and are often followed by winter wheat which benefits from the residual nitrogen. To reduce the risk of building up persistent soil-borne diseases such as footrots, field beans and related crops should not be grown within five years of themselves or each other. For spring beans, in addition, stem rot (*Sclerotinia sclerotiorum*) may be a problem and host crops such as linseed and oilseed rape should therefore have the same restrictions.

Seedbeds
Requirements are similar to those for cereals but ideally a little deeper. As a result of soil type and establishment method, the autumn seedbeds are fairly rough and may impair the activity of soil-acting herbicides.

Time of sowing
Winter beans are sown mid-October to mid-November. If sown too early in a mild autumn, the resultant soft growth is easily damaged by hard frosts and may be attacked by an early infestation of chocolate spot disease. This early spread of the disease can subsequently be difficult to control. Spring beans may be sown as soon as soil conditions permit after early February. This early sowing may reduce the effect of potential drought and may result in earlier harvesting.

Method of sowing
Winter beans may be ploughed-in at 10–15cm deep or may be broadcast on the ploughed surface and covered by harrowing. The spring crop in usually drilled up to a depth of 7.5cm, the required depth being influenced by choice of herbicide.

Varieties
These are detailed in the *NIAB Pulse Variety Handbook*. Winter beans are generally large seeded and are marketed for animal feed. Spring beans are usually smaller seeded and also used for animal feed but some varieties may attract a premium if suitable for specialist markets. *Victor* is the preferred variety for export for human consumption whereas *Maris Bead* has small rounded seed suitable for the pigeon trade. The newer white-flowered tannin-free varieties are more attractive to some feed compounders and for on-farm use for non-ruminants but may be lower yielding.
    Examples of recommended varieties:

	Winter	Spring
Coloured flower	*Target*	*Meli*
	Clipper	*Quattro*
	Striker	*Compass*
White flower	*Silver*	*Alpine*
		Lobo

Seed rates

The usual range for winter beans is 160–200 kg/ha and 160–250 kg/ha for spring beans. A number of factors will affect the number of seed sown and therefore the seed rate:

1 Thousand seed weight (TSW) (range 330–570 g).
2 Seedbed conditions and time of sowing, e.g. higher seed rate when sowing late in the autumn or early in the spring.
3 Plant population. Winter beans require an established plant population of 18–20 plants/m^2 post-winter and the chosen seed rate should allow for 20–25 % field loss. For spring beans the required population is about 40 plants/m^2 with a field loss of 0–5 %. Dense populations tend to increase disease problems whereas thin crops are less competitive to weeds. In some areas rooks may reduce plant populations.

Seed rate may be calculated as follows:

$$\frac{TSWg \times target\ population/m^2}{\%\ germination} \times \frac{100}{(100 - field\ loss)} = seed\ rate\ kg/ha$$

Fertilisers

To replace the nutrients removed from the soil 11.0 kg of P_2O_5 and 12.0 kg K_2O should be applied for each tonne of expected grain yield (e.g. at 3.5 t/ha, 38.5 kg of P_2O_5 and 42 kg of K_2O will be needed). However, allowance should be made for nutrients applied as organic manures. No nitrogen fertiliser will be required, but a magnesium fertiliser will be required at soil index 0 and 1.

Seed treatment

Certified seed is advised but if using home-saved seed it should be tested for germination and seed-borne leaf and pod spot (*Ascochyta fabae*) infection. Low levels of this disease may be treated with thiram and thiabendazole. It is recommended that all field bean seed is tested for stem and bulb nematode (*Ditylenchus dipsaci*) and should not be used for seed if nematodes are present.

Weeds

A sound rotation with good weed control in the previous crops particularly of grass weeds and perennial broad-leaved weeds will reduce the need and cost of weed control in field beans. A range of herbicides is available but there is only one approved post-emergence herbicide. Therefore, the pre-emergence herbicide should be chosen with care and there may be soil type or sowing depth restrictions and the field may need to be ploughed before sowing the next crop (Table 4 in Appendix 9).

Pests

Slugs may be a serious problem for winter beans especially where the crop is late sown in a cloddy seedbed and methiocarb or equivalent should be used. Rooks and pigeons may also damage the crop and bird scarers or other methods of control may be required.

Bean seed beetle (Bruchid beetle) may be a particular problem in crops for seed or human consumption and may be controlled by deltamethrin. This insecticide may also be used for the control of pea and bean weevil which tends to be more of a problem in spring beans and a monitoring system is available to forecast spray timing. Black bean aphid is also mainly a problem of spring beans and may be controlled by pirimicarb (Table 6.1). However, if insecticides are to be used when the crop is flowering then consideration should be given to protecting the bees which are useful in cross-pollinating the flowers.

Diseases
Wide rotations and the use of healthy seed will reduce the risk of some diseases whereas others are more influenced by weather conditions and for chocolate spot, mainly a problem in winter beans, early drilling and high populations may predispose the crop to infection. A range of approved products are available such as chlorothalonil plus cyproconazole for chocolate spot control and early infection by the disease will require treatment. There are no products approved for the specific control of leaf and pod spot (Table 7.1).

Harvesting
Winter beans are usually ready for harvesting in August or September; spring beans from late August on. They ripen unevenly, lower pods first, and although the leaves wither and fall off early combining has to be delayed until the stems lose their green colour. If required, a desiccant such as diquat (Reglone) may be used when at least 90% of pods are dry and black, particularly in weedy crops. Loss of shed seed when combining may be reduced by combining in dull weather or in the morning or evening. Beans for seed and specialist markets such as human consumption or pigeons must be carefully harvested and handled.

Drying and storing
Field bean seeds are large and must be dried carefully to ensure the moisture is removed from the centre of the seed and to avoid cracking. In continuous dryers, seed of over 20% moisture should be dried in two stages. On-floor dryers may be successfully used but if the ducts are widely spaced some beans near the floor may not dry properly as the air escapes easily. For long-term storage, beans should be dried to 14% moisture content and the marketing standard is normally 14% moisture and 2% impurities or a combination of the two up to 16%.

Yield
The yield in t/ha is as follows:

	Average	High
Winter beans	3.5	5.0
Spring beans	3.25	4.5

16.8.2   Dry harvested peas

Soils and climate
Peas grow best on well-drained loams and lighter soils. The pH should be about 6–6.5; if it is too high, manganese deficiency is likely and may reduce the quality of the harvested sample. Lodging may be a problem where fertility is high and so suitable varieties should be chosen.

Compared to the other pulses, peas are more tolerant of drought stress and high rainfall may make combining difficult and result in stained produce which may be less marketable. In dry situations irrigation may be of benefit at flowering and again at the pod-swelling stage.

Place in rotation
Peas should not be grown more often than 1 year in 5 and should be kept 4 years apart from other pulses, oilseed rape and linseed to avoid the build-up of soil-borne diseases and pests.

Seedbed
Timely ploughing will allow the preparation of a seedbed with a minimum number of cultivations because peas are sensitive to over-compaction and do not require a too fine tilth.

Time of sowing
As soon as possible after mid-February but seedbed condition and soil temperature are important considerations. Winter peas should be sown in early November.

Method of sowing
Narrow rows are advisable as this tends to result in better competition with weeds, higher yields and easier combining. Seed should be covered with about 3 cm of settled soil after rolling.

Varieties
The main market for combining peas is for animal feed and high yield is important. For the specialist markets such as human consumption, pigeons and micronising for pet food, choice of an appropriate variety is the main consideration and the premium will need to compensate for the lower yield. Disease resistance,

standing ability and ease of combining are particularly important in a wet season and the newer semi-leafless and tare-leaved varieties tend to be better for these characteristics. Varieties are detailed in the NIAB *Pulse Variety Handbook.*

Types of peas

- *Marrowfat.* This is the main type used for human consumption, the seeds are large, dimpled with a blue-green seed coat. A good colour and freedom from waste and stains are required quality factors so timeliness of harvest is important.
- *White peas.* These have round, smooth seeds with white/yellow seed coat. They are mainly used for animal feed but are also sold for use in soups and as split peas.
- *Small blues.* These have small, round, smooth, seeds with a blue-green seed coat. There is limited use for canning as small processed peas.
- *Large blues.* These have large, round, smooth seeds with a blue-green seed coat. Samples with good colour may also be used for micronising to be used for pet food or in packets for human consumption.
- *Maple peas.* These have purple flowers and round brown-speckled seeds. These varieties are grown as whole-crop forage or for the seed to be sold for pigeons.

Examples of recommended varieties:

	Spring		Winter
White peas	*Croma*	*Arrow*	*Blizzard*
Large blue peas	*Espace*	*Nitouche*	
Marrowfat	*Princess*	*Maro*	
Small blue peas	*Flare*		*Froidure*
Pigeon pea	*Minerva*		

Seed rates
The usual range is 200–250 kg/ha or a little higher. However, an adequate plant population is important and TSW and field losses should be taken into account.

Suggested target populations:

				Plants/m^2
Marrowfats e.g.	*Princess*	*Maro*		65
Others e.g.	*Arrow*	*Espace*	*Nitouche*	70
Winter peas				80

Field losses may vary from 20% down to 5% depending on sowing time and seedbed conditions.

The seed rate may be calculated as follows:

$$\frac{\text{TSW (g)} \times \text{target population/m}^2}{\% \text{ germination}} \times \frac{100}{(100 - \text{field loss})} = \text{seed rate kg/ha}$$

Thousand seed weight and germination percentage may be obtained from the seed merchant.

Fertilisers
To replace the nutrients removed from the soil 8.8 kg of P_2O_5 and 10.0 kg of K_2O should be applied for each tonne of grain/ha expected yield (e.g. at 4 t/ha, 35.2 kg/ha of P_2O_5 and 40 kg of K_2O should be applied). No nitrogen fertiliser will be required but a magnesium fertiliser should be applied at soil index 0 and 1. In areas of sulphur deficiency 25 kg/ha SO_3 (10 kg of elemental sulphur) should be applied.

If manganese deficiency occurs, symptoms being yellowing of leaves between veins which remain green, manganese sulphate should be applied, repeated at full flower and again 10–14 days later.

Seed treatment
Certified seed is advisable but there is no certification standard for *Ascochyta* and disease-free seed or thiram plus thiabendazole should be used. Pre-emergence damping-off is prevented by using thiram which should be used on all pea seed. Many varieties are resistant to downy mildew but if this is not the case then fosetyl aluminium may be used (Table 7.1).

Pests
Pea weevil may cause the characteristic 'U' shaped notches around the edges of the leaves but the main damage is as a result of the larvae feeding on the root nodules. A monitoring system is available to predict the need for sprays such as deltametherin. Pea aphids if present in large numbers may result in a severe reduction in yield. A predictive model is available and an approved pyrethroid should be applied when 15 % or more of plants are affected.

Pea moth larvae feed upon the developing seeds reducing the yield and value of the crop particularly for use as seed or for human consumption. A pea moth monitoring service is available and growers may use pheromone traps to aid spray decision making. One or more sprays of an approved pyrethroid will be required (Table 6.1).

Diseases
Crop rotation, varietal resistance and clean or treated seed are all vital for the pea crop particularly for diseases such as *Ascochyta* spp. which cannot be satisfactorily controlled in the growing crop. Some varieties are resistant to pea wilt and some have good resistance to downy mildew, otherwise an appropriate seed treatment should be used. Diseases such as *Botrytis* and *Mycosphaerella* may be a problem during flowering particularly if the weather is wet and may

be treated with an appropriate fungicide such as vinclozolin or azoxystrobin with or without chlorothalonil (Table 7.1).

Weeds

Good weed control is essential in peas which are not a very competitive crop and as well as increasing yield will aid combining and drying of the grain. A range of pre-emergence and foliar-applied herbicides is available but a few varieties are sensitive to herbicides and this should be checked. Where soil type, seed bed conditions and soil moisture are appropriate a pre-emergence herbicide such as cyanazine or pendimethalin is the preferred option. Once the crop is at the recommended growth stage post-emergence herbicides, for example, bentazone plus MCPB and cyanazine should be applied as soon as possible but it is advisable firstly to ensure the pea leaves are well-waxed by using the crystal violet dye test. Volunteer cereal, or grass weeds such as wild oats may be controlled with herbicides such as cycloxydim but if the grass weeds may be herbicide resistant then a specialist should be consulted (Table 4 in Appendix 9).

Growth stages

A knowledge of the growth stages of the pea plant will be helpful in deciding the best time to treat the crop for a specific problem (Table 16.2).

Harvesting

The price of combining peas for human consumption is largely dependent on quality and great care should be taken at harvest. A clean crop may be combined direct whereas if weedy a desiccant may be helpful but first consult the processor if the crop is for human consumption. Peas for animal feed may be combined a little later and drier than for human consumption and if the crop is lodged the combine must be fitted with efficient lifters and it may be necessary to combine one-way.

Drying and storage

As with beans, peas must be dried and stored with care. The relatively large size of the seed makes drying more difficult and low temperature dryers are safer. Peas for human consumption are dried at a lower temperature than peas for animal feed and if moisture content is over 24% a lower temperature should be used and two passes may be more appropriate. For long-term storage, peas should be dried to 14% moisture content and the normal marketing standard is 14% moisture and 2% impurities or, a combination of the two up to 16%.

Yield

The yields in t/ha are as follows:

Average	High
3.75	4.75

Table 16.2   Growth stages in peas
Definitions and codes for stages of development of the pea
These definitions and codes refer to the main stem of an individual plant. Definitions and codes for nodal development refer to all cultivars; only nodes where a stipule and leaf stalk develop are recorded; descriptions in brackets refer to conventional leaved cultivars only.

Code	Definition	Description
Germination and emergence		
000	Dry seed	
001	Imbibed seed	
002	Radicle apparent	
003	Plumule and radicle apparent	
004	Emergence	

Vegetative stage refers to main stem and recorded node. Two small-scale leaves appear first and the nodes where these occur are not recorded.

101	First node	(leaf fully unfolded, with one pair leaflets, no tendrils present)
102	Second node	(leaf fully unfolded, with one pair leaflets, simple tendril)
103	Third node	(leaf fully unfolded, with one pair leaflets, complex tendril)
10×	× node	(leaf fully unfolded, with more than one pair of leaflets, complex tendril found on later nodes)
	Last recorded node	(any number of nodes on the main stem with fully unfolded leaves according to cultivar)

Reproductive stage refers to main stem, and first flowers or pods apparent.

201	Enclosed buds	small flower buds enclosed in terminal shoot
202	Visible buds	flower buds visible outside terminal shoot
203	First open flower	
204	Pod set	a small immature pod
205	Fat pod	
206	Pod swell	pods swollen, but still with small immature seeds
207	Pod fill	green seeds fill the pod cavity
208	Green wrinkled pod	
209	Yellow wrinkled pod	seed 'rubbery'
210	Dry seed	pods dry and brown, seed dry and hard

Senescence stage refers to lower, middle and upper pods on whole plant.

301	Desiccant application stage. Lower pods dry and brown, seed dry, middle pods yellow and wrinkled, seed 'rubbery', upper pods green and wrinkled
302	Pre-harvest stage. Lower and middle pods dry and brown, seed dry, upper pods yellow and wrinkled, seed 'rubbery'
303	Dry harvest stage. All pods dry and brown, seed dry

16.8.3   Lupins

Lupin seed has a protein content of 30–40% with about 12% oil and therefore if the crop could be successfully grown in the UK would compete with imported soya. Modern varieties have been bred for low alkaloid content (sweet lupins) and new more determinate autumn- and spring-sown varieties are available with a yield potential of up to 4 t/ha.

The main commercial interest is in three species:

- White lupin (*Lupinus albus*) with winter and spring varieties.
- Blue lupin (*Lupinus angustifolius*) with spring varieties.
- Yellow lupin (*Lupinus luteus*) with spring-sown varieties.

Lupins require a pH below 7 and the seed should be inoculated with *Rhizobium* bacteria to encourage nodulation. It is important to establish a good plant population and time of sowing is critical for winter varieties. The crop does not require nitrogen and on fertile soils has little requirement for phosphate and potash. Crop protection products are available but generally with off-label approval and advice should be obtained.

Research has been conducted by the Marches Alternative Protein Group (co-ordinated by ADAS Rosemaund) and at IACR (Rothamsted). PGRO are now funding autumn- and spring-sown lupin trials by NIAB and this should supply more detailed agronomic information and response to variations in seasonal weather.

16.8.4   Dried 'navy' beans

These are the beans used for canned baked beans and if successful would have the potential to replace substantial imports. A breeding programme at Cambridge University produced varieties which are well adapted to UK conditions and agronomic requirements have been tested and developed in commercial crops. However, seasonal weather has caused problems and commercial arrangements have not been satisfactory – particularly in that the crop does not qualify for arable area payments.

16.9   Further reading

HGCA, Project reports and topic sheets for oilseed rape.
MAFF, Fertiliser Recommendations RB 209, The Stationery Office, 2000.
HGCA, *UK Recommended Lists for Oilseeds*.
NIAB, *Pulse Variety Handbook* (updated annually).
PGRO, Information and advisory service for pulse growers and users.
PGRO, Notes on growing combining peas.
PGRO, Notes on growing field beans.
CABI, *The UK Pesticide Guide* (updated annually).
Weiss E A, *Oilseed Crops,* 2nd ed, Blackwell Science, Oxford, 2000.

Part 4

Grassland

17

Characteristics of grassland and the important species

17.1 Types of grassland

Grass is the United Kingdom's most important crop and covers about 75% of the land surface. It should be treated as an important natural resource, not just as something which grows in a field! Recent MAFF statistics show the approximate areas of the main types of grassland. They are summarised in Table 17.1. For comparison, the total area of arable crops and set-aside in the UK is about 5 300 000 ha.

Table 17.1 Areas of grassland in the UK

	'000 hectares
Uncultivated Grassland	
'Sole right' rough grazing	4600
Commons	1200
Cultivated Grassland	
Permanent pasture	5450
Temporary grassland or 'leys'	1200

Grassland can be broadly classified as described below:

17.1.1 Uncultivated grasslands
These represent about 47% of the total grassland area. They consist of:

Rough mountain and hill grazings
The plants making up this type of grassland are not of great value. They consist mainly of fescues, bents, *Nardus* (mat grass) and *Molinia* (purple moor grass) as well as cotton grass, heather, bracken and gorse. Burning on a regular planned basis at between 10 and 15 year intervals can have beneficial results, both for grazing stock and wildlife. Traditionally, sheep and beef cattle and ponies were farmed in these areas with grouse shooting and deer stalking increasing in importance from the mid-nineteenth century. More recently, the payment of ewe premium together with the supplements for 'less favoured areas' (LFAs) and the Hill Livestock Compensatory Allowances (HLCAs) have encouraged overstocking, particularly with sheep, and a decline in sward quality. (Area payments replacing the HLCAs were announced in 2000.) In some areas, where the soil is extremely acid, conifer afforestation has been carried out.

Lowland heaths
Heather, bracken and grasses are the main species to be found here. Sheep's fescue is often the dominant grass on the predominantly acid soils. These heaths are to be found in south and east England and some of them have been reseeded. Traditional management involved regular burning and grazing by cattle and ponies, but nowadays many heaths are not grazed at all.

Calcareous downland
This occurs predominantly in southern England. Apart from herbs and broad-leaved flowering plants, grasses such as sheep's fescue and erect brome are found on these chalk and limestone soils. The management of these grasslands involved predominantly grazing by both cattle and sheep. Unimproved areas of calcareous downland are now restricted to the steeper slopes, inaccessible for arable cultivations.

Wetland and fen areas in east and south west England
Any unreclaimed areas are mostly poorly drained and are dominated by water-loving plants such as cotton grass, rushes and sedges. Fens in eastern England, when drained, are associated with intensive arable cropping. In south west England areas such as the Somerset levels have been partly improved by reseeding but are frequently inundated by floodwater.

Maritime swards
These consist of salt marshes and coastal dune areas with marram and cord grasses. The 'machair' areas of coastal Scotland and the Western Isles contain a wide variety of rare plants and offer a unique and valuable habitat for wildlife.

17.1.2   Cultivated grasslands
These represent about 53% of the total grassland area. They consist of:

Permanent pasture

The statutory definition of permanent pasture is grassland which is more than five years old. An alternative one would be 'grassland not normally included in an arable rotation'. The quality of a permanent pasture is dependent mainly on its content of perennial ryegrass and white clover. A sward containing more than 30% of its annual dry matter production as perennial ryegrass would be very productive and would not normally be thought to require much improvement. Other, less productive grasses, such as bents (*Agrostis* spp) and meadow-grasses (*Poa* spp) and a great many more, make up the balance. Although much less productive than the ryegrasses, these constituents perform a useful function in that they create a 'thatch' over the surface of the soil which, if well developed, has the capacity to carry stock physically and reduces the severity of the well-known phenomenon of treading or 'poaching' which can occur in wet conditions. In addition to clover and other legumes, a large number of broad-leaved plants also occur. Many of these, such as buttercup, dock, thistles and ragwort are weeds and ragwort is extremely poisonous. Some others, however, such as yarrow, burnet and ribwort are termed herbs and thought to have beneficial effects on grazing livestock.

Significant areas of species-rich permanent pasture are the subject of statutory protection, e.g. as Sites of Special Scientific Interest (SSSIs). Others have been targeted by one of a number of agri-environmental schemes, such as the Countryside Stewardship Scheme in England. Management strategies for such pastures are normally designed to maintain soil fertility at low or decreasing levels, in order to maintain or increase the species diversity.

Temporary grassland or 'leys'

These are temporary swards which have been sown to grass or grass/clover mixtures for a period of up to five years. A less precise definition is 'grass in an arable rotation'. Leys represent an important opportunity for the restoration of fertility to arable land. Fields which have been grazed, or which have included significant quantities of forage legumes, will contribute substantial amounts of available nitrogen to the following crops. Opportunities exist also for the reduction of some of the weed, pest and disease problems associated with arable farming. Winter wheat grown after a ley will normally yield very well indeed. In organic systems clover leys are one of the main ways in which soil fertility can be enhanced for subsequent arable cropping.

17.2   The nutritive value of grassland herbage

The great value of grass and indeed of other forages is their potential to provide cheap (often the cheapest) sources of energy and protein for farm livestock. Many calculations have been carried out to demonstrate their value and, in particular, the way in which good grassland management, and the achievement of high yields of digestible nutrients, can bring about worthwhile improvements in livestock performance, and a reduction (or complete elimination) of the need to buy in other feeds. Data published recently by Kingshay Farming Trust and by the Milk Development Council clearly demonstrate this and, in particular, point to the very low costs associated with grazing.

17.2.1   The effects of grass maturity

Short leafy grass is rich in protein and highly digestible. As it matures the proportion of yield represented by cell walls increases whereas the proportion of cell contents decreases. This results in an increase in the percentage of fibre and a decrease in the percentage of crude protein with maturity. As a result of the much greater leaf and stem area of the mature plant, and the greater opportunity for photosynthesis, the sugar (or water soluble carbohydrate) percentage also increases with maturity. This explains why it is usually easier to make a well-fermented silage from mature grass and why it is more difficult to do so from young leafy grass.

17.2.2   Digestibility ('D value')

Digestibility is one of the main characteristics which determines the feeding value of herbage plants, i.e. the amount of the plant which is actually digested by the animal. In the past, the digestibility of forages was only determined by *in vivo* animal feeding trials. With the development of the laboratory *in vitro* technique which simulates rumen digestion, it is now possible for the digestibility of any species of herbage plant to be assessed very much more easily and cheaply without using animals directly.

The digestibility of herbage plants is expressed as the 'D value' or DOMD. It is the digestible organic matter expressed as a percentage of the dry matter. High digestibility values are normally desirable as it means that animals are able to obtain large amounts of nutrients from the herbage being fed. It is now accepted that digestibility is a major factor affecting intake. Animals usually eat more of a highly digestible feed. However, other factors, such as sward height and density in a grazing situation and the pH of silages are also important. Dry matter yield increases as the plant grows; the proportion of leaf (the most nutritious and digestible part of the plant) decreases and the proportion of stem (more fibrous, lignified and less digestible) increases. This causes a gradual decline in the D value of the whole plant.

The rate of decline depends on the grass species and the variety. Cocksfoot, for example, always has a lower D value than the other grasses. The D value of particular varieties on any date is usually determined by the earliness or lateness of flowering. The D value of first cuts normally declines at a rate of between 3 and 5 units per week. The rates of decline in Timothy and Italian ryegrasses are slower than in the perennial ryegrasses. D values also differ between grass and legume species. For example, white clover, when young, has a D value of about 80 and the rate of decline is only about 0.8 units per week. Red clover and lucerne are substantially less digestible and their rates of decline are about 2.5 and 2.8 units per week respectively.

When cutting for silage, higher yields will result when the crop is cut at a more mature stage of growth. Many farmers aim to cut grass at or before it achieves a D value of 67. When making field-cured hay, a D value of 60 or even less is a more realistic expectation. This reflects the more mature nature of the crop which can be expected by the time the weather becomes suitable for hay making (if it ever does!).

The conversion of D values to ME (metabolisable energy) data for rationing purposes can be accomplished by the application of the following simple equations:

for fresh and dried grass and hay ME (Mj/kgDM) = 0.15 × DOMD   [17.1]

for grass silage ME (Mj/kgDM) = 0.16 × DOMD   [17.2]

17.3   Identification of grasses

17.3.1   Grass species of economic value

Over 150 different species of grasses can be found growing in this country, but only a few are of any importance to the farmer. They are: short duration ryegrasses, perennial ryegrass, Timothy, cocksfoot and meadow fescue. With the exception of the short duration ryegrasses most of the grasses described in Table 17.2 may be found in permanent pastures, either as indigenous types or bred varieties which were sown when the pasture was established. Before it is possible to recognise plants in a grass field, it is necessary to know something about the parts which make up the plant.

17.3.2   Identification of vegetative parts

Stems

There are two types of stems found on grass plants. They are:

- The *flowering stem or culm*. This grows erect and produces the flower. Most stems of annual grasses are culms.
- The *vegetative stem*. This does not produce a flower, and has not such an erect habit of growth as the culm. Perennial grasses have both flowering and vegetative stems.

Leaves

These are arranged in two alternate rows on the stem and are attached to the stem at a node. Each leaf consists of two parts (Fig. 17.1):

- The *sheath* which is attached to the stem.
- The *blade* which diverges from the stem.

Table 17.2 Recognition of grasses in the vegetative stage

	Short duration ryegrasses	Perennial ryegrasses	Meadow fescue	Cocksfoot	Timothy
Leaf sheath	Definitely split Pink at base Rolled in shoot*	Split or entire Pink at base Folded in shoot	Split Rolled in shoot	Entire at first, later split Folded in shoot	Split Pale at base Rolled in shoot
Blade	Broad Margin smooth Dark green	Narrow Margin smooth Dark green	Narrow Margin rough Lighter green	Broad Margin rough Light green	Broad Margin smooth Light green
Lower side of blade	Shiny	Shiny	Shiny	Dull	Dull
Ligule	Blunt	Short and blunt	Small, blunt, greenish-white	Long and transparent	Prominent and membranous
Auricles	Medium size and spreading	Small, clasping the stem	Small, narrow and spreading	Absent	Absent
General	Veins indistinct when held to light Not hairy	Not hairy	Veins appear as white lines when held to light Not hairy	Not hairy	Base of shoots may be swollen Not hairy

*Identification of hybrid ryegrass is more difficult; it can show a mixture of vegetative features similar to both Italian and perennial ryegrasses. Round tillers predominate.

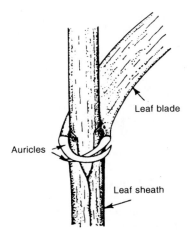

Fig. 17.1   Parts of the grass leaf.

The leaf sheath encloses the buds and younger leaves. Its edges may be joined (entire) or they may overlap each other (split) (Fig. 17.2). If the leaves are rolled in the leaf sheath, the shoots will be round (Fig. 17.3), but when folded the shoots will be flattened (Fig. 17.4).

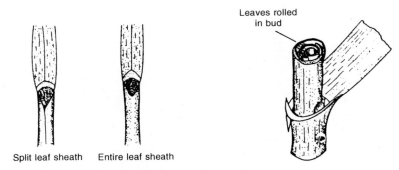

Fig. 17.2   Parts of the grass leaf.   **Fig. 17.3**   Parts of the grass leaf.

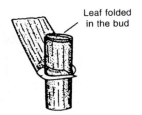

Fig. 17.4   Parts of the grass leaf.

Other structures

At the junction between the leaf blade and leaf sheath is the *ligule*. This is an outgrowth from the inner lining of the sheath (Fig. 17.5). *Auricles* may also be seen on some grasses where the blade joins the sheath. These are a pair of claw-like outgrowths (see Fig. 17.1).

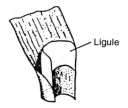

Fig. 17.5   Parts of the grass leaf.

In some species, the leaf blade will show distinct veins when held against the light and, according to the variety, the underside may be shiny or dull.

17.3.3   The inflorescence or flower head
The inflorescence consists of a number of branches called spikelets which carry the flowers. There are two types of grass inflorescence:

* *Spikes,* where the spikelets are attached to the main stem without a stalk (Fig. 17.6).
* *Panicles,* where the spikelets are attached to the main stem by a stalk (Fig. 17.7).

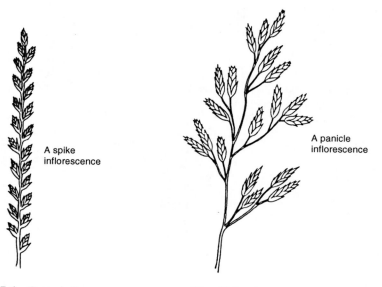

Fig. 17.6   Grass inflorescence.   **Fig. 17.7**   Grass inflorescence.

In some grasses the spikelets are attached to the main stem with very short stalks to form a dense type of inflorescence termed 'spike-like' (Fig. 17.8). The spikelet is normally made up of an *axis*, bearing at its base the *upper* and *lower glumes* (Fig. 17.9). Most grasses have two glumes. Above the glumes, and arranged in the same way, are the *outer* and *inner pales*. In some species these pales may carry *awns* which are usually extensions from the pales (Fig. 17.10).

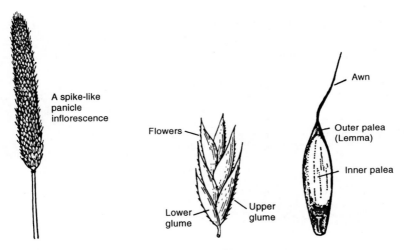

Fig. 17.8   Grass inflorescence.   **Fig. 17.9**   The spikelet.   **Fig. 17.10**   The paleas.

Within the paleas is the flower. The flower consists of three parts (Fig. 17.11):

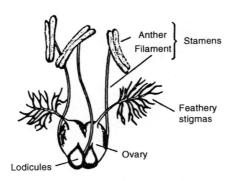

Fig. 17.11   The flower of the grasss.

- The male organs – the *stamens*.
- The female organ – the rounded *ovary* from which arise the feathery *stigmas*.

- A pair of *lodicules* – at the base of the ovary. They are indirectly concerned with the fertilisation process, which is basically the same in all species of plants (Fig. 1.22).

17.4   Identification of legumes

The important perennial forage legumes used by farmers in the UK are red and white clovers and lucerne. Sainfoin, birdsfoot trefoil and alsike are less important. A detailed description of each is given in Table 17.3.

Table 17.3   How to recognize the important perennial forage legumes

	Leaves, etc.	Stipule	General	Species
Mucronate tip	Centre leaflet with prominent stalk	Broad, serrated and sharply pointed	May be hairy	Lucerne
	Leaflets serrated at tip 6–12 pairs of leaflets, plus a terminal one	Thin, finely pointed	Stems 30–60 cm high Slightly hairy	Sainfoin
	Trifoliate (or pentafoliate if including the stipules)	Leaflet-like	Not hairy	Birdsfoot trefoil (yellow and orange/red flowers)
No mucronate tip	Trifoliate, dark green with white half-moon markings on upper surface	Membranous with greenish purple veins Pointed	Hairy	Red clover
	Trifoliate, serrated edge with or without markings on upper surface	Small and pointed	Not hairy	White clover
	Trifoliate, serrated edge, no leaf markings	Long tapering point, never red-veined.	Not hairy	Alsike clover (pink or white flowers)

Leaves
With the exception of the first leaf (which may be simple), all leaves are compound. In some species the mid-rib is extended slightly to form a mucronate tip. Other features on the leaf may be serrated margins, the presence or absence of marks, colour and hairiness (Fig. 17.12 and 17.13). The leaves are arranged alternately

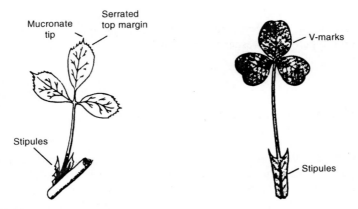

Fig. 17.12   Parts of the legume.   **Fig. 17.13**   Parts of the legume.

on the stem, and they consist of a stalk which bears two or more leaflets according to the species (Fig. 17.14).

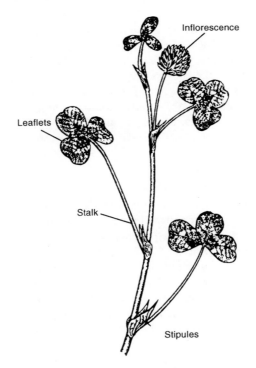

Fig. 17.14   Parts of the legume.

Stipules
These sheath-like structures are attached to the base of the leaf stalk. They vary in shape and colour (Fig. 17.12).

Flowers
The flowers are brightly coloured and, being arranged on a central axis, form an indefinite type of inflorescence (page 401).

17.5   Grasses of economic importance

17.5.1   Types of grass varieties
Details of good quality, reliable varieties from UK and European plant breeders can be found in the *Grasses and Herbage Legumes Variety* leaflet which is published annually by the NIAB. Details of grasses which have been bred for sports surfaces are provided by the Sports Turf Research Institute (STRI) also on an annual basis.

17.5.1.1   Early and late varieties
NIAB classifies perennial ryegrass and Timothy varieties according to date of heading as 'early', 'intermediate' and 'late'. Early heading varieties are usually associated with early spring growth and a less densely tillered plant with an erect habit of growth. These varieties are not very persistent. In the past, they were referred to as hay types. Late heading is associated with densely tillered plants which generally means good persistency. They have a prostrate growth habit and are more suitable for grazing. These varieties used to be called pasture or grazing types. Intermediate varieties combine the advantages of both the previous groups of varieties, namely relatively early spring growth but later flowering and good persistency. They were previously referred to as dual-purpose types.

17.5.1.2   Diploids and tetraploids
The majority of ryegrass varieties are diploids, i.e. they have the normal two sets of chromosomes in their cell nuclei. However, plant breeders have managed to breed some varieties with twice the normal number of chromosomes. These are known as tetraploid varieties (see page 239). NIAB gives details of whether varieties are diploid or tetraploid. Seeds merchants' catalogues usually provide the same information. Tetraploid varieties are often included in mixtures to improve palatability because of their high sugar contents. The dry matter contents of tetraploids are usually lower than those of diploids and their persistency and winter hardiness are also often inferior. As a result of being less 'aggressive' with fewer tillers, tetraploids make excellent companion grasses for legumes, especially white clover. Their high sugar content also makes them ideal for conservation as silage, particularly when the grass is wilted. They are often the first choice too for the very high dry matter (< 65% DM) 'haylage' which has become so popular as a roughage feed for horses. However, when a ley is being established specifically for making hay, it is important to

stipulate diploid varieties, since tetraploids are very difficult to dry well enough to make good hay in UK conditions.

Further information given by NIAB concerns the yields under a cutting or simulated grazing regime, the ground cover index which gives an indication of persistency, and the approximate date for the first cut at '67D'. Useful details are also given about the disease resistance of individual varieties.

17.5.2   Details of individual species
17.5.2.1   Short duration ryegrasses (Fig. 17.15 and Table 17.2)
This group consists of varieties of three species: westerwolds, Italian and hybrid ryegrasses.

Fig. 17.15   Short duration ryegrass.

Westerwolds ryegrass
This is an annual and the quickest growing of all grasses. A good crop can often be obtained within 12–14 weeks of sowing. It should not be undersown. It performs best when direct sown in the spring and summer, as it is not at all winter-hardy, except in very mild districts. NIAB does not recommend any varieties but a currently used one is *Lifloria*.

Italian ryegrass
This is short-lived (most varieties persist for 18–24 months) and very quick to establish. Sown in the spring, Italian ryegrass can produce good growth in its seeding year and an early grazing the following year. For optimum production it is best sown in summer or early autumn and will then produce a full crop

in the following year. It does well under most conditions, but responds best to fertile soils and plenty of nitrogen. Like all ryegrasses, its winter-hardiness is improved when surplus growth is removed in the autumn. Although stemmy, it is palatable with a high digestibility. *Atalja, Trajan* and *Bartissimo* are examples of NIAB-recommended diploid varieties. *Roberta, Danergo* and *Macho* are examples of NIAB-recommended tetraploid varieties.

Hybrid ryegrass
Some varieties in this group are similar to the less persistent Italian ryegrass varieties and others to the perennial ryegrasses. A feature of hybrid ryegrasses is their increased longevity (3–4 years) compared with Italians and good resistance to diseases. Most hybrid varieties are tetraploids. However, annual yields are usually lower than for the Italians.

   Molisto and *Aber Linnet* are examples of currently recommended varieties. *Barsilo, AberExcel* and *Twyblade* are recent introductions with annual yields comparable to those of Italian varieties.

17.5.2.2   *Perennial ryegrass (Fig. 17.16 and Table 17.2)*
Although it forms the basis of the majority of long leys and is the most important grass found in good permanent pasture, perennial ryegrass is, depending on the variety, also used in medium term leys. It is quick to establish and yields well in the spring, early summer and autumn. It does best under fertile conditions and responds well to nitrogen. Crown rust disease is a problem – particularly in the south west of England. The choice of a resistant variety is the best method of control (Table 7.1).

Fig. 17.16   Perennial ryegrass.

Examples:

–Early varieties:

- *Moy* and *Sherwood* are examples of recommended diploid varieties.
- *Sovereign* has good resistance to crown rust.
- *Anaconda* and *Labrador* are examples of recommended tetraploids.

–Intermediate varieties:

- *Fennema* and *AberElan* are examples of recommended diploid varieties;
- *AberSilo* has good resistance to crown rust.
- *Fetione, Rosalin* and *Napolean* are examples of recommended tetraploids.

–Late varieties:

- *Profit, Portstewart, Dromore* and *Twystar* are examples of recommended diploid varieties.
- *Condesa* and *Tivoli* are examples of recommended tetraploids;
- *Elgon* has good resistance to crown rust.

Details of perennial ryegrass varieties suitable for amenity and for sports turf use can be found in the annual recommended list of the Sports Turf Research Institute.

17.5.2.3   *Timothy (Fig. 17.17 and Table 17.2)*

Timothy is fairly slow to establish and is not particularly early in the spring. It is less productive than the other commonly used grasses, but is very palatable, although its digestibility is not as good as that of the ryegrasses. It is often included in grazing leys with perennial ryegrass. Timothy is winter-hardy and does well under a wide range of conditions, except on very light dry soils.

Fig. 17.17   Timothy.

Examples:

- Early varieties: *Promesse, Goliath.*
- Intermediate varieties: *Motim, Comtal.*
- Late varieties: *S48, Intenso, Farol.* (These varieties are becoming outclassed.)

17.5.2.4   Cocksfoot (Fig. 17.18 and Table 17.2)

This is quick to establish and fairly early in the spring. It is a moderately high yielding grass, but unless heavily stocked, most of the existing varieties soon become coarse and unpalatable. Cocksfoot is a deep-rooting grass and therefore often used in dry or drought-prone situations.

Fig. 17.18   Cocksfoot.

Cocksfoot does not need fertile soil conditions and produces well in late summer and autumn. Most varieties show good winter-hardiness. Cocksfoot should be considered as a special-purpose grass on drier, lighter soils in areas of low rainfall. Cocksfoot varieties are not as digestible as the other important grasses.

Example varieties: *Jesper, Sparta.*

17.5.2.5   Meadow fescue (Fig. 17.19 and Table 17.2)

Meadow fescue is not used very much now. This is because the once famous Timothy/meadow fescue mixture is no longer used; it is not so productive as the perennial ryegrass-based sward. Meadow fescue is slow to establish, but fairly early to start growth in the spring and has a high digestibility.

NIAB does not recommend any varieties at present. Examples of those currently available are *Bundy* and *Senu.*

Fig. 17.19   Meadow fescue.

17.5.2.6   Other grasses

Tall fescue is a perennial. Once it is well established, varieties such as *Dovey* and *Barcel* are useful for early grass in the spring, but production after this is mediocre. It is very hardy and it can be grazed in the winter. Because of its long growing season it is sometimes used for green crop drying enterprises.

The small fescue grasses, such as *creeping red fescue* and *sheep's fescue*, are useful under hill and marginal land conditions, and in some situations they will produce more than perennial ryegrass. They have no practical value under lowland farm conditions. They are typical 'bottom' grasses, and they produce a close and well-knit sward. *Chewings fescue* is often included in seed mixtures for lawns and playing fields. *Strong creeping red fescue* as its name implies is a very resilient grass and a frequent constituent of grass horse gallops. It is also sometimes included in grass mixtures for horse paddocks because of its hard wearing capability. *Boreal* is a frequently used variety.

The brome grasses occur in both permanent pasture (e.g. *meadow brome*) or as weeds of arable land (e.g. *sterile* or *barren brome*). Improved varieties of bromes (e.g. *Grasslands Matua*) have been introduced in the past from New Zealand but have proved unsuccessful due to lack of persistency under UK conditions. Some success is currently being achieved in southern England with the variety *Meribel*.

Rough stalked meadow-grass is indigenous to the majority of soils, but it prefers more moist conditions. In the later years of a long ley, it can make a useful contribution to the total production of a sward if white clover has been suppressed by heavy nitrogen fertilising. Rough stalked meadow grass has become an important arable weed on some farms. *Smooth stalked meadow-grass* (Kentucky blue grass) spreads by rhizomes, and can withstand quite dry conditions. It is very persistent and hard wearing and is sometimes included in mixtures for horse paddocks or gallops.

Crested dog's tail is not very palatable because of its wiry inflorescence. These days it is only used in seed mixtures for lawns, playing fields and sometimes for horse paddocks.

The bents. These are very unproductive and unpalatable grasses and, under most conditions, they can be considered as weeds. *Browntop bent* is often included in mixtures for lawn and amenity turf and *Highland bent* is a useful constituent of grass gallops, racecourses and amenity areas.

17.5.2.7   Weed grasses

The majority of other grasses growing naturally in this country are of little agricultural value but some of them are extremely persistent and are able to grow under very poor conditions where the more valuable grasses would not thrive. However, their production is always low, they are usually unpalatable and, under most conditions, they can be considered as weeds. The following are well-known examples of the less productive weed grasses.

Yorkshire fog is an extremely unpalatable grass except when very young. It is especially prevalent under acid conditions where fields have been repeatedly cut for hay. *Barley grass* is a common weed of permanent pastures in southern England. It is also extremely unpalatable. *Tall oat grass* is another very common grass of permanent pastures and hay meadows, as *onion couch* it is also an important weed of arable land. *Couch grass, blackgrass (slender foxtail), wild oat,* and *annual meadow-grass* are all important arable weeds. Detailed descriptions of these grasses are not possible here (see Table 1 in Appendix 9).

17.6   Forage legumes of economic importance

The clovers of agricultural importance are the red and white clovers. Red clovers are used in the short ley for conservation. White clovers are used for the longer duration grazing ley where they act as 'bottom plants'; with their creeping habit of growth they knit the sward together and help to keep out weeds. Although the majority of clovers are palatable, with a high feeding value, they are not so productive as the grasses and they should not be allowed to dominate the sward (aim for 30% clover).

The digestibility of red clover and most other legumes, except white clover, is not as high as that of the grasses, but their voluntary intake at equal digestibility is higher. Because of the ability of legumes to fix atmospheric nitrogen through the action of the *Rhizobium* bacteria in their root nodules, a balanced mixture of grasses and clover enables farmers to economise on fertiliser nitrogen inputs, and will leave a useful legacy of nitrogen for the following crop. This is particularly important in the case of organic farms where clover-based leys are one of the most important ways of naturally enhancing the fertility of the soil prior to arable cropping. Estimates of the quantity of nitrogen supplied to a mixed sward by a strong stand of clover, under UK conditions, vary from 150–200 kg N/ha/year.

17.6.1   Red clovers (Fig. 17.20 and Table 17.3)
Red clovers are usually a constituent of short term leys for cutting for hay or silage. Normally the mixture would also contain Italian or hybrid ryegrasses or Timothy although red clover can be sown alone. Although used more for conservation, some of the more persistent improved varieties are useful for aftermath grazing. In recent years red clover leys have become very popular for fattening store lambs. No breeding work has been undertaken in the UK for many years but the Institute of Crops and Environment Research (IGER) at Aberystwyth has recently started a new programme. There are two main types:

- *Early red clover. Merviot* is the most commonly used variety. It has good

Fig. 17.20   Red clover.

resistance to *Sclerotinia* (clover rot, Table 7.1), but is susceptible to stem
eelworm (Table 6.1). *Milvus* is a recent introduction from Europe which is
showing great promise.

- *Late red clover.* These varieties are later to start growth in the spring than
  the early red clovers. *Grasslands Pawera* and *Britta* yield about 10% less
  than the early varieties but are more persistent, lasting anything up to four
  years. They also demonstrate good resistance to *Sclerotinia* and stem
  eelworm.

17.6.2   White clovers (Fig. 17.21 and Table 17.3)

These should be regarded as the foundation of the grazing ley. They are not
so productive as the red clovers but are much more persistent. There are four
types, classified according to leaf size, and also a blend.

Fig. 17.21   White clover.

- *Very large-leaved white clovers.* The only clover in this group is *Aran*. It
  has become popular in recent years due to its ability to survive in conditions
  of high fertiliser nitrogen application and is suitable for inclusion in mixtures
  for dairy cow grazing and cutting. *Aran* is resistant to slug damage but shows
  poor resistance to *Sclerotinia* infection (Table 7.1).
- *Large-leaved white clovers. Alice* is one of the most popular varieties from
  this group and it is useful in a similar context to *Aran*.

- *Medium-leaved white clovers.* These are normally included in leys where a variety of different managements are likely. *AberHerald* is a variety which claims earlier spring growth than many others. *Menna* is one of the most persistent varieties.
- *Small-leaved white clovers.* These are used in the long ley and are particularly favoured for sheep grazing. They are rather slow to establish but can become dominant. *Kent wild white* and *S184* are the most popular varieties in this category.
- *Clover blends.* Since the management applied to leys containing white clovers can vary so much (sheep/cattle grazing, silage/hay cutting) it has become common practice to include blends of varieties of white clover, from several of the categories described above, in ley mixtures. This practice is designed to ensure that one or more varieties can contribute well under each of the different management systems imposed.

17.6.3 Lucerne (or alfalfa) (Fig. 17.22 and Table 17.3)
This is a very deep-rooted legume and it is therefore useful on dry soils, although it can be grown successfully under a wide range of soil conditions provided drainage is good. For best establishment lucerne should be sown in the spring (May is the ideal month). Undersowing or direct sowing after winter barley harvest are less satisfactory. Lucerne can be sown alone or with a companion

Fig. 17.22 Lucerne.

grass such as cocksfoot or meadow fescue. It yields best, however, as a single species stand. In the UK lucerne is mainly used for silage or haymaking but it is also very suitable for zero grazing in dry summers. Lucerne also forms the basis of several large-scale green-crop drying ventures. The main problems associated with lucerne growing in the UK concern *Verticillium* wilt, a soil-borne fungal disease (Table 7.1) and stem nematode. A variety which shows good resistance to both is *Vertus*. French varieties such as *Daisy* are being introduced on farms where stem eelworm is known to occur.

17.6.4   Sainfoin (Fig. 17.23 and Table 17.3)

The growth characteristics of this most traditional of forage legumes are similar to those of lucerne. It yields less well than lucerne or red clover but can do reasonably well on calcareous soils. Traditionally it was used for hay (especially valued for horses) and the aftermath growth used for fattening lambs. Its most important characteristic is that of being a non-bloating legume. The most suitable companion grass is meadow fescue.

- *English giant.* This is high yielding but short lived, usually for just one harvest year.
- *Common sainfoin.* There are several traditional varieties, e.g. *Hampshire common* and *Cotswold common*. A new introduction *Emyr* is showing great promise and yields almost as well as lucerne.

Fig. 17.23   Sainfoin.

17.6.5   Birdsfoot trefoil or lotus (Table 17.3)

Although indigenous in many lowland pastures *Birdsfoot trefoil* is becoming increasingly popular as a constituent of grazing mixtures alongside clovers. In common with sainfoin it is a non-bloating legume and also has some anthelmintic properties. Examples of current varieties are *Leo* and *Maku*.

17.6.6   Alsike (Table 17.3)

Alsike is sometimes included as a substitute for part of the red clover content of a ley. It grows better in poor soils than red clover. It should never be used for horse hay as it can cause photosensitivity and even liver damage.

17.6.7   Other forage legumes

Wild red clover and *Yellow suckling clover* are indigenous species in many pastures but of low productivity. *Crimson clover* and *Black medick* (also known as *Trefoil* or *Yellow trefoil* but not to be confused with the perennial *Birdsfoot trefoil*) are sometimes sown as 'winter annuals' for subsequent hay or silage crops. Sometimes these species are mixed with Italian ryegrass, or even, in an organic situation especially, sown as a green manure crop to increase the soil nitrogen supply for a spring-sown crop.

17.7   Herbs

These are deep-rooting plants which are generally beneficial to pastures. However, to be of any value they should be palatable, in no way harmful to stock, and they should not compete with other species in the sward. They have a high mineral content which may benefit the grazing animal.

    Yarrow, chicory, rib grass and *burnet* (Fig. 17.24–17.27) are the most useful of the many herbs which exist. They can be included in a seed mixture for a grazing type of long ley. They are not cheap, however, and one or more of them may well establish in a sward of its own accord. *Chicory* is occasionally sown as a crop specifically for fattening lambs and a New Zealand variety *Grasslands puna* has recently been introduced into this country.

17.8   Grass and legume seeds mixtures

17.8.1   Traditional mixtures

Grasses and legumes may be sown as single species or even single variety swards but it has become normal (sometimes for poorly defined reasons) to

Fig. 17.24   Yarrow.

Fig. 17.25   Chicory.

sow mixtures. It is of course possible to create a rationale for this phenomenon. So, for example, the mixing of species or varieties which exhibit desirable

Fig. 17.26   Rib grass.   **Fig. 17.27**   Burnet.

factors such as high yields, early spring growth, palatability, persistency and winter hardiness all in the same field would seem to offer the farmer the combined benefits of all of them. Add to this further factors such as drought tolerance, suitability for cutting or grazing by a variety of different stock, to say nothing of the desirability of clover and herbs in a mixed sward and it is possible to devise a mixture of such complexity, and with so many constituents, that it is likely that none of the desired characteristics would be demonstrated to any worthwhile degree. However, this was quite common practice for many years and so, for example, the traditions of the 'Cockle Park' and 'Clifton Park' mixtures survive even to this day. Indeed, the need to re-establish some of the species-rich grasslands so rashly discarded over the last 50 years has created a resurgence of interest in them.

17.8.2   Modern seeds mixtures

The following sections contain examples of mixtures which would be suitable for mainstream agricultural and organic use. Examples of good quality varieties recommended by NIAB for general use have already been given in sections 17.5 and 17.6. The reader should not feel restricted by these however, and it is strongly recommended that the current issue of the NIAB *Grasses and Herbage Legumes Variety* leaflet together with information from reputable seeds houses, be consulted for the most up-to-date information. Organic producers should also check for the availability of organically produced seed, and use it if available.

Short-term mixtures

Westerwolds ryegrass is only suitable for a one-year crop and is normally sown in the spring for grazing or cutting in the same year. The majority of short-term (one or two year) mixtures incorporate Italian ryegrass varieties because of their very high yield potential. They are suitable for early grazing and for cutting for hay or silage. The inclusion of early perennials or hybrids is often practised in order to extend the potential of the ley into a third year. However, yields will almost certainly suffer in year three as the Italians will have died out. Tetraploid varieties with high sugar contents are favoured for silage and haylage production. Where it is intended to make hay regularly only diploid varieties should be chosen. The most suitable types of clover are the large- or very large- leaved white clovers and/or red clover. Example mixtures are shown in Table 17.4.

Table 17.4 Examples of short-term leys (1–2 years)

Plant	Ley characteristics
35 kg/ha westerwolds ryegrass or 25 kg/ha westerwolds ryegrass 10 kg/ha Italian ryegrass ———— 35 kg/ha (14 kg/acre)	A one-year ley ideal for sowing in spring; not suitable for undersowing; it should produce two good grazings or silage cuts within the year.
32–40 kg/ha Italian ryegrass* or 16–20 kg/ha Italian ryegrass* 16–20 kg/ha hybrid ryegrass* ———— 32–40 kg/ha (13–16 kg/acre)	These leys can be productive for up to two years; usually autumn-sown after cereals or maize; suitable for early bite grazing and subsequent silage cuts.
16–20 kg/ha hybrid or Italian ryegrass* 10 kg/ha early red clover ———— 26–30 kg/ha (10.5–12 kg/acre)	Suitable for conventional or organic use over two years; suitable for silage, hay and aftermath grazing; select diploid ryegrass varieties for haymaking.

*Blends of diploids and tetraploids; where tetraploid varieties are predominant use the higher seed rate.

Medium-term cutting/grazing mixtures

This type of ley typically will be taken for one or more silage cuts and followed by grazing. Early, intermediate and late perennial ryegrasses will all be suitable but obviously early heading perennials would be favoured for early cutting. These mixtures also contain late perennial ryegrasses which are most suited to grazing. It is normal to include both tetraploid (more palatable and better for making silage) together with diploid (more persistent) varieties. A white clover blend including both large- and very large-leaved and medium-leaved varieties would be suitable. Example mixtures are given in Table 17.5.

Table 17.5   Examples of medium-term leys (up to 5 years)

Plant	Ley characteristics
15 kg/ha early perennial ryegrass* 10 kg/ha intermediate perennial ryegrass* 5 kg/ha late perennial ryegrass* 5 kg/ha Timothy 2.5 kg/ha large-/very large-leaved white clover 37.5 kg/ha (15 kg/acre)	A 'dual-purpose' type mixture suitable for one or two cuts of silage followed by aftermath grazing. For later cutting the proportion of early perennial ryegrass varieties could be reduced and replaced with intermediate and late varieties.
12 kg/ha tetraploid hybrid ryegrass 12 kg/ha intermediate perennial ryegrass 10 kg/ha late perennial ryegrass 2.5 kg/ha white clover blend 5 kg/ha early red clover 41.5 kg/ha (17 kg/acre)	A ley suitable for inclusion in an organic rotation with a high proportion of both red and white clovers. This ley could be cut for silage and then grazed. Alsike clover could be substituted for part of the clover content. It can grow well in low fertility situations. Alsike must not be used in leys intended for horse hay.

*Blends of diploid and tetraploid varieties.

Long-term grazing mixtures
Intermediate and late perennial ryegrasses predominate in such mixtures and some tetraploids and Timothy can be included for improved palatability. However, the lower persistency of the tetraploids precludes their large-scale use in these mixtures. A white clover blend with small- medium- and large-leaved varieties should be used especially where grazing may be carried out by sheep as well as cattle. Example mixtures are given in Table 17.6.

Mixtures including lucerne and sainfoin
Lucerne and sainfoin may be sown with a grass companion but there is some evidence that lucerne, in particular, yields better as a single species sward. Red clover, cocksfoot and meadow fescue are suggested as suitable species for inclusion. Examples of suitable mixtures are shown in Table 17.7.

Table 17.6   Examples of long-term leys (over 5 years)

Plant	Ley characteristics
10 kg/ha intermediate perennial ryegrass* 20 kg/ha late perennial ryegrass* 5 kg/ha Timothy 2.5 kg/ha medium-/large-leaved white clover blend 37.5 kg/ha (15 kg/acre)	A long-term mixture suitable for intensive grazing by dairy cows.
30 kg/ha late perennial ryegrass* 3 kg/ha small-leaved white clover 33 kg/ha (13.5 kg/acre)	An extreme long-term pasture type ley for grazing by sheep.
7.5 kg/ha intermediate perennial ryegrass 20 kg/ha late perennial ryegrass 10 kg/ha strong creeping red fescue 2.5 kg/ha smooth stalked meadow-grass 40 kg/ha (16 kg/acre)	A long-term mixture suitable for horse paddocks. White clover has not been included as it can become dominant and indigenous clover often establishes in hard grazed areas.
5 kg/ha early perennial ryegrass 5 kg/ha intermediate perennial ryegrass 15 kg/ha late perennial ryegrass* 3.75 kg/ha late red clover 3.75 kg/ha white clover blend 2.5 kg/ha birdsfoot trefoil 35 kg/ha (14 kg/acre)	A longer term mixture suitable for organic use. Birdsfoot trefoil is included to reduce the problem of bloat. It may also have some anthelminthic benefit.

*Blends of diploid and tetraploid varieties.

Table 17.7   Examples of mixtures incorporating lucerne and sainfoin

Plant	Characteristics of mixture
90 kg/ha sainfoin (unmilled seed) 6 kg/ha cocksfoot or meadow fescue 96 kg/ha (39 kg/acre)	Where milled seed is available a substantial reduction in the weight of sainfoin seed sown would be possible. This mixture could last for about 3 years.
20 kg/ha lucerne (8 kg/acre)	Timothy, cocksfoot or meadow fescue are suitable companion grasses for lucerne. 5 kg/ha of lucerne could be substituted by 10 kg/ha of grass seed. Lucerne will persist for up to 5 years in UK conditions.
12 kg/ha lucerne 7 kg/ha early red clover 19 kg/ha (7.75 kg/acre)	Sowing these two legumes together helps to compensate if pest or disease damage reduces the contribution of either. Red clover will die out after about 2 years.

All of these mixtures would be suitable for use on organic farms

17.9   Further reading

Andrews J and Rebane M, *Farming and Wildlife: a Practical Management Handbook*, Royal Society for the Protection of Birds, 1994.

Frame J, *Improved Grassland Management*, Farming Press, 2000.

Frame J, *Temperate Forage Legumes*, CAB International, 1998.

Hague J and Hutchinson M, *Forage Costings for Home Grown Crops*, Kingshay Farming Trust, 1999.

Hopkins A (ed.), *Grass: its Production and Utilization*, 3rd ed., Blackwell Science, 2000.

Hubbard C E, *Grasses: a Guide to their Structure, Identification, Uses and Distribution in the British Isles*, 3rd ed. (revised by J C E Hubbard), Penguin Science, 1984.

Kingshay Farming Trust, *Grass and Clover Mixtures for Grazing and Silage*, Kingshay Farming Trust, 1998.

Milk Development Council, *Winter Sward Management*, Milk Development Council, 2000.

Monsanto, *Renewed Grass for Improved Milk and Meat Production*, 1998.

NIAB, *Grasses and Herbage Legumes Variety* leaflet, (Annual).

Nature Conservancy Council, *Effects of Agricultural Land Use Change on the Flora of Three Grazing Marsh Areas*, 1989.

Palmer J, *Future for Wildlife on Commons. Part I: The report; Part II: The case studies*, 1989.

Sheldrick R D, Newman G and Roberts D J, *Legumes for Meat and Milk*, Chalcombe Publishing, 1995.

Sheldrick R D (ed.), 'Grassland management in the environmentally sensitive areas'. *Proceedings: Symposium*, Lancaster, September 1997.

Sports Turf Research Institute, *Turfgrass Seed*, STRI (Annual).

18

Establishing and improving grassland

18.1 Establishing leys

Newly established leys containing almost 100% of improved grass and legume varieties constitute the most productive form of grassland and so it is appropriate to commence this chapter with a section on their establishment.

18.1.1 Terminology
The following refer to the various practices associated with sowing grass seeds:

- *Direct sowing*. Sowing grass seeds in spring or autumn without the cover crop used for undersowing.
- *Undersowing*. Sowing grass seeds with a cover crop, e.g. spring cereals or peas, at a reduced seed rate, which may be taken for grain harvest or wholecrop silage, leaving the undersown ley established in the stubble. Sometimes a light cover crop of cereals or forage rape may be grazed.
- *Direct reseeding*. Sowing grass seeds again in a field which has previously contained grass with no arable break.
- *Sod seeding*. A form of direct drilling often associated with the improvement of an old ley or permanent pasture by the introduction of new species without cultivation.
- *Oversowing*. The introduction of new species by surface broadcasting, in some cases associated with raking (for example using the 'Vertikator' or 'Einbock' machines).

18.1.2 Spring sowing

If direct sowing without a cover crop, the maiden seeds can give valuable production in the summer. Establishment can be enhanced by stock being able to graze the developing sward within a few weeks of sowing. This is not possible to the same extent with autumn sowing. A limiting factor with spring sowing may be moisture, and in the drier districts the seeds should be sown in March if possible. The plant should thus establish itself sufficiently well to withstand a possible dry period in early summer. Undersowing in a cereal crop in the spring means that the greatest possible use is being made of the field, although with the slower growing grasses, establishment may not be so good, especially in a dry year.

18.1.3 Late summer/early autumn sowing

Leys are normally sown at this time of year and provided satisfactory establishment is achieved it ensures top yields in the following full harvest year. There is usually some rain at this time; heavy dews have started again, and the soil is warm. When clovers are included in the seed mixture, sowing in August is preferred before the onset of frosts. Whether earlier sowing is possible depends on when the preceding crop in the field is harvested. Winter barley is the ideal crop to follow. Sowing later than September is not advised.

18.1.4 Direct sowing

Reference has already been made (page 196) to the cultivations necessary for preparing the right type of seedbed for grass seeds. But it must be re-emphasised that grass and clover seeds are small, and should be sown shallow, and that a fine, firm seedbed is necessary. With ample moisture the seed can be broadcast. This should be on to a ribbed-rolled surface, so that the seeds fall into the small furrows made by the roller. Most fertiliser distributors can be used for broadcasting, and seeds and fertiliser are sometimes sown together. Seeds may then be covered with a light harrow or chain harrow or simply rolled in.

In the drier areas, and on light soils, drilling to about 2 cm is safer. The seed is then in much closer contact with the soil. The 10 cm coulter spacing of an ordinary cereal seed drill should give a satisfactory cover of seeds, but for an improved result, cross drilling could be considered. After broadcasting or drilling, rolling will complete the whole operation. For those with the time to spare and seeking the best possible result, drilling the grass seeds and subsequently broadcasting the clover seeds should give the best chance of good establishment to both. At soil index 2 50 kg/ha each of phosphate and potash can be broadcast (up to 120 kg/ha at index 0) and worked in during the final seedbed preparation. For spring sowing, 60 kg/ha of nitrogen is also advised. Some farmers also use some nitrogen (<40 kg/ha) on seeds sown in August but this may not always be necessary and may adversely affect the establishment of clovers. For full details of fertiliser recommendations for ley establishment see Tables 18.1 and 18.4.

18.1.5 Direct drilling

Grassland can be direct reseeded by direct drilling. Although it is no cheaper than conventional reseeding, the practice does allow a much quicker turn-round from the old grass to the new sward. It also reduces poaching. Paraquat or, more frequently, glyphosate, are desiccant herbicides which can be used to kill off or suppress the existing vegetation prior to the introduction of the new seeds. Cropping with spring-sown direct drilled kale, turnips or forage rape prior to direct drilling with grass seeds in late summer or autumn ensures the best possible establishment from this technique. Another option, particularly if the old ley is thick and matted, is to spray in the autumn and allow the old sward to die off during the winter months prior to direct drilling in the spring. Rolling immediately following drilling is particularly beneficial and control of slugs, leatherjackets and frit fly is almost always necessary.

'Sod seeding' sometimes also referred to as 'strip seeding' usually refers to the direct drilling of new seeds without the complete kill of the old sward. Specialist drills have been evolved for this purpose which in some instances incorporate a 'band' spraying facility sufficient to kill off or suppress the old sward adjacent to the strip where the new seeds have been sown. Sod seeding is a useful way of introducing white clover into an established perennial ryegrass pasture.

18.1.6 Undersowing

Good establishment can be achieved when undersowing spring cereals. The cereal is sown first, usually at about three-quarters of the normal seed rate and followed immediately by the seed mixture drilled or broadcast, followed by light harrowing or harrow combing and rolling. This is the best sequence of operations for slower-establishing long-term leys. With more vigorous species such as Italian ryegrass and red clover, it is often preferable to broadcast the seed after the cereal is established and at about the three-leaf stage. This helps to preclude excessive growth of the undersown species which, particularly in a wet season, may seriously interfere with the harvesting of the cereal.

Another useful technique is to take the cover crop of cereals for wholecrop cereal silage for which the arable area payment may still be claimed on eligible land. An alternative which has been used in recent years is undersowing in peas (grain or forage varieties are both suitable) which may also be taken for silage. Weed control in undersown crops is sometimes difficult, especially if clovers have been included in the mixture, and it is essential to use a legume-safe herbicide. If peas are being used as the cover crop then a stale seedbed prior to sowing is the preferred option. Undersowing in autumn-sown cereals is not normally advised due to the competitive nature of the cover crop. However, in an organic cropping situation with a less densely tillered cereal crop, undersowing in spring has been shown to be successful in conjunction with operations such as harrow combing for weed control which can also rake in and cover the grass seeds.

18.1.7 Weed, pest and disease control in establishing leys

Seedling diseases such as 'damping off' (*Pythium*) may adversely affect the establishment of some grass varieties. Some seed merchants offer seed treated with a fungicidal dressing which may be well worth while, particularly when establishing grasses late in the season or under less than ideal conditions.

Particularly important pests are slugs, which can cause damage at any time of year, but particularly in wet conditions, and almost always when sowing in the autumn or direct drilling. Seeds can be supplied with slug pellets already mixed in or slug pellets can be applied to the surface of affected fields after test baiting. Leatherjackets, larvae of the crane fly or 'daddy longlegs', can cause massive damage and complete crop failure especially when direct reseeding (grass after grass) is being practised in the spring. Frit fly larvae can also be present in old turf in enormous numbers and can seriously damage grass seeds sown at any time of year. Both of these pests can also be present in significant numbers in arable situations, particularly in grassy stubbles. Control of both with chlorpyrifos spray is currently approved.

Control of annual broad-leaved weeds in new grass leys is relatively straightforward. The most important point, of course, is to remember to use a legume-safe herbicide when spraying grass/clover mixtures. The preferred material for lucerne is 2,4-DB but it is sometimes difficult to obtain. The most important weed to control is chickweed, particularly in autumn-sown leys as this can grow extremely vigorously if left unchecked and causes significant loss of plant through competition and smothering.

18.1.8 Early management

Leys can often be grazed about eight weeks after sowing if conditions are suitable. Grazing an August-sown ley with sheep in October can be very beneficial as it encourages the grasses to tiller. Grazing a spring-sown ley in June or July with cattle should be possible if conditions are dry enough. Grazing in wet conditions should be avoided, however, as it will inevitably lead to poaching, a loss of sown species and the early ingress of weeds. When weed growth is not severe a light topping will often control annual weeds adequately and precludes the need to use a herbicide.

18.2 Grassland improvement and renovation

18.2.1 Problem identification

This section will deal mainly with the improvement and renovation of permanent pastures which have deteriorated, although this obviously does not preclude operations designed to improve the composition of an old ley. Direct reseeding, as described above, is an expensive operation and should only be carried out when absolutely necessary. Evidence of deterioration should be sought from the species composition and from the condition of the soil. The problems may be obvious (e.g. infestation with such plants as docks, thistles, nettles and ragwort) or slightly more difficult to discern (e.g. a general lack of productivity) in comparison with other pastures.

The first step should be to undertake a detailed botanical analysis of the species present in the field so far as is possible. The obvious points to note are the presence or absence of significant quantities of perennial ryegrass. If it is possible to identify, say 25% + of the grass tillers in the sward as perennial ryegrass then it should be possible to improve the sward by changing the management. If, on the other hand, there is hardly any perennial ryegrass present, then some form of complete sward renewal by direct reseeding or, at least, a renovation operation, may be necessary. However, before any operations are undertaken it is necessary to check on a few of the important 'basics'.

18.2.2   Drainage

Evidence of poor drainage should be fairly easy to find. Obviously standing water and rushy patches are important indicators; also blocked ditches and drainage outfalls which are failing to run. Digging a fairly shallow pit for soil examination may be worthwhile. The grey and rusty red mottling in the topsoil known as 'gleying' will indicate seasonal waterlogging and a probable need for improved drainage although it may not be necessary to lay a complete new pipe system. Frequently there are drains in a field already which just need unblocking or renewing. In other cases, where the soil is clayey it may be possible to set out a 'mole' system (page 188), leading the mole channels into main drains covered with permeable backfill. Improving the drainage of a wet grass field has been shown to be capable of improving dry matter yields by up to 30% and increases the length of the grazing season by up to a month.

18.2.3   Soil pH

Acidity can also have a major effect on the productivity of grassland. In the 'Park Grass' classical experiment at the Institute for Arable Crops Research (IACR) Rothamstead it has been shown that maintaining a soil pH of between 6.5 and 7.5 increases grass dry matter yields by up to 30% when compared to a pH of 5. Furthermore, in the most acid areas the species composition has deteriorated to almost 100% Yorkshire fog. Correcting the soil pH by the application of agricultural lime is very straightforward and cost-effective.

18.2.4   Phosphate and potash

Routine soil samples should be taken to check on soil phosphorus and potassium as well as pH. Visual evidence may include grass with very poor stunted growth and purple discolouration in the leaves. This is indicative of a deficiency of phosphate. Potash deficiency is often well-demonstrated by the deep green colour and vigorous growth of grass growing in dung and urine patches compared to an otherwise poor sward with a pale green colour. Correction of low (index 0 or 1) phosphorus and potassium levels should be carried out as recommended by MAFF (2000) in Reference Book 209 *Fertiliser Recommendations*. These recommendations are summarised in Table 5 in Appendix 9.

18.2.5 Control of perennial weeds (see also page 107)

The control of perennial weeds such as docks, thistles, nettles and ragwort by the use of an effective herbicide can have a markedly beneficial effect on the productivity of the sward. It can also (in theory) be enforced under the provisions of the Weeds Act (1959). The information contained in Table 5 in Appendix 9 indicates the appropriate herbicides for use on established grassland. Particular care should be taken where clovers or other perennial forage legumes are a valued sward component, to select a suitable legume-safe herbicide. Regular topping to prevent weeds setting seed is also important (a mature dock plant can set up to 60 000 viable seeds in a year). In the case of ragwort there is often no alternative to pulling by hand, removing and burning, since it is so poisonous and becomes very attractive to animals when it has been sprayed or mown. Sheep grazing during the autumn and winter periods can also help to control ragwort plants in their first winter 'rosette' stage. For badly infested fields a machine (the 'Eco-pull') has been designed to ease the burden of hand-pulling.

18.3 Improving a sward by changing the management

In the case of a sward which contains significant quantities of perennial ryegrass it should be possible to improve the species composition by management. The first step is to ensure that the basic good management practices, as set out above, are in train. Then increase the annual application of nitrogen fertiliser to near optimum with a corresponding increase in the intensity of utilisation, either by intensive grazing or by cutting early for silage or, preferably, both. Avoid cutting late for hay as this enables the poorer grasses to come to maturity and to set seed. Autumn grazing by sheep, although it may to some extent reduce the earliness of the spring flush in the following year, will certainly reduce the incidence of the poorer bents and meadow-grasses. It is important to avoid both undergrazing, overgrazing and excessive poaching. Regular defoliation on a rotational basis to a sward height of 6–8 cm, plus judicious inputs of nitrogen fertiliser, should substantially favour the perennial ryegrass component of the sward over time and lead to a gradual improvement.

18.4   Improving a sward by renovation

In the case of a sward which contains few species of value it will be necessary to introduce new seeds, specifically of perennial ryegrass and white clover, and to encourage them to establish and thrive. The advantage of this technique is that it is cheaper than complete sward renewal through a direct reseeding operation. Also it is possible to retain much of the 'thatch' supplied by the old turf which is so valuable in physically carrying stock and reducing poaching. Such operations are usually most successful in August. September- sowing can lead to disappointing establishments of both grass and clover.

There are two basic approaches. One is to introduce the new seeds by way of one of the sod-seeding machines described above under the section on direct drilling. Examples of these machines currently in use are the Hunter rotary strip seeder, the Vreedo slit seeder and the Opico seeding rake. A degree of sward suppression may be achieved by an application of glyphosate, or the sward may be severely grazed prior to the operation. The application of a suitable compound fertiliser, rolling, and the application of pesticides to control slugs, leatherjackets and frit fly larvae as previously described should complete the operation.

The other approach involves a very severe grazing and/or vigorous surface cultivation to expose soil and obtain a tilth. This can most conveniently be undertaken with disc harrows. Broadcasting seeds and fertilisers then takes place and the operation is concluded by rolling and pesticide application as previously described.

18.5   Fertilisers and manures for grassland

Grass, like all other crops, needs mineral nutrients for its establishment, maintenance and production; nitrogen, phosphorus and potassium are the most important. However, before discussing these major nutrients it is essential to remind the reader of the importance of lime.

18.5.1   Liming grassland

The importance of liming has already been referred to in the section on grassland improvement and renovation. Soil acidity is probably the biggest single factor adversely affecting the productivity of grassland and it is one of the most simple to correct. The ideal pH for permanent pastures is around 6.0 whilst for establishing new seeds the soil should be limed to a pH of 6.5. The ideal pH for lucerne is 6.5–7. Agricultural limes are relatively cheap and extremely cost-effective. The usual dressing is about 5 t/ha, and the materials that are used on arable land are equally suited to grassland. Magnesian or dolomitic lime can, in addition, have a useful effect on the soil magnesium level.

Other materials sometimes favoured for use on grassland include calcified seaweed and granular lime. Like any other liming materials they should be evaluated solely on the basis of their neutralising value (NV). Such an evaluation may make them appear expensive; however, availability in small bags and the ability to apply through a farm fertiliser spreader may make them attractive for very small areas such as horse paddocks. When seeding a new ley it is always advisable to have the soil tested for lime and, should it be necessary, the best advice is to lime on the ploughed furrow and work it into the seedbed before the seeds are sown. When direct drilling into an old sward it is advisable to check the pH of the top 2–3 cm of soil as this is frequently more acid than the rest of the soil profile.

18.5.2 Nitrogen

18.5.2.1 Optimum nitrogen

Nitrogen is the most important nutrient required by grasses. When applied in the form of organic manures or inorganic fertilisers, nitrogen has an immediate and very positive effect on grass growth provided there are no other limiting factors. It is often possible to observe very strong correlations between the application of nitrogen fertilisers, farm stocking rates and ultimate profitability. Nitrogen has been the central pivot to the development of intensive grass-based livestock systems and has also been responsible for the creation of some environmental problems.

A great many trials have been carried out to observe the extent to which grasses respond to nitrogen. In the absence of any other limiting factors such as poor drainage, soil acidity or lack of one of the other important nutrients, it is probably true to say that summer rainfall and soil type have the major impact on this response. In that most readable (still relevant, but, sadly, out of print) book *Milk from Grass* (Thomas *et al*, 1991) the authors set out various hypotheses based on the 'GM 20' series of trials. One of the main suggestions was the subdivision of the UK into five 'site classes' according to their theoretical ability to grow grass. These are set out in Table 18.2. Responses to nitrogen fertiliser inputs for the five site classes are summarised in Fig. 18.1. This gives very general indications of the likely optimum annual levels of nitrogen fertiliser input under different soil and climatic conditions. It also provides useful pointers as to the potential dry matter yields of perennial ryegrass based swards under these conditions. The 2000 edition of MAFF Reference Book 209 gives slightly different definitions of grass growing conditions, but the same basic analysis and a very similar range of annual application rates for nitrogen fertiliser.

It is, of course, true to say that there are many more factors impinging on the farmers' decisions about the optimum rate of nitrogen fertiliser in particular circumstances. A full discussion of the soil nitrogen status, the contribution of clover, sward management (cutting or grazing) and the use of FYM or slurry, could fill the rest of this book. However, decision support systems are being

Table 18.1   Fertiliser nitrogen recommendations for grazed swards and swards cut for silage

Date	Grazing swards	Swards cut for silage
February/ March or T° sum 200	Apply 60 kg/N/ha; only apply in February where March grazing is possible; if grazing is carried out in March apply a further 60 kg N/ha.	Apply 40–60 kg N/ha; if an 'early bite' grazing is taken follow this with a further 40 kg N/ha.
April/May	Continue to apply nitrogen monthly at 60 kg N/ha or up to 75 kg N/ha where soil N supply levels are low; apply after each grazing on a rotational system; where set stocking is practised either apply monthly to the whole area or weekly to 25% of the area.	Apply the balance of N to bring the total up to 120 kg N/ha (150 kg N/ha where soil N supply levels are low); this application *must* be carried out more than 6 weeks before the planned date of cutting; as soon as the first cut is cleared apply a further 100 kg N/ha for the second out.
June/ July/ August	Extra grazing is normally introduced at this time after 1st cut silage; continue to apply 40–60 kg N/ha monthly to the whole area until the target N level is reached; in very dry conditions (typical of site classes 4 & 5 in some years) nitrogen application should be discontinued; it is not necessary to apply nitrogen fertiliser after mid-August.	For 3rd and 4th cuts apply a further 80 kg N/ha for each cut; in very dry conditions further significant growth is unlikely before September and these applications should be omitted.

Notes
(1) Urea is an acceptable material to use in the period February–May. After May ammonium nitrate usually gives better results.
(2) For beef and sheep grazing the optimum N levels are usually about 50 kg/ha below those given in Table 18.1. At low stocking rates or where significant contributions from clover are expected nitrogen applications may be as low as 100–150 kg N/ha.
(3) For hay cuts nitrogen levels of 60–75 kg N/ha are usually sufficient. Higher levels will give rise to heavy cuts which will be difficult to dry prior to baling.
(4) Slurry is best applied to the cutting fields and care is needed to avoid contamination of the crop. Always take account of the likely contribution of nutrients from this source when deciding on individual applications.
(5) For ley establishment between April and mid-August apply up to 60 kg N/ha. For autumn establishment nitrogen is not usually necessary.
(6) Where the clover content of a sward is high then applications of nitrogen should be restricted, in order to maintain it, to no more than 50 kg N/ha in March followed by a further 50 kg N/ha in August.
Based on MAFF RB 209 (2000)

Table 18.2 Definition of grass growth site classes

Soil texture	Average April–September rainfall (mm)			
	> 500	425–500	350–425	< 350
All soils except shallow soils over chalk/rock or gravelly and coarse sandy soils	1	2	2	3
Shallow soils over chalk or rock and gravelly and coarse sandy soils	2	3	4	5

Based on Thomas *et al., Milk from Grass*, 1991.

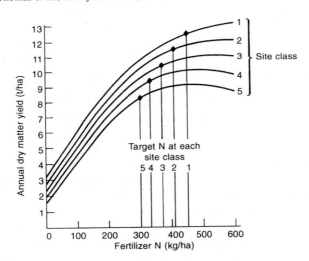

Fig. 18.1 Responses of grass to nitrogen at five site classes. (From Thomas *et al, Milk from Grass*, 1991.)

developed in order to remove some of the guesswork from this most important of calculations (in particular about the soil nitrogen status). Soil mineral nitrogen (SMN) levels can be easily estimated and such information is being increasingly used by farmers to inform their decisions.

18.5.2.2 Clover
In the absence of fertiliser nitrogen, clover will make a useful contribution to the requirements of the grass crop. The *Rhizobium* bacteria in the root nodules fix atmospheric nitrogen and, as they die and are replaced, some of this nitrogen is made available for the companion grass plants. The amount of nitrogen fixed will be proportional to the amount of clover in the sward, as well as to pH, soil temperature and rainfall. With about 30% of a mixed perennial ryegrass sward present as white clover, a reasonable expectation is for a contribution

of up to 200 kg nitrogen/ha. If nitrogen fertiliser is applied, then this contribution may well decrease although it is common for farmers with clover rich swards to apply some nitrogen (say 50 kg/ha) early in the season, before the growth of clover commences, and a similar quantity in August. The vigorous large- and very large-leaved white clover varieties such as *Alice* and *Aran* will survive well in a nitrogen-fertilised sward and continue to supply the nutritional benefits for which they were also sown.

18.5.2.3   Organic manures and slurry

On most farms, with grass-based livestock enterprises, it is necessary from time to time to apply organic manures or slurry to the grass area. These are best applied to the conservation (hay or silage) areas where their contents of phosphate and potash will be most valuable in replenishing that removed by cutting. They may also be applied to fields intended for forage maize, kale, reseeding, etc. It is important, from an environmental as well as an economic point of view, to take account of the nutrient content of these materials when calculating the optimum inputs of inorganic fertilisers. Organic manures and slurry are some of the most important sources of nitrate contamination of surface and ground water. Within Nitrate Vulnerable Zones (NVZs) there are specific regulations governing the amounts of nitrogen in organic form which can be applied to agricultural land and the times of year when they may be applied. An important reference for this situation is the DEFRA publication, *Guidance for Farmers in Nitrate-Vulnerable Zones* (PB3277).

Nitrogen losses from organic sources can be minimised by application in the spring and summer rather than the autumn and winter, and such practices as soil injection will reduce volatilisation losses. Nitrification inhibitors have been investigated but their effects are often variable, and their widespread use is precluded by high costs. Dilution, acidification and separation are also techniques which have been investigated and which show some promise.

18.5.2.4   Timing of nitrogen applications

Grass growth is very seasonal with the main period of rapid growth in the UK falling between April and June. It is sensible therefore for the main applications of nitrogen fertiliser to be made during this period. Two-thirds of the target nitrogen application before the end of May is a good rule of thumb. The usual principle is for fairly large or frequent applications early in the season followed by smaller applications after each cut or grazing. Some suggested targets and programmes are set out in Tables 18.1 and 18.3.

The timing of the first nitrogen application is always an area for discussion. The UK Farming Press has taken it upon itself to foster the 'T° sum' calculation each spring. This is simply a calculation of positive day degrees Celsius based on air temperature and running from the 1 January each year. The most advantageous time to apply nitrogen is believed to be that at which the sum reaches 200. Scientific evaluation of the technique in the UK has proved very

Table 18.3 Target grass dry matter yields and optimum nitrogen rates in different grass growing conditions

Grass growth site class	Optimum annual application rates of N fertiliser (kg/ha) assuming moderate levels of soil N supply	Target dry matter yield from perennial ryegrass (t DM/ha)
1	420	12.7
2	380	11.6
3	340	10.5
4	300	9.5
5	300	8.4

In conditions of low soil nitrogen supply, responses may be obtained up to a *maximum* of 460 kg N/ha for site class 1 and 420 kg N/ha for site class 2.
Based on Thomas *et al, Milk from Grass,* 1991 and MAFF RB 209, 2000.

inconclusive. However, since the date of T° sum 200 often coincides with a period in February or March when soil conditions are suitable for fertiliser application, many farmers adhere to it. Monitoring soil temperatures is also quite straightforward, and application when the soil temperature at 10 cm reaches 5 °C is advocated by some.

The date advised for the latest application of nitrogen is also a matter under current discussion. The advice regarding extending the grazing season into the autumn and early winter frequently also involves some fairly late applications which are potential environmental hazards. The latest safe date recommended for the application of nitrogen fertiliser is mid-August. In most seasons further applications are completely unnecessary since warm soils and plentiful autumn rainfall frequently result in large quantities of nitrogen being mineralised from the large organic reservoirs within soils.

18.5.3 Phosphorus and potassium

Phosphate and potash fertilising of grassland is quite well understood. In a grazing situation there is frequently no need to apply any because of the frequent returns of dung and urine. Only in deficient soils will any response be observed. In the case of potash there is an important connection with the incidence of hypomagnesaemia or grass staggers. High levels of potassium in the herbage impedes the uptake of magnesium. This is a very serious condition affecting all classes of stock and which can cause sudden death or substantial loss of production. The main periods of risk are in the spring, especially during cold wet spells after turnout. Suckler cows are particularly at risk in the autumn. The best advice is to avoid any applications of potash fertilisers in the spring months except where the soil is seriously deficient. Where needed, potash can usually be applied safely in the summer. The other remedy (apart from the attentions of the vet) is to feed a suitable magnesium supplement throughout the risk period. This can be variously given as a feed supplement, lick, added

Table 18.4 Phosphorus and potassium fertiliser recommendations for grassland

Management system		Phosphorus soil index level					Potassium soil index level				
		0	1	2	3	3+	0	1	2	3	3+
Ley establishment (seedbed)		120	80	50	30	0	120	80	50	0	0
Grazing (annual application)		60	40	20	0	0	60	30	0	0	0
Cutting for silage	1st cut	90	65	40	20	0	140	110	70	30	0
	2nd cut	25	25	25	0	0	120	100	75	40	0
	3rd cut	15	15	15	0	0	80	60	60	20	0
	4th cut	10	10	10	0	0	70	70	55	20	0

Notes
(1) N P&K can be applied as 'straights' or as compound or blended fertilisers of a suitable analysis.
(2) Excessive potash applications on grazing fields can give rise to *hypomagnesaemia* (grass staggers). Except in the case of serious potash deficiency, applications to grazing areas should be made in the June/July period.
(3) Where individual potash applications in excess of 100 kg/ha are required application in the previous autumn is advised.
(4) On fields with inherently low potash levels (index 0–2) an additional 60 kg/ha K_2O should be applied in the autumn after the last cut.
(5) Where applications of FYM or slurry have taken place account should always be taken on the nutrient contribution from these sources.
(6) Sulphur deficiency can be a problem in second and subsequent silage cuts. Compound fertilisers should be used to apply about 40 kg/ha SO_3 if deficiency is confirmed by herbage analysis.
Based on MAFF RB209 (2000)

to the water supply or as an ingested bolus. Another option is to dust the pasture regularly with feed-grade calcined magnesite but this can be readily washed away by rainfall. A magnesium based pasture spray is another popular option.

On cut swards phosphorus and potassium are both required in large amounts, especially potassium, where optimum annual dressings for frequently cut swards may total 200–300 kg/ha K_2O. Recommendations are set out in Table 18.4. Table 3.5 sets out the phosphorus and potassium contents of average quality organic manures and slurries. As with nitrogen, these values should be taken into account when calculating the supplementary applications of inorganic materials.

18.5.4   Other nutrients

18.5.4.1   Magnesium (Mg)
It is possible to obtain a response to up to 85 kg/ha of MgO applied as calcined magnesite where the soil index level for magnesium is 0. Another avenue for the application of magnesium is the use of magnesian or dolomitic lime for correcting soil pH.

18.5.4.2   Sodium (Na)
Sodium (in compound form) is sometimes applied to grazed swards in an attempt to improve palatability and intake. Sodium application can also reduce the risk from grass staggers in some situations.

18.5.4.3   Sulphur (S)
ADAS trials in the 1990s have shown yield increases from cut grass of 20–30% where sulphur deficiency has occurred. This is mainly as a result of a marked decline in the level of industrial pollution in recent years. Deficiencies tend to occur in areas removed from the main centres of population. Grass growing on free draining soils is vulnerable and situations where winter rainfall is unusually low have also shown this deficiency. Farms where little or no FYM or slurry is applied to grass, may also be vulnerable. Sulphur, as up to 40 kg/ha of SO_3, should be applied in deficient situations. Second and third cut silage crops in fairly dry situations have shown the most frequent deficiencies of sulphur. This deficiency can best be assessed by an analysis of herbage.

18.6   Irrigation of grassland

In dry areas with low summer rainfall (e.g. grass growth site classes 4 or 5) it is the summer months of June to August which give rise to the poorest rates of grass growth. Soil moisture deficits (SMDs) during this period can frequently reach 100 mm or more in the south and east of England. Grass growth is usually severely impaired when the SMD reaches 50. Irrigation, if it is available at a reasonable cost (e.g. £5/ha mm), can be a way of overcoming such severe summer droughts and (in effect) transforming the farm from site class 5 to 3, for example. Only a relatively few farms have installed irrigation equipment sufficient to make a worthwhile impact on grass growth. Most frequently, when grass is irrigated, it is through the marginal use of equipment which is already being justified by its use on a high value arable crop such as potatoes, salads or vegetables. A 20 year study carried out in the south east of England and reported in the 1980s, referred to an increase in grass dry matter yields of 25 %, as the result of the sustained use of irrigation.

18.7 Further reading

DEFRA, *Guidance for Farmers in Nitrate Vulnerable Zones* (PB3277), DEFRA, 2001.
DEFRA, *Manure Planning in Nitrate Vulnerable Zones* (PB3577), DEFRA, 2001.
Frame J, *Improved Grassland Management,* Farming Press, 2000.
Hopkins A (ed.), *Grass: its Production and Utilization*, 3rd ed., Blackwell Science, 2000.
Institute of Arable Crops Research, *Guide to the Classical Field Experiments*, Rothamsted Experimental Station, AFRC, 1991.
MAFF *Fertiliser Recommendations for Agricultural and Horticultural Crops* (RB209), The Stationery Office, 2000.
Monsanto, *Renewed Grass for Improved Milk and Meat Production*, 1998.
Simpson N A, *Survey Review of Information on the Autecology and Control of six Grassland Weed Species*, English Nature Research Report no 44, 1993.
Smallwood P, *The Horse at Grass,* British Horse Society, 1988.
Thomas C, Reeve A and Fisher G E J, *Milk from Grass*, The British Grassland Society, 1991.
Younie D (ed.), *Legumes in Sustainable Farming Systems*, BGS Occasional Symposium no 30, British Grassland Society, 1996.

19

Grazing

The utilisation of grass by grazing is, for dairy cattle, beef cattle and sheep, undoubtedly the most economical means of production. Reference has already been made (page 393) to the low cost of energy as grazing compared with conserved grass. Estimates vary, according to the way the calculations have been completed, to between 30% and 50% of the costs of conserved feeds and even lower when compared with the costs of energy as bought-in feed. There is little point here in trying to describe just how cheap energy as grazed grass will be as the calculation will vary substantially from farm to farm. Most are agreed, however, that maximising the contribution from grazing, whatever the system of production, is a sensible approach to reducing costs and maintaining an acceptable degree of profitability, even in (or perhaps especially in) a time of depressed prices.

19.1   Stocking rate or density

By convention, stocking rates are usually expressed as 'Grazing Livestock Units' (GLU) per hectare; a unit being a 500 kg Friesian/Holstein cow. Stocking rates should be accurately calculated from the following information, and based on a monthly reconciliation of actual stock numbers in the various categories on a farm and the total forage hectares. There is often a strong correlation between stocking rate, the application of nitrogen fertilisers and farm profitability. High stocking rates (e.g. > 3 GLU/ha) would normally only be associated with the very best grass growing conditions (site class 1 page 431) and with near optimum applications of nitrogen fertiliser. Lower rates (1.5–2.0 GLU/ha) might be associated with organic farms or with the poorer grass growing conditions (site class 5 page 431). Stocking rate calculations are an integral part of the applications for EU livestock subsidy payments (such as 'Extensification Premium'). NB Calculations for subsidy purposes should be made accurately, using the livestock unit equivalents supplied by MAFF and not those supplied here which are for farm management calculations only.

The following Grazing Livestock Unit figures are those currently in use for farm management calculations:

Cattle	GLU	Sheep	GLU
Dairy cows (Friesian/Holstein)	1	Lowland ewes	0.11
Beef cows (excluding calves)	0.75	Upland ewes	0.08
In-calf heifers	0.8	Hill ewes	0.06
Bulls	0.65	Breeding ewe hoggs	0.06
Other cattle 0–1 years old	0.34	Other sheep over 1 year	0.08
Other cattle 1–2 years old	0.65	Store lambs under 1 year	0.04
Other cattle over 2 years old	0.8	Rams	0.08

(Taken from the Farm Management Pocketbook, Nix 2001)

19.2   Principles of grazing management

Grazing management is a balancing act. Balancing the requirements of stock against the potential of grass requires a good knowledge of both and is strongly influenced by the situation of the farm and the actual livestock production system. In fact, the choice of production system (especially the dates of calving or lambing) is often strongly influenced by the grass production characteristics of a particular farm. Such considerations would include 'earliness' of grass growth and the potential date of turnout, the likelihood of a mid-season drought and the potential for extended grazing into the autumn and early winter. Consequently, the synchronisation of lambing or calving with the expected growth of grass so as to turn out stock to grass ready to maximise utilisation from grazing seems a good approach to minimising production costs. But of course other considerations obtain, such as the need to maintain milk supply throughout the year and to market fatstock at the most advantageous time as well.

19.2.1   Animal potential
So far as milk production is concerned, calving in late winter/early spring should ensure that cows can be turned out to grass at peak lactation and in a position to maximise the contribution of grazed grass to their production. Autumn calvers too can achieve good levels of production from grazing, both from autumn

grass prior to housing, and again in the spring towards the latter half of the lactation. The susceptibility of a farm to summer drought should influence the decisions about calving date, and it may be necessary to consider making special provisions for summer feeding in droughty conditions. The potential is for grazing to provide sufficient energy and protein for a daily milk yield of up to 30 l from animals capable of producing it. This figure would fall to about 25 l in late summer.

In the case of fattening stock (young cattle or lambs) the potential is for grazing to provide sufficient nutrients for individual daily liveweight gains of up to 1.25 kg for beef cattle and up to 300 g for lambs. There would be falls in potential gains of similar proportions to milk production with the advancing season. In these cases further complication of management decisions are presented by the combined grazing of adults and juveniles prior to weaning, the inevitable build-up of parasitic worm populations both pre- and post-weaning, and the need to integrate parasite control measures together with the provision of sufficient quantities of nutritious grass at the correct stage of growth throughout the grazing season.

19.2.2 Grass potential

The factors which affect grass yields such as sward species and varieties, site class and fertiliser (especially nitrogen) application, have been dealt with at length in Chapters 17 and 18. Suffice it here to remind the reader of them, and of the value of the more leafy and persistent intermediate and late heading perennial ryegrasses for grazing, of the excellent palatability of tetraploid ryegrasses and Timothy and of the suitability of the different types of white clover for grazing by various classes of stock. Seeds mixtures suitable for grazing

Table 19.1 Suitability of seed mixtures

Purpose	Varieties which should be used
For early grazing, i.e. early bite	Mainly Italian ryegrass and hybrid ryegrass
For optimum production throughout the season	Ryegrass (supported by either non-ryegrass or permanent pasture)
For winter-grazing cattle – foggage (provided conditions allow)	Non-ryegrass
For the grazing block	Herbage plants which produce a closely knit sward
For the cutting block	Herbage plants which produce a relatively tall habit of growth

Note: The heading dates will generally coincide if the above points are borne in mind when deciding on a seeds mixture. As far as possible, varieties should correspond because, with only a few exceptions, the digestibility of a plant starts to decline when the ear emerges. A grass crop cannot be so valuable if its various plant components head at different times. If, for example, the crop is cut or grazed at the average date for the plants making up the sward, the earliest heading varieties will have declined in quality, whereas the latest will not have produced any sort of yield. Compatibility is important.

in various situations have also been described in detail on page 415 and in Tables 19.1, 17.4, 17.5 and 17.6. Other site factors such as aspect and soil factors such as texture and drainage may affect the suitability for grazing, and, in particular, the suitability of a field for early grazing in spring or extended autumn grazing. On some farms specific crops such as rye, rye plus Italian ryegrass or triticale are sown for early grazing. These crops are described in more detail on page 359. Winter wheat and oats, sown as grain crops, may also be grazed in early spring so long as soil conditions are suitable and the crop has not reached the 'stem extension' growth stage. Although there may be some suppression of yield the strategic gain in terms of cost saving to the livestock enterprise may well outweigh this. Another extremely important factor is the susceptibility of a farm to summer drought, and the need to make special provision for feeding stock during times of shortage of grazing. These might include sowing an area of quick-growing turnips for summer grazing, or the inclusion of a field of lucerne with its potential for zero grazing (cutting and carting) during the summer months. A further provision might simply be making available a stock of silage specifically for feeding during the summer in times of grass shortage.

19.2.3   Sward height

Estimates of sward height are important in order to maintain good performance at grass from all classes of stock. As a sward is grazed closer to the ground herbage intake and performance will be progressively reduced. The ideals are represented in Table 19.2 which gives information about set stocked and

Table 19.2   Suggested target range of sward heights for set stocked animals or post-grazing sward heights for animals kept on rotational grazing systems

	Target sward height
SHEEP	
In spring and summer	
Ewes and lambs (medium growth rates)	4–5 cm
Ewes and lambs (high growth rate)	5–6 cm
Dry ewes	3–4 cm
In autumn	
Flushing ewes	6–8 cm
Finishing lambs	6–8 cm
Store lambs	4–6 cm
CATTLE	
Dairy cows	7–10 cm
Finishing cattle	7–9 cm
Beef cows and calves	7–9 cm
Store cattle	6–8 cm
Dairy replacements	6–8 cm
Dry cows	6–8 cm

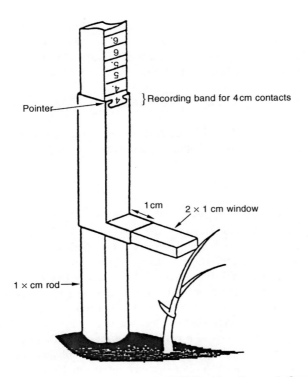

Pointer

} Recording band for 4 cm contacts

1 cm   2 × 1 cm window

1 × cm rod

Fig. 19.1   The sward stick designed by the Hill Farming Research Organization.

rotationally grazed pastures. Estimates of sward height are best made with a sward stick such as the one shown in Fig. 19.1, but in the absence of anything so sophisticated, a ruler or even a mark on the side of a wellington boot are more than adequate substitutes! To get a reliable estimate of average sward height, at least 20 measurements should be taken in a field at random from both grazed and rejected areas. Gateways, rough areas close to trees and water troughs should be avoided. Sward height estimates are likely to be of most use to farmers grazing cattle or sheep on a set stocked basis.

19.2.4   Sward density and dry matter yield estimates

Rising plate meters (e.g. the Ashgrove Pasture Meter) have become popular in recent years and provide reasonably reliable estimates of sward dry matter yields. Again, it is important to walk the paddocks in a random pattern and make 30 compressions. The readings recorded can then be compared with a set of standards to give an estimated yield of dry matter per hectare. Readings taken before and after grazing will give a good indication of the average level of dry matter intake per head as set out in the example below. This technique is likely to be mainly of use to dairy farmers grazing on a rotational paddock system.

Example
A herd of 120 cows graze a 1 hectare paddock for one day. Estimates using the plate meter show a pre-grazing yield of 3800 kg DM/ha. The post-grazing residue is measured as 1800 kg DM/ha.

$$\text{DM intake per cow} = \frac{2000}{120} = 16.6 \text{ kg DM} \qquad [19.1]$$

Assuming about 11.5 MJ of ME/kg DM for leafy grass the daily intake of ME = 190 MJ. This would be sufficient for maintenance and 24 l of average quality milk.

19.2.5   Stocking rates at grass

The output of particular grazing systems is strongly correlated with the stocking rate sustained during the grazing season. As a result of the well-understood changes in the rate of growth of grass as the season progresses it is usual to find that the area devoted to grazing expands during this period. This usually happens as a result of the progressive introduction of silage or hay aftermath areas into the grazing block. However, it is normal to find that the stocking rates possible in the early part of the grazing season are the highest. On intensively managed grassland it would be normal to find dairy cows allocated about 0.2–0.25 ha per head at this time. Ewes with lambs would normally start the grazing season at about 18–24 ewes with lambs/ha. However, if spring grass growth is poor for any reason, it is normal to allocate increased grazing to ewes with lambs at the expense of the conservation area in the expectation that a hay or silage cut may become available later in the season when grass growth improves. With young cattle or dairy replacements at grass, it is normal to express the stocking rate in terms of the liveweight of animals per hectare since individual weights can vary so much. On well-fertilised grassland, about 2.5 t of liveweight/ha would be a good target in the early season falling subsequently to 1–2 t/ha later in the season.

19.3   Grazing systems

19.3.1   Strip grazing

Strip grazing (Fig. 19.2) (also known sometimes as 'flexible block' grazing) is particularly suitable for the smaller dairy herd. The system allows the animals access to a limited flexible area of fresh grass either twice daily, daily, or for longer intervals. Wherever possible a back fence should be used, whereby the area just grazed is almost immediately fenced off. This is to protect the recovering sward from constantly being nibbled over. Without a back fence the recovery rate is slower, although few strip grazing systems use it these days. Strip grazing is also time-consuming and requires a daily decision as to

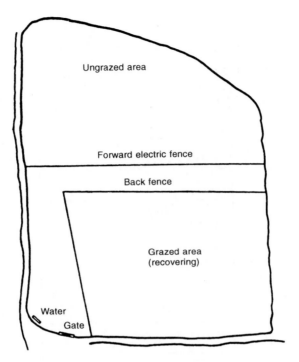

Fig. 19.2 Strip grazing.

how much grass is needed for the cows. Wet soil conditions can lead to serious poaching along the line of the fence. It is always important to maintain good access to water. Strip grazing is also a suitable system for utilising crops such as forage rye for early bite, or turnips, rape and kale later in the season. Beef cattle can be strip grazed although they may take some time to get used to the routine. Some farmers leave round-baled straw in the fields together with direct drilled turnips to facilitate feeding to store cattle while strip grazing the turnips. Strip grazing is not a suitable system for suckler herds. Ewes or store lambs grazing rape and turnips during the autumn and early winter are usually rationed in this way, with blocks of forage being introduced for periods of 3–4 days with double strand electric fencing.

19.3.2 Rotational or paddock grazing

The principle of this system is rotational grazing alternating with rest periods (Fig. 19.3). The grazing area is divided into equal-sized paddocks. Good access tracks and a reliable water supply to each paddock are essential features of the system. On a dairy farm the grazing paddocks will normally occupy fields close to the milking parlour.

There is huge variation over the number of paddocks considered to be 'ideal', but the common feature is a system which allows animals to graze for

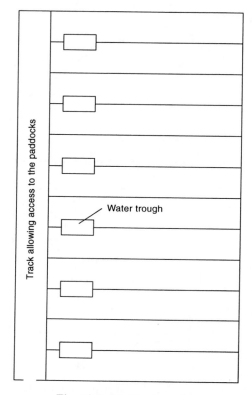

Fig. 19.3   Paddock grazing.

a limited period (between half a day and one week) followed by a period of recovery. Rotational grazing can be used for all forms of farm livestock and is beneficial for horses as well. The more intensive systems with one-day paddocks for example are obviously most suited to the dairy herd. At the peak of spring growth it would be expected that animals would rotate around a paddock system on a 21-day cycle. It is quite possible that some of the paddock areas may become available for an early cut of silage. As the season progresses the rate at which grass recovers after grazing will decline. At this time the grazing obtained from part of the area which has already been cut can be introduced. Fertilising with nitrogen after each grazing (Table 18.1) and regular 'topping' of rejected material and weeds will maintain the quality of the grazing throughout the season. 'Leader/follower' grazing is in operation on some dairy farms with high yielding cows grazing ahead of a low yielding group. Such a system can lead to very high levels of pasture utilisation.

Young cattle and sheep, when managed on rotational grazing systems, usually have access to fewer paddocks, 4–6 would be normal, but the length of the recovery periods would be roughly the same as for dairy cows. Leader/follower systems where young calves graze the same set of paddocks ahead of older animals have been practised in the past. 'Forward creep' grazing, where lambs graze ahead of ewes through a creep gate can also reduce the levels of parasitism. However, these systems are demanding in terms of labour and expensive to set up and, bearing in mind the recent difficulties of the livestock industry, it is not surprising to find that they have become less popular. The one sector to which they may still be applicable is organic production where the precluding of routine dosing with anthelminthics necessitates the use of management strategies which minimise the likelihood of parasite infestations.

19.3.3 Set stocking

The majority of beef cattle and sheep in the UK are managed on what amounts to set stocked grazing systems where animals remain in the same area for grazing for extended periods. The obvious appeal of set stocking is the simplicity of management, but it is extremely important to ensure that the swards are of suitable quality (short, dense and leafy) and, most of all, that the grazing pressure, as determined by the stocking rate at grass, is at the correct level to maintain the optimum sward height. Optimum sward height recommendations are set out in Table 19.2. They can have a substantial effect on daily dry matter intakes and animal performance. Sward heights should be checked regularly (e.g. weekly) and adjustments made to the stocking rate by removing animals to other fields or introducing extra grazing (e.g. silage or hay aftermaths) if the need arises. Dairy cows too can be managed by set stocking, and they often are, but it is generally observed that better performance at grass is obtained from rotational or strip grazing. Set stocking would be a suitable grazing system for cows coming to the end of their lactations (e.g. those that calve June–August) during the grazing period.

Fertiliser application to set stocked areas can follow one of two patterns, either fertiliser is applied to the whole area once a month or to a part of the area (e.g. a quarter) on a regular (e.g. weekly) basis. It has been observed, with sheep in particular, that animals tend to avoid the most recently fertilised areas for several days and this can lead to a kind of rotational grazing without fences. In the past the injection of up to 250 kg N/ha as aqueous or anhydrous ammonia into set stocked pastures was popular, but this involved contractor costs and precluded the individual operator applying more or less fertiliser according to seasonal conditions. A further modification of the same principle recently has seen a whole season's nitrogen fertiliser applied to set stocked areas at the start of the grazing season in liquid form with the addition of a chemical nitrification inhibitor. In both cases since large quantities of nitrogen are being applied in a single application there have been concerns about the possible consequences of extra nitrate leaching into ground water.

19.3.4 Zero grazing

Cutting and carting green forage to farm animals obviously involves substantial extra costs as well as wear and tear on machinery. On the majority of UK farms it would only be regarded as a technique for use in particular situations. One of these might be green feeding of forage crops like lucerne, kale or forage maize where either the nature of the crop or conditions underfoot precluded the use of strip or block grazing. Another situation might be the need to cut green forage from outlying fields in times of shortage of grazing.

19.4   Strategies to minimise parasitism at grass

The value of 'clean' (i.e. relatively parasite free) grazing for young cattle and lambs is obvious. Grazing young cattle or ewes with lambs on the same fields year after year will encourage the development of massive populations of parasitic nematodes, require the routine administration of anthelminthics and the probable development of resistance to them. Strategies for minimising this problem include alternating sheep and cattle on different fields for grazing, year on year. The 'leader/follower' and 'forward creep' systems have already been mentioned. A reduction in parasitism can be brought about in this way by reducing the extent to which the young cattle or lambs are forced to graze heavily contaminated material at the base of the sward. The regular introduction of new leys into the system has obvious value, and moving young cattle or weaned lambs on to silage or hay aftermaths after worming in mid-season is also extremely effective. The '1/3 : 2/3' system for young cattle, illustrated in Fig. 19.4, incorporates this latter principle. Improving the drainage of liver fluke-infected pastures and the consequent reduction in habitat for the mud snails which act as an intermediate host can be an important factor in minimising the incidence of this important parasite. There is some interest currently, especially on organic farms, in the increased use of legumes such as sainfoin and birdsfoot trefoil, which have some anthelminthic properties, for grazing young cattle and lambs.

19.5   The energy yield from grass and forage–the UME calculation

Assessment of the real value of the output from grass and forage on livestock farms is difficult for practical reasons. The measurement of the yield of a crop of wheat or potatoes is easy as there is a quantifiable, physical and marketable yield. With grass and forage crops although the yield of dry matter is obviously present and is sometimes measured directly (e.g. loads of grass or numbers of bales), it has to be utilised by farm animals before any financial or marketable output is created. Where a high proportion of the animal production is obtained from grazing these estimates are particularly difficult.

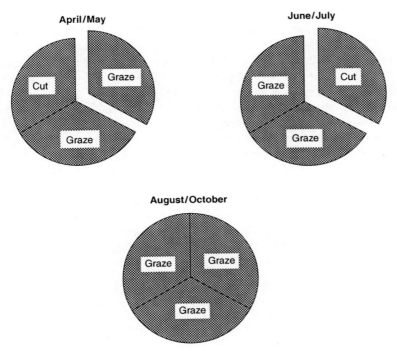

Fig. 19.4 The 1/3:2/3 system for grazing young cattle.

19.5.1 The Utilised Metabolisable Energy (UME) system

This system takes account of the whole of the theoretical energy requirements of forage-based farm livestock and estimates the proportion of that energy which has been produced on the farm by the forage areas. Energy estimates are made as megajoules (MJ), or gigajoules (GJ), where 1 GJ = 1000 MJ.

The calculations assume that if the total annual ME requirements for the livestock on the farm are known, as well as the annual amount of ME fed to the stock from bought-in feed and concentrates, then the remaining ME must come from home-produced forage. This would include all the forage area on the farm, grazing, silage, hay, and any other forage crops.

Example
To record UME production from grass for the dairy herd on a dairy farm, the following information would be needed:

1 The annual energy (ME) requirement per cow.
   For maintenance this would be about 25 000 MJ and for the production of average quality milk, 5.3 MJ/litre for (say) 6471 l. (For extra accuracy an allowance might also be made for pregnancy, different breeds and actual animal liveweights and milk of varying quality–however, the above will suffice for a rough calculation.)

2 The annual ME purchased as:
   (a) Concentrates – say 1.6 t per cow with a dry matter of 85% and an ME
   value 13 MJ/kg DM.
   (b) Other purchased feeds – say 100 kg of hay per cow with a dry matter of
   84% and an ME of 9 MJ/kg DM plus 1 tonne of wet brewers grains per cow
   with a dry matter of 20% and an ME of 11 MJ/kg DM.
   The total ME purchased per cow would then be the total of a + b above.
3 UME obtained from grass and forage per cow = the annual ME requirement
   per cow minus the total ME purchased per cow (a + b).
4 The stocking rate in GLUs per forage hectare – say 1.97.
5 The annual UME per grass and forage hectare for the dairy herd = UME
   per cow × stocking rate. The complete calculation is shown in Table 19.3.
6 The grass growth site class (see Table 18.4). The UME targets suggested in
   Milk from Grass are also given in Table 19.3. The reader can then substitute
   actual farm figures for this calculation, use the appropriate site class for
   comparison and assess the efficiency with which forage is being produced
   and utilised on a particular farm.

Table 19.3   The UME calculation from the example on page 445

Maintenance	+	Milk production	=	Total annual ME requirements	
		6471×5.3			
25 000	+	= 34 296 MJ	=	59 296 MJ	
Supplied by:					MJ of ME
Concentrates		Hay		Brewers grains	purchased/cow
$1665 \times 85\% \times 13$		$100 \times 84\% \times 9$		$1000 \times 20\% \times 11$	
= 18 398 MJ	+	= 756 MJ	+	= 2200 MJ	= 21 354 MJ
Total ME	–	ME		= UME/cow	
requirements		purchased/cow			
59 296 MJ	–	21 354 MJ		= 37 942 MJ	
UME/cow	×	Stocking rate GLU/ha		= UME/hectare	
37 942 MJ	×	1.97		= 74 745 MJ/ha	= 74.7 GJ/ha

Target UMEs taken from *Milk from Grass*

Site Class (see Table 18.2)	Target GJ/ha
1	126
2	115
3	105
4	93
5	83

The calculation of UME figures for a dairy enterprise is quite simple. The
total milk output is always very well documented. The calculation for beef and
sheep enterprises is more complex in that the output figures required are total

annual liveweight gains which are obviously more difficult to obtain reliably and are subject to a great deal more error.

19.6   Further reading

Crofts A and Jefferson R G (eds), *The Lowland Grassland Management Handbook*, English Nature, 1999.

Frame J, *Improved Grassland Management*, Farming Press, 2000.

Hodgson J, *Grazing Management: Science into Practice*, Longman Scientific and Technical, 1990.

Hodgson J and Willius A, *Ecology and Management of Grazing Systems*, CAB International, 1996.

Hopkins A, *Clovers and Other Grazed Legumes in UK Pasture Land*, IGER, 1994.

Hopkins, A (ed.), *Grass: Its Production and Utilization*, 3rd ed., Blackwell Science, 2000.

Newton J, *Organic Grassland and Sheep Husbandry: 15 Questions Anwered*, Newton Publishing, 1999.

Nix J, *Farm Management Pocketbook*, Imperial College at Wye, 2001.

Pearson C J and Ison R L, *Agronomy of Grassland Systems*, 2nd ed., Cambridge University Press, 1997.

Thomas C, Reeve A and Fisher G E J, *Milk from Grass,* The British Grassland Society, 1991.

Vallentine J F, *Grazing Management*, 2nd ed., Academic Press, 2001.

Voisin A, *Better Grassland Sward: Ecology, Botany, Management*, Crosby Lockwood, 1960.

20

Conservation of winter feed

20.1 Silage

20.1.1 The change from hay to silage

20.1.1.1 Reasons for the change

Grass and a range of other forage crops are conserved as winter feed, mainly as silage in the UK. Hay was the traditional method for conserving grass in western Europe. However, as a result of the prolonged drying period necessary for the reduction of the moisture content to a safe level for storage and feeding and the vagaries of the climate of the UK in particular, it is not difficult to establish the reasons for the dramatic decline in haymaking and the rise in the popularity of silage. In fact the quantity of hay conserved virtually halved between 1970 and 1994 whilst, during the same period, the popularity of silage making increased dramatically from about 2 million tonnes of dry matter to an estimated 13 million tonnes.

Weather was not the only reason for these dramatic events. They mirror complete changes in the ways in which livestock has been managed and fed on British farms. Grassland improvement, increases in intensity of stocking and, in particular, increases in the use of nitrogen fertiliser virtually forced farmers into silage making, since the earlier cutting of very heavy, highly digestible crops completely precluded haymaking. Other important factors have included increases in herd and flock sizes, improvement in ensiling techniques and the advent of reliable systems for mechanically handling and feeding silage. Field machinery too has improved dramatically, and the advent of the widespread use of contractor teams for silage making has meant that most British farmers who want it now have access to some of the most sophisticated and efficient machinery for making silage anywhere in the world. Speed of operation has become very important since declines in the D value of grass are now well appreciated. It is possible for a dairy farmer, with a farm of average size for example, to expect an efficient contract team to make an entire cut of first cut silage within two days, a job which in the 1970s with farm machinery, would have taken weeks!

20.1.1.2 Losses from silage making

However, all these developments have brought with them associated problems. The costs of silage making have rocketed and although it can be argued that growing heavy crops of grass can minimise the costs per tonne of dry matter conserved, the depressed livestock and milk prices at the turn of the millennium brought home to farmers the true costs of conservation and the need to minimise them whilst maximising the use of grazing. There is also a high potential for losses if silage making is not carried out well, and although modern machinery and practice have reduced such problems it is still routinely possible to lose 15–25% of the dry matter ensiled. An ADAS survey carried out in the 1980s indicated that the true level of losses from silage making (field, storage, effluent and feed-out) ranged from 25–45% of crop dry matter. Precision chopping, good consolidation and clamp sealing can all help to minimise losses, but the greatest potential for losses and for environmental pollution comes from silage effluent.

It is possible, from very wet, direct cut material, to produce as much as 200 litres of effluent per tonne. Silage effluent has a very heavy biochemical oxygen demand (BOD) which makes it about 200 times as polluting as raw domestic sewage. Pollution of groundwater or watercourses by silage effluent has resulted in many prosecutions and some very heavy fines under the Control of Pollution Act 1992. All silos must have perimeter drains and a sealed effluent pit which is emptied regularly, especially during the period immediately after ensiling when effluent production is greatest.

Silage effluent can be conserved and fed to livestock. One system is to add an absorbent material (e.g. dried sugar beet pulp) at ensiling; another is to conserve the effluent and preserve it either with the addition of a material such as formalin or by enclosing it in an airtight container (e.g. the Eff system). However, the simplest expedient is to wilt silage grass (or other materials to be ensiled) in the field to 25–30% dry matter in which case the production of effluent is minimised. In the case of forage maize, harvesting of the standing crop is delayed until the dry matter reaches about 30%.

20.1.2 Crops for silage making

20.1.2.1 Grass

All types of grass can be ensiled. Feed value will be very dependent on the stage of growth and digestibility (D value) of the crops. The fermentation quality will depend on a great many other factors discussed later, but the dry matter percentage and the content of water soluble carbohydrates (sugars) are two of the most important. Grass for silage should be mown at a height of 7.5–10 cm in order to minimise soil contamination and to allow good recovery.

20.1.2.2 Legumes

Red and white clovers are frequently made into silage, usually in mixtures with grasses of various types although occasionally red clover may be sown and harvested as a pure stand. Lucerne is grown on farms in the drier parts of southern and eastern England often as a pure stand, but also mixed with red clover or with grass. D values of red clover and lucerne when cut for silage are often lower than for grasses but protein content is much higher (up to 20–22 % crude protein at early flowering).

Annual legumes such as peas (forage or grain varieties are both suitable) and vetches can be made into high protein silage as well. These crops are also often mixed with spring barley or oats as 'arable silage mixtures' in order to improve the fermentation characteristics of the silage and to increase the starch percentage of the final product. Such mixtures can be harvested at any stage from when the grain is 'milky' up to the 'cheesey' stage and when the peas or vetches have pods partly filled. Recently interest has developed into using forage varieties of lupin as an arable silage crop, and trials are underway too with forage varieties of soya bean. Both crops are fairly late maturing and have potential in their own right or as 'bi-crops' with forage maize.

Most legume plants are more difficult to make into well-fermented silage than is grass. This is because they usually contain quite low levels of sugar. Hence the need arises for mixing with grasses or cereals, which have much higher levels of sugar or starch, to improve the fermentation of the silage. Other possibilities, particularly with lucerne and clovers, are to wilt to at least 30 % dry matter or to use some sort of acid additive at an effective rate.

20.1.2.3 Forage maize

Maize as a silage crop has increased substantially in popularity in recent years and detailed information about its agronomy is given on page 359. According to the area where it is being grown, a variety of maize should be selected on the basis of its maturity group, which determines its ability to yield ripe cobs which contain starch. At harvest the cob yield should be about 50 % and the starch should comprise about 25–30 % of the dry matter. This situation usually obtains when the dry matter percentage of the whole crop reaches about 30 %. Another test is to examine the grains on a number of cobs. If it is just possible to make a dent with a finger or thumbnail then the crop is ready to harvest.

Very fine (1–2 cm) chopping and grain cracking are essential for making good maize silage. The crop must be well consolidated (some contractor teams have taken this to the extreme of employing a road roller!) in order to avoid over-heating and the growth of moulds (particularly at feed out). Since there is so much starch and some sugar still in maize when it is harvested, additives should not normally be needed and a satisfactory fermentation should be easy to obtain. 'Ground ear maize' (GEM) and 'corn cob mix' (CCM) are the products of cob-only harvesting options using a maize snapper header either with a forage harvester (for GEM) or a combine harvester (for CCM) to make very high energy silages. In the case of corn cob mix, feeding it to pigs has become popular, especially in other parts of Europe.

20.1.2.4 Other cereals for 'wholecrop' silage

Any type of cereal can be taken for silage, but the best yields and quality are usually obtained from winter wheat. Advantages are that it is an excellent source of cereal starch (although maize is favoured by livestock farmers) and it substitutes for forage maize in the north of England, Scotland and Northern Ireland where maize cob and starch yields may not be so good. Advantages over maize are that the full cereal arable aid payment can be claimed on eligible land (this aspect could change, however, and readers should always check the current situation with their regional DEFRA centre). Any surplus can be left for combining and harvest takes place in July/August leaving many more post-harvest options for the fields in question.

There are two basic systems for making wholecrop cereal silage. One is to cut the crop when it is still fairly green at about 35% dry matter in July for normal fermented wholecrop silage. The other is to leave the crop until it is nearly ripe for combining (about 50% dry matter) and to use feed-grade urea as an additive at the clamp. When the clamp is sealed (good sealing is absolutely essential for this technique) the heat of the initial fermentation releases ammonia gas from the urea. This has the effect then of preserving the grain and straw as a moist relatively non-fermented high pH product. The product known as 'alkalage' is made in the same way, but from a crop at about 70% dry matter and requires the use of a forage harvester with a processing mill.

Other systems exist for ensiling moist grain after crimping (partial grain cracking through fluted rollers) and with the use of preserving agents such as propionic acid. The use of brewers grains for ensiling has also become commonplace either as layers within grass clamps, on its own or mixed with an absorbent such as dried sugar beet pulp (grainbeet).

20.1.2.5 Kale for silage 'kaleage'

This is a system pioneered by farmers whereby leafy kale varieties are ensiled after mowing and wilting for 24–48 hours. Sowing at a high seed rate gives a thick crop which can be wilted on a fairly high stubble. Kale can then be harvested for clamping or, more frequently, for round baling and wrapping (six times to give a really good seal). Other options investigated at IGER and elsewhere have involved growing kale as a bi-crop with spring cereals for making into clamp silage.

20.1.2.6 Fodder beet

This crop has been popular for ensiling in some European countries such as Denmark, but has recently been eclipsed by the rise in popularity of forage maize. In the UK some pioneering work was done at ADAS Rosemaund. The system involves growing varieties (e.g. *Kyros*) which can be top lifted in reasonably dry field conditions in September to give a moderate yield of fairly clean beet. A special harvester then chops the whole crop, tops and all, and ensiling takes place between alternate layers of chopped straw (sometimes ammonia treated) and an absorbent such as dried sugar beet pulp. An alternative is to clamp whole beet in the normal way and to make silage from the tops. In both cases excessive soil contamination is a potential problem to the quality of the fermentation of the silage.

20.2   The silage-making process

20.2.1   Silage fermentation

Most silage is made from grass. Therefore most of the information in this section concerns the making of grass silage. Silage making or ensilage is a process of anaerobic fermentation (i.e. fermentation without air). The speed with which air can be excluded from the clamp and prevented from re-entering it subsequently will determine the success of the whole operation. During fermentation the carbohydrates and proteins present in the crop are acted upon by bacteria and plant enzymes. Some of the bacteria will be those naturally occurring on the crop before ensiling. Others may have been added in the form of an inoculant additive. The most desirable outcome of this initial fermentation is a rapid fall in the pH of the silage to about 4.0. This is caused by an accumulation of lactic acid and of other organic acids which are a by-product of fermentation by a group of bacteria of which the genus *Lactobacillus* is the most common, and one of those which has been cultured to produce inoculant additives. In a well-fermented silage the concentration of acids reaches a peak after less than a week and the pH will stabilise at about 4.0. Such a silage will have a light brown colour, a sharp smell and an acid taste (for those who fancy it!).

In a poorly fermented silage, which may result from very wet grass or where contamination with soil or faecal material has taken place, the bacteria which dominate the fermentation may be from another group known as *Clostridia*. In this case the end result is an accumulation of a weaker acid called butyric acid. The pH of such a silage may well be in the region of 5.0, and it will appear an olive green colour with a foul smell. One of the unpleasant smells will be that of ammonia which will almost certainly be present in large quantities as a result of the inefficient preservation of proteins and their subsequent breakdown. Apart from the dry matter percentage and the pH, one of the most useful indicators of silage quality by analysis is the percentage of the total nitrogen in the silage which is present as ammonia. This should be as low as possible (3–5%); if it is high (20–30%) this will almost certainly be as a result of a poor fermentation and the predominance of butyric acid.

20.2.2 Factors affecting silage fermentation

20.2.2.1 Type of crop

Crops with high levels of fermentable carbohydrates (sugars or starch) are usually the easiest from which to make good silage. So the Italian and tetraploid ryegrasses with their high sugar contents and maize and cereals with high starch contents usually ferment well and there is no need for additive application. Low sugar crops such as short leafy grass and legumes at most stages of growth are more difficult to ferment and need some extra treatment such as wilting, the use of an effective additive, or both.

20.2.2.2 Fertiliser treatment

This can affect the sugar content of grasses in particular. It is important to stress that the nitrogen recommendations and timings set out in Table 18.1 should be adhered to. Late or excessive applications of fertiliser can obviously lead to the ensiling of grass containing large quantities of nitrogen in the ammoniacal or nitrate form. This in turn can lead to the depression of grass sugar percentage and the presence of excessive levels of ammonia in the final silage. Where fertiliser application to silage crops is delayed because of poor weather or soil conditions, rates should be appropriately reduced.

20.2.2.3 Weather

Dry, sunny weather is obviously to be preferred for silage making. Wilting conditions will be improved and the possibility of soil contamination is minimised. Grass sugar levels are also likely to be highest in sunny weather. Many farmers prefer to have grass mown in the afternoon or evening in an attempt to maximise grass sugar levels.

20.2.2.4 Minimising contamination

The control of moles and the rolling of silage fields are important preparatory tasks in the spring. On most farms they are standard practice. Slurry application, although common on silage fields, should be avoided if possible. Where there is no alternative, it should be applied as early as possible relative to the proposed cutting date. The heavy contamination of silage grass with slurry can lead to poor fermentation and subsequent losses in feed value. Other fairly obvious ways in which contamination can be minimised include power washing the clamp prior to ensiling, tipping the crop on a clean concreted area and ensuring that the wheels of the vehicles undertaking the buckraking are as clean as possible. Levels of contamination can most easily be assessed by reference to the percentage of 'ash' in the silage analysis report. A figure of less than 10% is acceptable whereas 15% would indicate that significant levels of contamination have taken place.

20.2.2.5 Wilting

Wilting in the field after cutting has become a standard treatment when making

Table 20.1 The Star System

		Example 1	Example 2
Grass species (sugar content)	Timothy/meadow fescue	*		
	perennial ryegrasses	**	**	
	Italian/hybrid ryegrasses	***		***
Growth stage	leafy silage	0	0	0
	stemmy mature	*		
Fertiliser nitrogen	heavy (125 kg/ha +)	minus*	minus*	minus*
	average (40–125 kg/ha)	0		
	light (below 40kg/ha)	*		
Weather conditions	dull, wet	minus*		
	dry, clear	0	0	0
	brilliant, sunny	*		
Wilting	none (15% DM)	minus*		
	light (20% DM)	0	0	
	good (25% DM)	*		*
	heavy (30% DM)	**		
Chopping and/or bruising	disc/drum cutting	0		
	flail cutting	*		
	double chop	**	**	
	meter/twin chop	***		***

The Star System will help to decide on the need or otherwise for an additive. Low levels of grass sugars provide insufficient material at too low a concentration for rapid and effective acid production. The need for additives is lessened when the sugars in the plant are either concentrated by wilting or their levels are improved by sunshine. The Star System simply assesses the likelihood of an acceptable fermentation by taking into account the species of grass; the level of nitrogen fertilising; the weather prior to, and at, cutting; the amount of wilting; and the length of the crop to be ensiled. **Each factor is given a star rating and if five stars are obtained an additive should not be necessary. But for every star short of five, a recommended rate of additive should be applied.**

Example 1 A perennial ryegrass sward (**), in leafy silage growth stage (0) and heavily manured (minus*), being ensiled in dry weather (0) and only lightly wilted (0) with pick-up by double chop (**), gives a total score of *** and will show a benefit from 2.25 litres formic acid (**). Thus the final score is brought to *****.

Example 2 An Italian ryegrass sward (***), in a leafy stage (0), which is heavily manured (minus*), in average weather (0) but well wilted, 25% DM (*) and meter chopped (***) will give a total score of ****** and will not require an additive.

grass or legume crops into silage. However, there is more to wilting than simply leaving the mown swath in the field. In fact doing just that can be counter-productive and lead to a degree of aerobic fermentation (heating) of the swath which is very undesirable. In most cases, thick swaths mown by very wide high performance mowers are subsequently spread by a tined implement and may be tedded as well, prior to rowing up for harvesting with a precision chop harvester. Whereas this process will undoubtedly lead to an increase in dry matter percentage, it can also lead to quite high levels of field losses (up to 15%

for very dry material), and also to a degree of contamination with soil. 'Conditioning' the crop (i.e. physically damaging it with nylon brushes, tines or rollers) can also be beneficial, and increases the rate of wilting substantially so that in good conditions it is easily possible to achieve dry matters of 25% or more from 24 hours wilt. New high performance mower/conditioners mow, condition and spread in one operation, minimising wilting time in the field and soil contamination since all that is required subsequently is rowing up prior to final harvesting. Wilting to much higher dry matter levels 30–40% is required for ensiling in towers so that the unloading equipment will work well. This normally requires a 48 hour wilt and some additional tedding.

Kale for ensiling should also be wilted for about 48 hours. The best technique is to mow the crop with a mower/conditioner and leave it on a fairly high stubble (about 10 cm) prior to baling direct from the swath. Lucerne and red clover benefit from the action of roller/crimpers (if available) which crack the thick stems of these plants and improve wilting rate in this way. A dry matter of at least 30% is very desirable for ensiling lucerne and red clover. Maize and wholecrop cereals are 'self wilting' crops since harvesting does not normally take place until they have started the natural dehydration process associated with ripening.

20.2.2.6 Harvesting

On the majority of farms grass silage is now harvested using high output metered-chop machines. Wilted grass is usually chopped to between 20 and 50 mm whereas maize benefits from even shorter chopping to between 10 and 20 mm. In addition, when harvesting maize, it is beneficial for the grains to be cracked to aid their fermentation and subsequent digestion by farm animals feeding the silage. Finely chopped material usually consolidates better and it is possible to achieve true anaerobic conditions in the clamp more quickly. Fine chopping also results in the more rapid release of sugars from grass for fermentation.

20.2.2.7 Baled silage

About 20% of the silage made on UK farms at the turn of the millennium was harvested by big balers. This is a system of production which has substituted for hay on many of the smaller stock farms. The saving in the large capital expense of a clamp silo has obvious attractions for the smaller farmer. Grass is normally wilted up to about 40% dry matter prior to baling in high density round or square bales. As soon as the grass has been baled it should be wrapped. Silage bags are now very infrequently used and the advent of very efficient bale wrapping machines, using polythene stretch film, has meant that this system of covering is now almost universal. The more wrapping takes place the more efficient will be the exclusion of air and the quality of the final product. All possible measures should be taken to avoid puncturing the wrap after it has been completed, and stacks should be fenced so that stock do not damage them, and netted to avoid bird damage as well.

20.2.2.8   Baled 'haylage'

Crops of grass for very high dry matter baled 'haylage' for horses are normally made from quite mature grass and are wilted and tedded for up to three days to achieve a dry matter of about 65%. Tetraploid Italian or hybrid ryegrasses ensure a high sugar content and a satisfactory fermentation, and wrapping with six layers of high quality stretch film ensures that the bales remain airtight. The best quality material is normally made from first cuts when the level of soil contamination is likely to be lowest. Haylage has become popular mainly as a result of its dust-free characteristics and very palatable nature.

20.2.2.9   Filling a clamp silo

Firstly it should be ensured that the clamp is as airtight as possible and, if necessary (e.g. with a sleeper sided clamp) a side sheet should be used. Filling a clamp silo should then take place as rapidly as possible. Buckraking and subsequent consolidation should aim at making a wedge-shaped clamp in the initial stages. Rolling and consolidation should be a continuous process during clamp filling and special attention should be paid to consolidating the edges of the clamp. If possible they should be built a little higher than the centre of the clamp. If the process of filling the clamp is likely to take several days it is advisable (although not always popular) to sheet the clamp each night. This will help to minimise the intake of air into the clamp.

20.2.2.10   Final sheeting and sealing

Ideally, clamps should not be opened after ensiling until feeding starts. Sealing should take place as soon as filling and consolidation have finished. The majority of clamps will be sealed with two thicknesses of plastic sheet and the 'shoulders' of the clamp may well be covered by a third layer, from the side sheet. Weight should be placed evenly on top of the sheet all over the clamp. The ideal materials for doing this are the ubiquitous used car tyres or, better still, small straw bales placed tightly together.

20.2.2.11   The use of silage additives

Silage makers have used additives for many years. Some of the earliest, developed in Scandinavia, included dilute hydrochloric and sulphuric acids (the 'AIV' process) and molasses have also been used for many years. In spite of quite different modes of action both of these approaches can give rise to improved fermentation either by direct acidification or by increasing the supply of sugar leading to an increased rate of lactic acid production in the clamp. Similar materials are still in use today and formic acid marketed as 'Add-F' is one of the longest standing and most effective of all additives. Molasses are also still in use and extremely effective in the ensiling of low sugar crops such as lucerne. However, the inconvenience of applying large quantities of liquid molasses, and the extremely unpleasant and corrosive nature of acids or acid/formalin mixtures (particularly unwelcome on expensive contractor-operated machinery) have led farmers to consider alternatives.

The most widely used category today is the inoculant additive which works by the application of very large numbers (to be effective about 1 million per gramme of grass) of live, lactic acid producing bacteria, at the time of ensiling. Although not always so effective as direct acidification the inoculants have achieved a high degree of popularity mainly because of their total safety.

There is a very large array of commercially available additives of all types. The interested reader is best referred to the annual review undertaken in the UK by ADAS on behalf of the United Kingdom Agricultural Supply Trade Association (UKASTA) and known as the UK Forage Additive Approval Scheme (FAAS). The results of this are usually published in the farming press in November of each year. Additives are rated according to the beneficial effects which they can be shown to produce as a result of scientific trial evidence. These effects are either rated as having benefits on the performance of livestock, or as being beneficial to the fermentation characteristics of silage. Stability of silage at feed out and reductions in the production of effluent are also useful effects shown by some products.

Additives and inoculants are most frequently applied as liquids via applicators on the forage harvester, although granular or powder applicators have also now been developed. It would be true to say that both forms of application frequently leave a great deal to be desired in terms of accuracy. Additives can cost between 50 pence and £5 per tonne of grass and so can become quite significant items of expenditure on livestock farms. The 'star' system for determining the need for an additive was originally developed at the one time ADAS Experimental Husbandry Farm at Liscombe in Somerset. It still has relevance today and is reproduced for readers' interest as Table 20.1.

20.2.2.12 Secondary fermentation

Secondary fermentation or 'aerobic spoilage' can occur when the silage is being fed. This is particularly the case if the rate at which a clamp is being used is fairly slow, or if the size of the face exposed is large. Aerobic bacteria, yeasts and moulds can affect the silage and if the face is left exposed for a long period of time there can be significant deterioration. Bacteria of the *Listeria* family can be particularly damaging in the case of baled silage where the wrapping has been punctured, and in some cases this can lead to a serious brain infection (listeriosis) of stock feeding the silage. Maize silage is particularly susceptible to secondary fermentation at feed out because of its high carbohydrate content. Really good consolidation at ensiling coupled with ensiling in a long narrow clamp are ways in which such deterioration can be minimised. Some silage additives also claim benefits in improving the stability of silage at feed out. The use of mechanical block cutters when feeding all types of silage is also beneficial since they minimise the extent to which the face of the clamp is disturbed.

20.3   Hay

With the intensification of grassland management has come a marked reduction in the popularity of hay. On many smaller or more traditional farms the more predictable big bale silage system has replaced hay. However, from a feeding point of view it is probably still true to say that 'good hay hath no fellow' and many farmers still prefer to feed small quantities of hay as a diet 'conditioner' with high D value silage.

20.3.1   Traditional hay production

Making hay traditionally is obviously totally weather-dependent and a very risky method of feed conservation in UK conditions. The target is to reduce the moisture content of mown grass from about 80–85% to 12–15% for field conditioning and barn storage. This can frequently mean that the hay is in the field for a week or more and subject to prolonged respiration losses, as well as the physical and leaching losses that will inevitably result from frequent tedding, rowing up and the action of dews or rain showers. Such losses from material which was already of fairly low feed value (e.g. a D value of about 60 when cut) frequently result in hay with a feeding value which is little better than straw. Further hazards can arise of course as a result of baling and carrying the hay at too high a moisture content. The heating of bales with the consequent growth of bacteria and moulds, at best results in a further deterioration of feeding value and at worst in the spontaneous combustion of the stack. Handling and feeding mouldy hay can be hazardous to stock as well as causing substantial hazards to workers. The condition known as 'farmers lung' can be caused in this way.

There are a few 'dos and don'ts', however, which may have a beneficial effect on hay quality. One of these is always to stipulate diploid varieties when selecting grasses for a ley mixture which may be used for hay. The very sappy, sugary tetraploids, whilst ideal for making haylage, are extremely difficult to dry sufficiently well to make good quality field hay. Another important point is not to condition severely and certainly not to mow with a flail any crop which is to be made into field-cured hay. Leaching losses in the event of rain showers from crops so treated will be excessive. Additives to restrict microbial activity in hay bales have been investigated. These have mainly been based on propionic acid or propionate salts but have not been widely taken up. For most farmers, field-cured hay will remain merely an opportunist method of conservation should the right crop and weather conditions be obtained.

Increases in the designations of traditional species-rich hay meadows for conservation purposes may well increase marginally the quantity of this type of material conserved in the future. The management agreements for many of these meadows usually stipulate very late (e.g. July) cutting to facilitate the setting of seeds from the desired species. By this time the feeding value of the hay will be very low indeed, and some potentially hazardous infections with fungi such as ergot may have occurred.

20.3.2 Barn-drying or conditioning

The obvious value of some form of artificial drying of hay in store was recognised in the 1950s and 1960s when barn-drying installations involving electric or diesel driven fans were popular to a limited extent. The obvious disadvantages of high fuel costs, and the very high demand for labour, have subsequently reduced these installations to a very few, most of which are now producing specialist high value products for the equine market. It is possible, by producing low density bales (about 100 kg/m³), to bale and carry hay with up to 40% moisture content and to dry to less than 20% moisture in store. This can also be achieved with loose chopped hay, and various systems exist for handling and drying in store relatively small tonnages of such material. Lucerne or sainfoin hay destined for the equine market can also be successfully barn-conditioned and the stemmy nature of both of these legumes when cut for hay facilitates air movement through the bales and facilitates drying. Hay should be blown continuously and drying can take up to 21 days.

20.4   Green-crop drying

Green-crop drying can no longer be considered as a farm enterprise but as an industrial operation using agricultural crops. Most of the output is milled and cubed as grass and lucerne nuts but in some cases the material is sold long as dried grass or lucerne or chopped and mixed with molasses (e.g. 'Alfa–A').

Crops for drying may be drawn from a single farm source or contracted out onto neighbouring farms as a break from arable cropping. Grass and lucerne constitute the main output. Tall fescue is favoured in some cases because of its resilience to mowing every five to six weeks and its long growing season. Occasionally, other crops such as wholecrop maize or cereals have been dried and sold as blends with the main output of grass or lucerne nuts. In most cases the nuts are sold on to feed compounders, but some farmers and horse keepers feed them direct to stock. One of the valuable constituents of grass meal is the range of pigments in it; this is one of the main reasons for its inclusion in poultry rations, for example, to improve the colour of egg yolks. A certain amount of dried grass and lucerne is sold to livestock producers and horse owners as very high quality hay.

The system of production is extremely simple and grass is cut and dried if possible on the same day. The driers were at one time oil fired with about 200–300 litres of oil being required per tonne of dried grass (depending on the dry matter content of the original material). Many driers have recently converted to using low grade coal as a more economical fuel source. The driers themselves are usually large rotating drums and the wet grass is introduced into a stream of very hot air (over 1000 °C). It spends about three minutes in the drier, emerging at a temperature of about 120 °C and is then milled and cubed at a moisture content of about 10%. Obviously the logistics of such an operation need to be very efficiently organised, and during the main production season the plants are run 24 hours a day and 7 days a week.

Crop agronomy and fertiliser inputs will be much the same as for a livestock based unit, except that the absence of any return of organic manures necessitates the substantial use of both nitrogen and potash fertilisers on grass and mainly potash fertiliser on lucerne. In the latter case potash (K_2O) application can be as much as 375 kg/ha in order to replace that removed by the crop.

In the post-BSE era the popularity of dried grass and lucerne has improved. However, in spite of the availability of EU aid for the industry, green-crop drying has never developed in the UK to the same extent as in the rest of Europe where co-operative ventures have been (as ever) very willing to accept the very generous levels of aid available.

20.5   Further reading

Baillie J, *Assessment of Silage Additives in High Quality Wilted First Cut Grass Silage*, Kingshay Farming Trust, 1997.

British Grassland Society, 'Big bale silage: the technology of making silage in big bales and its place in the general forage conservation scene'. *Proceedings: Conference*, Stoneleigh, February 1989.

Cooper M M and Morris D W, *Grass Farming*, 5th ed., Farming Press, 1983.

Hopkins A (ed.), *Grass: Its Production and Utilization*, 3rd ed., Blackwell Science, 2000.

Nash M J, *Crop Conservation and Storage in Cool Temperate Climates*, Pergamon Press, 1985.

National Economic Development Office, *Adoption of Technology in Agriculture: Silage*, 1986.

Raymond W F and Waltham R, *Forage Conservation and Feeding*, 5th ed., Farming Press, 1996.

Spedding A and Diekmahns T, *Grasses and Legumes in British Agriculture*, CAB, 1972.

Wilkinson J M, *Silage UK*, 6th ed., Chalcombe Publishing, 1990.

Wilkinson J M, Allen D M and Newman G, *Maize: Producing and Feeding Maize Silage*, Chalcombe Publishing, 1998.

Appendices

Appendix 1

Soil texture assessment in the field

The texture of a soil (i.e. the amount of sand, silt, clay and organic matter present) can be measured by mechanical analysis of a representative sample of the soil in the laboratory. But it is important for the farmer and his adviser to be able to assess the texture of the soil in the field, not only as a guide for cultivations and general management, but also because it affects the recommended application rate of soil-acting pesticides (mainly herbicides). Many of these chemicals are adsorbed by the clay and/or organic matter in the soil and so higher dose rates may be required on soils which are rich in these materials. On sandy soils, surface-applied residual herbicides may be washed into the root zone of the crop too easily and so cause damage.

With practice, it is possible to become reasonably skilled at assessing soil texture by feeling the soil in the following way:

Carefully moisten a handful of stone-free soil until the particles cling together (avoid excess water). Work it well in the hand until the structure breaks down; rub a small amount between the thumb and fingers to assess the texture according to how gritty, silky or sticky the sample feels. The handful of moist soil can also be assessed by the amount of polish it will take, and the ease or difficulty of moulding it into a ball and other shapes.

Sands feel gritty, but are not sticky when wet (loose when dry) and do not stain the fingers.

Clays (at the other extreme of particle size) take a high polish when rubbed, are very sticky, bind together very firmly, and need some pressure to mould into shapes.

Silty soils have a smooth silky feel and the more obvious this is, the greater is the amount of silt present. The amount of polish the sample takes, and its grittiness, are guides to the amount of clay and sand present.

Loams have a fairly even mixture of sand, clay and silt and, because these tend to balance each other, loam soils are not obviously gritty, silky or sticky and they take only a slight polish. A ball of moist loam soil is easily formed, and the particles bind together well.

These are the main texture grades, but a wide range of intermediate grades exist, each having different amounts of sand (of various sizes), silt and clay particles. All this can be complicated by the amount of organic matter present which has a soft silky feel and is usually dark brown or black in colour.

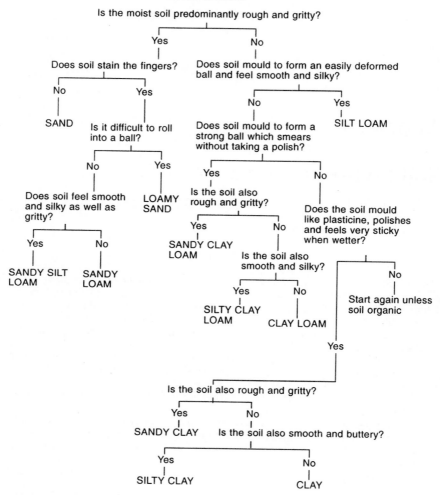

SOIL TEXTURE
MINERAL SOILS

Based on MAFF/DEFRA
RB209 (Crown copyright)

Once the particular qualities described are recognised, it should be possible to use the Texture Key to arrive at a single texture.

The prefix *organic* can be applied to the above classes if organic matter levels are relatively high, i.e. about 6% in sands or loamy sands, and about 10% in clays; prefix *peaty* when 20–35% organic matter; *peat soils* when more than 35% organic matter (peat has over 50% organic matter). The prefix *calcareous* (Calc) can be applied if more then 5% calcium carbonate is present.

The above textural groups are used in advisory work for making recommendations on dose rates for soil-acting herbicides and for assessing available-water capacity, suitability for mole drainage, workability and stability of soils.

Appendix 2

Nomenclature of crops

Crop names	Botanical names
Cereals – wheat	*Triticum aestivum (T. vulgare)*
durum	*T. durum (T. turgidum var. durum)*
barley	*Hordeum sativum*
oats	*Avena sativa*
rye	*Secale cereale*
triticale	*Triticale hexaploide*
maize	*Zea mays*
Potato	*Solanum tuberosum*
Sugar beet, mangel, fodder beet	*Beta vulgaris*
Cabbage group	*Brassica oleracea*
savoy	*B. oleracea* var. *bullata*
drumhead	*B. oleracea* var. *capitata*
red (white)	*B. oleracea* var. *rubra (alba)*
Brussels sprouts	*B. oleracea* var. *bullata* s. var. *gemmifera*
Cauliflower	*B. oleracea* var. *botrytis*
Sprouting broccoli, calabrese	*B. oleracea* var. *botrytis* s. var. *italica*
Kohlrabi	*B. oleracea* var. *gongyloides*
Kale – marrow stem	*B. oleracea* var. *acephala* s. var. *medullosa*
thousand head	*B. oleracea* var. *acephala* s. var. *millicapitata*
curly	*B. oleracea* var. *acephala* s. var. *laciniata*
Turnip group	*B. rapa* var. *rapa*
turnip oilseed rapes	*B. campestris* var. *oleifera*
Swede group	*B. napus*
swedes	*B. napus* var. *napobrassica*
swede forage rape	*B. napus* var. *oleifera*
swede oilseed rapes (colza)	*B. napus* var. *oleifera*
hungry gap and rape kale	*B. napus*
Mustard – brown	*B. juncea*
white	*Sinapsis alba (Brassica alba)*

Crop names	Botanical names
Fodder radish	*Raphanus sativus* var. *campestris*
Buckwheat	*Fagopyrum esculentum*
Carrot	*Daucus carota*
Parsnip	*Pastinaca sativa*
Celery	*Apium graveolens*
Onion	*Allium cepa*
Pea	*Pisum sativum*
Beans – field and broad	*Vicia faba*
green, dwarf, French	*Phaseolus vulgaris*
runner	*P. coccineeus*
soya	*Glycine max*
Vetch (tares)	*Vicia sativa*
Lupins – yellow (white)	*Lupinus luteus (L. albus)*
pearl (blue)	*L. mutabilis (L. augustifolius)*
Lucerne (alfalfa)	*Medicago sativa*
Sainfoin	*Onobrychis viciifolia*
Linseed and flax	*Linum usitatissimum*
Sunflower	*Helianthus annuus*
Grasses: ryegrass – Italian	*Lolium multiflorum*
hybrid	*L. (multiflorum × perenne)*
perennial	*L. perenne*
Grasses: cocksfoot	*Dactylis glomerata*
Timothy	*Phleum pratense*
meadow fescue	*Festuca pratense*
tall	*F. arundinacea*
red	*F. rubra*
Clovers: red	*Trifolium pratense*
white	*T. repens*
alsike	*T. hybridum*
crimson	*T. incarnatum*
Borage	*Borago officinalis*
Evening primrose	*Oenothera biennis*
Fenugreek	*Trigonella fenumgraecum*
Quinoa	*Chenopodium quinoa*
Fibre Hemp	*Cannabis sativa L.*
Elephant Grass	*Pennisetum purpureum* or *Miscanthus sinensis*
Kenaf	*Hibiscus cannabinus*

Appendix 3

Nomenclature of weeds

(A) annual, (B) biennial, (P) perennial

Common names	Botanical names
Barley – meadow (P)	*Hordeum secalinum*
– wall	*H. murinum*
Bent – black (P)	*Agrostis gigantea*
– common (P)	*A. capillaris*
– creeping (watergrass) (P)	*A. stolonifera*
Bindweed – black (A)	*Bilderdykia convolvulus*
– field (P)	*Convolvulus arvensis*
Birdsfoot trefoil – common (P)	*Lotus corniculatus*
Bistort amphibious (P)	*Persicaria bistorta*
Blackgrass (A)	*Alopecurus myosuroides*
Borage	*Borage officinalis*
Bracken (P)	*Pteridium aquilinum*
Bristly oxtongue (A) or (B)	*Picris echioides*
Brome – soft (A) or (B)	*Bromus hordeaceus* ssp *hordeaceus*
– barren (sterile) (A) or (B)	*B. sterilis (Anisantha sterilis)*
– rye (A) or (B)	*B. secalinus*
– great (A)	*B. diandrus*
– meadow (A) or (B)	*B. commutatus*
Broomrape – common	*Orobanche minor*
Burdock – greater (B)	*Arctium lappa*
Burnet – salad (P)	*Poterium sanguisorba*
– fodder (P)	*P. polygamum*
Buttercup – bulbous (P)	*Ranunculus bulbosus*
– corn (A)	*R. arvensis*
– creeping (P)	*R. repens*
– meadow (crowfoot) (P)	*R. acris*

Campion – red (B)	*Silene dioica*
– bladder (P)	*S. vulgaris*
Canary grass – awned (bristle-spiked) (A)	*Phalaris paradoxa*
Carrot – wild (A)	*Daucus carota*
Cat's ear (P)	*Hypochaeris radicata*
Chamomile – corn (A) or (B)	*Anthemis arvensis*
Charlock (yellow) (A)	*Sinapsis arvensis*
Chervil – rough (B)	*Chaerophyllum temulum*
Chickweed – common (A)	*Stellaria media*
– common mouse-ear (A)	*Cerastium fontanum*
Cleavers (A)	*Galium aparine*
Coltsfoot (P)	*Tussilago farfara*
Corncockle (A)	*Agrostemma githago*
Cornflower (A)	*Centaurea cyaneum*
Corn mint (P)	*Mentha arvensis*
Cornsalad – common (A)	*Valerianella locusta*
Couch – common (P)	*Elytrigia repens*
– onion (false oat grass) (P)	*Arrhenatherum elatius*
	(var. *bulbosum*)
Cowbane (water hemlock) (P)	*Cicuta virosa*
Cow parsley (wild chervil) (P)	*Anthriscus sylvestris*
Cow wheat – field (A)	*Melampyrum arvense*
Crane's – bill (A)	*Geranium* spp.
Cress – hairy bitter (A)	*Cardamine hirsuta*
– swine (A)	*Coronopus squamatus*
Cuckoo flower (P)	*Cardamine pratensis*
Daisy – common (P)	*Bellis perennis*
Dandelion (P)	*Taraxacum officinale*
Darnel (A)	*Lolium temulentum*
Dead nettle – red (A)	*Lamium purpureum*
– Henbit (A)	*L. amplexicaule*
Docks – broadleaved (P)	*Rumex obtusifolius*
– curled (P)	*R. crispus*
Duckweed – common	*Lemna minor*
– ivy-leaved	*L. trisulca*
Fat hen (A)	*Chenopodium album*
Fescue – red (P)	*Festuca rubra*
– sheep's (P)	*F. ovina*
Fiddleneck (Tarweed) (A)	*Amsinckia intermedia*
Fleabane – Canadian (A)	*Conyza canadensis*
Flixweed (A)	*Descurainia sophia*
Fool's parsley	*Aethusa cynapium*
Forget-me-not (field) (A)	*Myosotis arvensis*
Foxglove (B) or (P)	*Digitalis purpurea*
Foxtail – meadow (P)	*Alopecurus pratensis*
Fritillary (P)	*Fritillaria meleagris*
Fumitory – common (A)	*Fumaria officinalis*
Gallant soldier (A)	*Galinsoga parviflora*
Garlic – field (P)	*Allium oleraceum*
Gorse (whin, furze) (P)	*Ulex europaeus*
Goosefoot – many seeded (A)	*Chenopodium polyspermum*

Gromwell – field (A)   *Lithospermum arvense*
Ground elder (goutweed) (P)   *Aegopodium podagraria*
Ground ivy (P)   *Glechoma hederacea*
Groundsel (A)   *Senecio vulgaris*

Hawkbit – autumn (P)   *Leontodon autumnalis*
     – rough (P)   *L. hispidus*
Hawk's-beard – rough (A) or (B)   *Crepis biennis*
     – smooth (A) or (B)   *C. capillaris*
Heather (P)   *Calluna vulgaris*
     – bell (P)   *Erica cinerea*
Hedge mustard (A)   *Sisymbrium officinale*
Hedge parsley – upright (A)   *Torilis japonica*
Hemlock (A) or (B)   *Conium maculatum*
Hemp nettle – common (A)   *Galeopsis tetrahit*
Henbane (A)   *Hyoscyamus niger*
Hogweed (B)   *Heracleum sphondylium*
     – giant (P)   *H. mantegazzianum*
Horsetail – field (P)   *Equisetum arvense*
     – marsh (P)   *E. palustre*

Knapweed – common (P)   *Centaurea nigra*
Knawel – annual (A)   *Scleranthus annuus*
Knotgrass (A)   *Polygonum aviculare*
Knotweed – Japanese (P)   *Reynoutria japonica*

Loose silky bent (A)   *Apera spica-venti*

Marigold – corn (A)   *Chrysanthemum segetum*
Mayweed – scented (A)   *Matricaria recutita*
     – scentless (A)   *Tripleurospermium inodorum*
     – stinking (chamomile) (A)   *Anthemis cotula*
Meadow-grass – annual (A)   *Poa annua*
     – rough-stalked (P)   *P. trivialis*
     – smooth-stalked (P)   *P. pratensis*
Medick – black (A)   *Medicago lupulina*
Mercury – annual (A)   *Mercurialis annua*
     – dog's (P)   *M. perennis*
Mignonette – wild, cut-leaved (A) or (P)   *Reseda luta*
Mugwort (P)   *Artemisia vulgaris*
Mustard – black (A)   *Brassica nigra*
     – white (A)   *Sinapsis alba*
     – treacle (A)   *Erysimum cheiranthoides*

Nettle – common (P)   *Urtica dioica*
     – small, annual (A)   *U. urens*
Nightshade – black (A) or (B)   *Solanum nigrum*
     – deadly (P)   *Atropa belladonna*
Nipplewort (A)   *Lapsana communis*
Nutsedge – yellow   *Cyperus rotundus*

Oat – bristle (greys) (A)   *Avena strigosa*
     – spring, wild (A)   *A. fatua*
     – winter, wild (A)   *A. sterilis* spp. *ludoviciana*

Oat grass – downy (P)   *Helictotrichon pubescens*
         – false (onion couch) (P)   *Arrhenatherum elatius* (var. *bulbosum*)
Onion – wild (crow garlic) (P)   *Allium vineale*
Orache – common (A)   *Atriplex patula*

Pansy – field (A)   *Viola arvensis*
        – wild (A)   *V. tricolor*
Parsley – cow (B) or (P)   *Anthriscus sylvestris*
          – fool's (A)   *Aethusa cynapium*
Parsley-piert (A)   *Aphanes arvensis*
Pearlwort (A) or (P) – procumbent   *Sagina procumbens*
Pennycress – field (A)   *Thlaspi arvense*
Persicaria – pale (A)   *Persicaria lapathifolia*
Pineapple weed (rayless mayweed) (A)   *Chamomilla suaveolens*
Plantain – greater (P)   *Plantago major*
          – ribwort (narrow-leaved) (P)   *P. lanceolata*
Poppy – corn or common (A)   *Papaver rhoeas*
        – Californian (A)   *Eschscholzia californica*
Primrose (P)   *Primula vulgaris*

Radish – wild (sea) (A) or (B)   *Raphanus raphanistrum*
Ragwort – common (B) or (P)   *Senecio jacobia*
          – Oxford (B)   *S. squalidus*
Ramsons (P)   *Allium ursinum*
Red bartsia (P)   *Odontites verna*
Redshank (A)   *Persicaria maculosa*
Reed – common (P)   *Phragmites australis*
Restharrow – common (P)   *Ononis repens*
             – spiny (P)   *O. spinosa*
Rush – jointed (P)   *Juncus articulatus*
       – soft (common) (P)   *J. effusus*
       – hard (P)   *J. inflexus*
       – heath (P)   *J. squarrosus*

Saffron – meadow (P)   *Colchicum autumnale*
St. John's wort (P) – perforate   *Hypericum perforatum*
Scabious – field (P)   *Knautia arvensis*
Scarlet pimpernel (A)   *Anagallis arvensis*
Sedges (P)   *Carex* spp.
Selfheal (P)   *Prunella vulgaris*
Shepherd's-needle (A)   *Scandix pecten-veneris*
Shepherd's-purse (A)   *Capsella bursa-pastoris*
Silverweed (P)   *Potentilla anserina*
Soft grass – creeping (P)   *Holcus mollis*
Sorrel – common (P)   *Rumex acetosa*
         – sheep's (P)   *R. acetosella*
Sow-thistle – corn or perennial (P)   *Sonchus arvensis*
              – smooth, milk (A)   *S. oleraceus*
Speedwell – common, field (Buxbaum's) (A)   *Veronica persica*
             – germander (P)   *V. chamaedrys*
             – green field (A)   *V. agrestis*
             – ivy-leaved (A)   *V. hederifolia*
             – wall   *V. arvensis*
Spurge – sun (A)   *Euphorbia helioscopia*

– dwarf (A)	*E. exigua*
Spurrey – corn (A)	*Spergula arvensis*
Stork's-bill – common (A) or (B)	*Erodium cicutarium*
Thistle – creeping (field) (P)	*Cirsium arvense*
– spear (Scotch) (B)	*C. vulgare*
Toadflax – common (P)	*Linaria vulgaris*
Traveller's-joy (old man's beard)	*Clematis vitalba*
Trefoil hop (A)	*Trifolium campestre*
Tussock grass (P)	*Deschampsia cespitosa*
Venus's-looking-glass (A)	*Legousia hybrida*
Vetch – common (tares) (A) or (B)	*Vicia sativa*
– kidney (P)	*Anthyllis vulneraria*
Viper's bugloss (B)	*Echium vulgare*
Water-dropwort (hemlock) (P)	*Oenanthe crocata*
Willow-herb (rosebay or fireweed) (P)	*Chamerion angustifolium*
Woodrush – field (P)	*Luzula campestris*
Yarrow (P)	*Achillea millefolium*
Yellow iris or flag (P)	*Iris pseudacorus*
Yellow rattle (A)	*Rhinanthus minor*
Yorkshire fog (P)	*Holcus lanatus*

Appendix 4

Insect pests

Crop	Common names	Latin names
Cereals		
Wheat, barley, oats	cereal cyst nematode	*Heterodera avenae*
Oats, rye	stem nematode	*Ditylenchis dipsaci*
Barley	bird cherry aphid	*Rhopalosiphum padi*
Barley, oats	fescue aphid	*Metopolophium festucae*
Barley, oats, wheat	rose grain aphid	*Metopolophium dirhodum*
All	grain aphid	*Sitobion avenae*
All	leatherjackets	*Tipula* spp. and *Nephrotoma* spp.
All	wireworms	*Agriotes* spp.
All	slugs	*Deroceras reticulatum*
Wheat, barley, rye	wheat bulb fly	*Delia coarctata*
Wheat	yellow cereal fly	*Opomyza florum*
Barley, wheat	gout fly	*Chlorops pumilionis*
Oats, barley, wheat, maize	frit fly	*Oscinella frit*
All	saddle-gall midge	*Haplodiplosis marginata*
All	Thrips	*Limothrips* spp.
Barley	cereal ground beetle	*Zabrus tenebrioides*
Wheat	orange wheat blossom midge	*Sitodiplosis mosellana*
Wheat	yellow wheat blossom midge	*Contarinia tritici*
All	grain weevils	*Sitophilus* spp.
All	saw-toothed grain beetle	*Oryzephilus surinamensis*
All	mites	*Acarus siro*
All	swift moth	*Hepialus* spp.
Potatoes		
	Stubby root nematodes	{ *Trichodorus* spp. and *Paratrichodorus* spp.
		Globodera rostochiensis

Crop	Common names	Latin names
	potato cyst nematode-PCN	*Globodera pallida*
	yellow cyst nematode	*Myzus persicae, Aulacorthum*
	white cyst nematode	*solani, Macrosiphum euphorbiae*
	aphids	and *Aphis nasturtii*
	Colorado beetle	*Leptinotarsa decemlineata*
	Cutworms (noctuid moths)	
	e.g. turnip moth	*Agrotis segetum*
	large yellow underwing	*Noctura pronuba*
	garden dart moth	*Euxoa nigricans*
	slugs, leatherjackets, swift	
	moth, wireworms, see	
	Cereals	
	chafer grubs	
		Melolontha melolontha and
		Phyllopertha horticola
Sugar beet		
	beet cyst nematode	*Heterodera schachtii*
	aphid – peach potato	*Myzus persicae*
	aphid – black bean (blackfly)	*Aphis fabae*
	mangold fly	*Pegomya hyoscyami*
	millipedes	*Blaniulus guttulatus*
	docking disorder	*Longidorus* and *Trichodorus* spp.
	cutworms, slugs, wireworms,	
	chafer grubs, leatherjackets,	
	see Potatoes and Cereals	
Peas and Beans		
	pea and bean weevil	*Sitona lineatus*
	pea cyst nematode	*Heterodera gottingiana*
	stem nematode	*Ditylenchus dipsaci*
	black bean aphid (blackfly)	*Aphis fabae*
	bean beetle	*Bruchus rufimanus*
	pea moth	*Cydia nigricana*
	pea aphid	*Acyrthosiphon pisum*
	pea midge	*Contarinia pisi*
	flea beetle	*Phyllotreta* spp.
	cabbage root fly	*Delia radicurim*
	blossom (pollen) beetle	*Meligethes aeneus*
	cabbage seed weevil	*Ceuthorhynchus assimilis*
	cabbage aphid	*Brevicoryne brassicae*
	cabbage stem flea beetle	*Psylliodes chrysocephala*
	Brassicae pod midge	*Dasneura brassicae*
	small white butterflies	*Pieris rapae*
	diamond-back moth	*Plutella xylostella*
	cabbage stem weevil	*Ceatorhynchus pallidactylus*
Onions		
	onion fly	*Delia antiqua*
Carrots		
	carrot fly	*Psila rosae*

Crop	Common names	Latin names
	carrot willow aphid	*Cavariella aegopodii*
	nematodes	*Longidorus* spp.
		Trichodorus spp.
Grasses		
	frit fly, leatherjacket,	
	swift moth, wireworm, slugs,	
	see Cereals.	*Dilophus febrilis*
	bibionid flies	*Bibio marci*
Herbage legumes		
	stem nematode	*Ditylenchus dipsaci*
	sitona weevil	*Sitona* spp.

Appendix 5

Crop diseases

Crop names	Common names	Botanical names
Cereals		
All cereals	mildew$^\Delta$	*Erysiphe graminis*
Wheat, barley, rye	yellow rust	*Puccinia striiformis*
Barley	brown rust	*P. hordei*
Wheat	brown rust	*P. recondita*
Oats	crown rust	*P. coronata*
Barley	Rhynchosporium leaf blotch	*Rhynchosporium secalis*
Wheat	Septoria leaf + glume blotch	*Stagonospora nodorum 'Leptosphaeria nodorum'*
Wheat	Septoria leaf blotch	*Septoria tritici 'Mycosphaerella graminicola'*
Oats	dark leaf spot, speckle blotch	*S. avenae 'Leptosphaeria avenaria'*
Oats	stripe, leaf spot, seedling blight	*Pyrenophora avenae*
Barley	leaf stripe	*P. graminae*
Barley	net blotch	*P. teres*
Oats	halo blight	*Pseudomonas coronafaciens*
All cereals	black point	*Alternaria* spp.
Wheat, barley, rye	tan spot	*Pyrenophora tritici-repentis*
Barley	halo spot	*Selenophoma donacis*
All cereals	pink snow mould	*Microdochium nivale*
Rye, wheat, barley	ergot	*Claviceps purpurea*
Wheat, rye	bunt, covered, stinking smut	*Tilletia caries*
Barley, oats, rye	bunt, covered, smut$^\Delta$	*Ustilago hordei*
Oats	bunt, covered, smut	*U. avenae*
Barley	loose smut	*U. nuda*
Wheat, rye	loose smut different	*U. tritici*

Wheat, barley, oats	eyespot–rye type	*Tapesia acuformis*
Wheat, barley, oats	eyespot–wheat type	*T. yallundae*
Wheat, barley, oats, rye	sharp eyespot	*Rhizoctonia cerealis 'Cerato basidium cereale'*
Wheat, barley, rye	take-all	*Gaeumannomyces graminis*
Oats	take-all	*G. graminis* var. *avenae*
All cereals	brown foot rots (and ear blight)	*Fusarium* spp.
Wheat, barley	black (sooty) mould	*Cladosporium* spp.
Winter cereals (esp. barley)	snow rot	*Typhula incarnata*
Maize	bunt, covered, smut	*U. maydis*
Maize	stalk rot	*Fusarium* spp.
Maize	eyespot	*Kabatiella zeae*

Potatoes

	late blight	*Phytophthora infestans*
	pink rot	*P. erythroseptica*
	common scab	*Streptomyces scabies*
	powdery scab	*Spongospora subterranea*
	gangrene	*Phoma exigua* var. *foveata*
	watery wound rot	*Pythium ultimum*
	wart disease	*Synchytrium endobioticum*
	sclerotinia white mould	*Sclerotinia sclerotiorum*
	blackleg	*Erwinia carotovora*
	skin spot	*Polyscylatum pustulans*
	black scurf and stem canker	*Rhizoctonia solani 'Thanatephores cucumeris'*
	dry rot	*Fusarium* spp.
	brown rot	*Pseudomonas solanacearum*
	black dot	*Colletotrichum coccodes*
	silver scurf	*Helminthosporium solani*

Sugar beet

	blackleg	*Pleospora bjoerlingii*
	downy mildew	*Peronospora farinosa*
	powdery mildew	*Erysiphe betae*
	Ramularia leaf spot	*Ramularia beticola*
	beet rust	*Uromyces betae*
	violet root rot	*Helicobasidium purpureum*
	vector of rhizomania	*Polymyxa betae*

Legumes

Peas, beans	downy mildew	*Peronospora viciae*
Peas	leaf and pod spot	*Ascochyta pisi*
Beans	leaf and pod spot	*A. fabae 'Didymella fabae'*
Beans	bean rust	*Uromyces viciae-fabae*
Beans, peas	chocolate spot and grey mould	*Botrytis* spp. *'Botryotinia fuckeliana'*
Beans	stem rot (bean sickness)	*Sclerotinia trifoliorum*
Red clover	clover rot (sickness)	*S. trifolium*
White clover	phyllody virus	
All legumes	damping-off of seedlings	*Pythium* spp.
Peas	pea wilt	*Fusarium oxysporium*
Lucerne	verticillium wilt	*Verticillium albo atrum*

Lucerne	bacterial wilt	*Corynebacterium insidiosum*
Peas	powdery mildew	*Erysiphe polygoni*
Peas	bacterial halo blight	*Pseudomonas syringae*
Peas	anthracnose	*Colletotrichum lindemuthianum*
Peas	foot rot and leaf and pod spot	*Mycosphaerella pinodes*

Brassicae

	club root (finger and toe)	*Plasmodiophora brassicae*
	powdery mildew	*Erysiphe cruciferarum*
	downy mildew	*Peronospora parasitica*
	white blister	*Albugo candidus*
	light leaf spot	*Pyrenopeziza brassicae*
	dark leaf and pod spot	*Alternaria* spp.
	stem canker and leaf spot	*Phoma lingam 'Leptosphaeria maculans'*
	ring spot	*Mycosphaerella brassicicola*
	soft rot	*Pectobacterium cartovorum*
	stem rot	*Sclerotinia sclerotiorum*
	white leaf spot	*Mycosphaerella capsellae*

Linseed

	Alternariab light	*Alternaria linicola*
	pasmo	*Mycosphaerella linicola*
	powdery mildew	*Sphaerotheca lini*
	grey mould	*Botryotinia fuckeliana*

Onions

	white rot	*Sclerotium cepivorum*
	downy mildew	*Peronospora destructor*
	neck rot	*Botrytis allii*
	leaf rot	*Botrytis squamosa*

Grasses

	Drechslera leaf spot	*Drechslera* spp. 'Pyrenophora spp.'
	ergot	*Claviceps purpurea*
	powdery mildew	*Blumeria graminis*
	rusts	*Puccinia* spp.
	Rhynchosporium leaf spot	*Rhynchosporium* spp.
	crown rust	*Puccinia coronata*

Δ Different races found on different cereals

Note The sexual (teleomorph) stages for some diseases have only just been identified. These are given''. The anomorph, imperfect or asexual classification is still commonly used.

Appendix 6

Crop seeds

The following are average figures and are only intended as a general guide and for comparisons. Precision drilling of crops requires seed counts per kilogram to be known for the stock of seed being sown and merchants will usually supply these figures.

Crop	1000 seeds weight (g)	Seeds per kilogram (000's)	Seeds per m² for every 10 kg/ha sown	kg/hl
Cereals – wheat	48	21	21	75
barley	37	27	27	68
oats	32	31	31	52
maize	285	3.5	3.5	75
Grasses – ryegrass – Italian	2.2	455	455	29
ryegrass – Italian tetraploid	4	250	250	
ryegrass – hybrid	2.15	465	465	
ryegrass – perennial S 24	2	500	500	35
ryegrass – perennial tetraploid	3.3	303	303	
cocksfoot	1	1000	1000	33
Timothy	0.3	3333	3333	62
meadow fescue	2	500	500	37
tall fescue	2.5	400	400	30
Clovers – red	1.75	571	571	80
white – cultivated	0.62	1613	1613	82
wild	0.58	1724	1724	82
Lucerne (alfalfa)	2.35	425	425	77
Sainfoin – milled	21	48	48	
Peas – marrowfats	330	3	3	
large blues and whites	250	4	4	78
small blues	200	5	5	

Crop	1000 seeds weight (g)	Seeds per kilogram (000's)	Seeds per m² for every 10 kg/ha sown	kg/hl
Beans – broad	980	1	1	
winter and horse types	670	1.5	1.5	
tick	410	2.4	2.4	80
dwarf, green, French	500	2	2	
Vetches (tares)	53	19	19	77
Linseed and flax	10	100	100	68
Carrots	1.5	660	660	
Onions – natural seed	4	246	246	
mini-pellets	18	55	55	
Sugar beet – pelleted	62	16	16	
Cabbages	4.1	240	240	
Kale	4.5	220	220	
Swedes	3.6	280	280	
Turnips	3	330	330	
Oilseed rape	4.5	220	220	
Brussels sprouts	4.7	210	210	
Cauliflower	4.1	240	240	

Appendix 7

Metrication

The following factors are approximate conversions for easy mental calculations. They are correct to within 2% error, which is acceptable in agriculture! For accuracy, reference should be made to conversion tables. To convert metric to Imperial the multiplying factors should be inverted.

Imperial	Metric	Multiply by	Example
inches	to millimetres	$^{100}/_4$	12 in = 300 mm
feet	to metres	$^3/_{10}$	30 ft = 9 m
yards	to metres	$^9/_{10}$	100 yd = 90 m
square feet	to square metres (m^2)	$^1/_{11}$	55 sq ft = 5 m^2
square yards	to square metres (m^2)	$^{10}/_{12}$	60 sq yd = 50 m^2
cubic feet	to cubic metres (m^2)	$^{11}/_{400}$	800 cu ft = 22 m^3
cubic yards	to cubic metres (m^2)	$^3/_4$	12 cu yd = 9 m^3
nos. per foot	to nos. per metre	$^{10}/_3$	30 per ft = 100 per metre
nos. per sq ft	to nos. per sq metre	11	20 per sq ft = 220 per sq metre
lb per cu ft	to kg per cu metre	16(4 × 4)	50 lb/ft^3 = 800 kg/m^3
lb per bushel	to kg per hectolitre	$^{10}/_8$	60 lb bush = 75 kg/hl
pounds	to kilograms	$^9/_{20}$	100 lb = 45 kg
chain (22 yards)	= chain (20 metres)		
acres	to hectares	$^4/_{10}$	16 acres = 6.4 ha
nos. per acre	to nos. per hectare	$^{10}/_4$	40 per acre = 100 per ha
fert. units per acre	to kg per hectare	$^{10}/_8$	40 units/ac = 50 kg/ha
cost per unit	to cost per kilogram	2	9 p/unit = 18 p/kg
tons	to tonnes	1	for accuracy, × by 1.016
cwt	to kilograms	50	for accuracy, × by 50.8
tonnes per acre	to tonnes per hectare	$^{10}/_4$	2 tons/ac = 5 tonnes/ha
cwt per acre	to tonnes per hectare	$^1/_8$	32 cwt/ac = 4 tonnes/ha
cwt per acre	to kg per hectare	$^{1000}/_8$	2 cwt/ac = 250 kg/ha

Imperial	Metric	Multiply by	Example
pounds per acre	to kg per hectare	$^{11}/_{10}$	10 lb/ac = 11 kg/ha
pints per acre	to litres per hectare	$^{7}/_{5}$ or $^{11}/_{8}$	5 pints/ac = 7 litres/ha
gallons per acre	to litres per hectare	11	20 gal/ac = 220 litres/ha
tons per cubic yard	to tonnes per cubic metre	$^{4}/_{3}$	1 ton/yd^3 = 1.3 tonnes/m^3
pints	to litres	$^{4}/_{7}$	7 pints = 4 litres
gallons	to litres	$^{9}/_{2}$	1100 gal = 5000 litres
lb per gallon	to kg per litre	$^{1}/_{10}$	4 lb/gal = 0.4 kg/litre
pence per lb	to pence per kg	$^{20}/_{9}$	18 p/lb = 40 p/kg
horsepower	to kilowatts	$^{3}/_{4}$	100 h.p. = 75 kW
acres per hour	to hectares per hour	$^{4}/_{10}$	5 ac/hour = 2 ha/hour
lb ft in^2 (psi)	to kilopascals (kPa)	7	28 psi = 196 kPa (2 bars)
lb ft in^2 (psi)	to bars	$^{7}/_{100}$	200 psi = 14 bars
irrigation inch/acre	= cm/ha = 100 m^3		

Appendix 8

Agricultural land classification (ALC) in England and Wales

DEFRA land classification maps and reports

The 12 million hectares of agricultural land in England and Wales have been classified into five grades by the Department for Environment Food and Rural Affairs (DEFRA). These are used by planners when considering requests for planning permission which would take land out of agricultural use.

This survey work is published on coloured Ordnance Survey maps on three scales:

(i) 1:625000 covering the whole of England and Wales
(ii) 1:250000, There are six of these, each covering approximately one DEFRA region.
(iii) 1:63260 i.e. 1 inch to 1 mile

The details shown in (i) are more generalized than in (ii) while (iii) shows areas accurate to approx. 50 acres. The classification is based mainly on:

Climate – rainfall, accumulated temperature.
Relief – slope, surface irregularities, flood risk.
Soil – wetness, depth, texture, structure, stoniness and moisture balance.

These characteristics can affect the range of crops which can be grown, the level of yield, the consistency of and the cost of obtaining the yield. Land is graded into five grades under this classification with the third grade divided into two sub-grades.

Grade 1 – Excellent. Land with no, or very minor, limitations to agricultural use. A very wide range of agricultural and horticultural crops can be grown. Yields are high and consistent. This grade occupies 3% of agricultural land in England and Wales and is coloured dark blue on the maps.
Grade 2 – Very good. Land with minor limitations which affect crop yield, cultivations or harvesting. A wide range of crops can be grown, but there may be difficulties with very demanding crops. The level of yield is generally high, but it can be lower or more

variable than Grade 1. This grade occupies 14% of agricultural land and is coloured light blue.

Grade 3 – Good to moderate. Land with moderate limitations which affects choice of crop, timing and type of cultivations, harvesting or level of yield. Yields are generally lower or more variable than on land in Grades 1 and 2. Grade 3 is the largest group of soils, occupying 50% of agricultural land and is coloured green.

Sub-grade 3a – Moderate/high yields of a narrow range of crops (especially cereals) or moderate yields of a wide range of crops. Some good grassland.

Sub-grade 3b – Moderate yields of a narrow range of crops and lower yields of a wide range of crops.

Grade 4 – Poor. Land with severe limitations restricting range and/or yield levels. Suited to grass plus some low and variable yields of cereals. It also includes droughty arable land. It occupies 20% of agricultural land and is coloured yellow.

Grade 5 – Very poor. Land with very severe limitations. Permanent pasture or rough grazing. It occupies 14% of agricultural land and is coloured light brown.

Urban areas are coloured red and non-agricultural areas are coloured orange on the maps.

Additionally, DEFRA is undertaking a physical classification of the hill and upland areas of Grades 4 and 5 which will be divided into two main categories:

The Uplands – The enclosed and wholly or partially improved land; there will be five sub-grades.

The Hills – Unimproved areas of natural vegetation; there will be six sub-grades.

The physical factors to be taken into account will be vegetation, gradient, irregularity and wetness.

The object of this classification is to help the farmer plan possible improvement schemes when conditions permit. It is also intended to be a guide to theose concerned with conservation and amenity who are trying to preserve scenery and ecology. Booklets giving full details of all the classifications and grades and maps can be obtained from the DEFRA.

Soil Series and Soil Survey Maps

In the grading of land for the Land Classification Maps and Reports, use was made of the maps which show the Soil Series Classification of England and Wales. A 'soil series' is a soil formed from the same parent material, and having similar horizons (layers) in their profiles. Each soil series is given a name, usually where it was first recognized and described. The name is used for the soil series, however widespread, throughout the country; it is also used on the maps which are produced on 1:250000 scale. A few examples of the named soil series are:

Romney series. These soils are deep, very fine, sandy loams found in the silt areas of Romney Marsh, and also in parts of Cambridgeshire, Lincolnshire and Norfolk. They are potentially very fertile soils.

Bromyard series. Red-coloured, silt loam soils (red marls) found in Herefordshire and other parts of the West Midlands and in the south-west. They should not be worked when wet. Suitable crops are cereals, grass, fruit and hops.

Evesham series. Lime-rich soils formed from Lias (or similar) clays, found in parts of Warwickshire, Gloucestershire, Somerset, East and West Midlands. They are normally very heavy soils and are best suited to grass and cereals.

Sherborne series. Shallow (less than 25 cm deep), reddish-brown, loam-textured soils of variable depth and stoniness (and so subject to drought). They are mainly found on the soft oolitic limestones of the Cotswolds, and in parts of Northamptonshire, and the

Cliff region of Lincolnshire. The soils of this series are moderately fertile, easy to manage and mainly grow cereals and grass.

Worcester series. Red silt loam (or silty clay loam) formed from Keuper Marl and found in the East and West Midlands and in the south west. They are slow-draining, require subsoiling regularly and are best suited to grass and cereals.

Newport series. Free-draining, deep, easy-working, sandy loams over loamy sands with varying amounts of stones. They are formed from sands or gravels of glacial origin and are found in many areas of the Midlands and north. They are well suited to arable cropping.

Several different soil series may be found on the same farm and sometimes in the same field.

The Land Use Capability Classification (LUCC) in the United Kingdom has been modelled on the United States Department of Agriculture Classification Scheme. It uses the soil series groups in conjunction with limitations imposed by wetness (w), soils (s), gradient and soil pattern (g), erosion (e) and climate (c). The Classification divides all land into seven classes (1–4 classes are very similar to grades 1–4 of the Land Classification Maps prepared by DEFRA; Classes 6 and 7 are virtually useless for agriculture). Each class has sub-divisions and limitations as indicated by the letters w, s, g, e, or c, after the class number, e.g. Class 2w. This classification was carried out by the Soil Survey and Land Resource Center, Silsoe. Although a more accurate and useful system than ALC the LUCC is hardly ever used.

Land is *Not* graded on suitability for grass only. Wheat and potatoes at scales of 1:250000 for the whole country, and at 1:63360 or 1:25000 for seleted areas.

Appendix 9

Weed control

Table 1 Guide to the recognition of some common arable grass weeds by their vegetative characters

1. Shoots flattened. Leaves with boat-shaped
tips and 'tramlines' on upper surfaces

A. Young leaves with non-glossy under-surface;   annual meadowgrass
    non-stoloniferous; ligule tall and rounded.

B. Young leaves glossy on under-surface;   rough stalked meadow-
    stoloniferous; ligule pointed; heading   grass
    mid-May onwards.

2. Shoots cylindrical

A. Leaves and leaf sheaths hairy.
    (i) Auricles and rhizomes present; heading June.   common couch
    (ii) Auricles and rhizomes absent.
          Basal leaf sheath with red coloration restricted   Yorkshire fog
          to veins; dense short hairs; heading May.
          Basal leaf sheaths with red colour not restricted barren brome
          to veins; long hairs and ragged ligule; heading
          from mid-May.
    (iii) Auricles absent, bulbous base present;   onion couch
          slightly hairy leaf blades; tufted habit.
    (iv) Auricles absent; no stem-based coloration;   wild oats
          hairs present on leaf margin; long blunt ligule.

B. Leaves, leaf sheaths and stems non-hairy,
    no auricles.
    (i) Prostrate habit; stoloniferous; internodes with   creeping bent
          strong colour; ligule long; heading July.
    (ii) Upright habit; rhizomatous; ligule long   black bent
          and blunt; heading June.
    (iii) Upper surface of leaves with distinctly   blackgrass
          marked veins; no rhizomes or stolons; leaf
          sheaths purplish; ligule blunt; heading from May.
    (iv) Annual; ligule long and oblong; mainly   loose silky bent
          found in the east and south east of England.
    (v) Similar to blackgrass; leaf sheath base   awned canary grass
          pinkish; red sap oozes out of stem base
          when squeezed.

Table 2 Grass and broad-leaved weed control in autumn-sown cereals

Crop codes: w = wheat, b = barley, o = oats, r = rye, d = durum, t = triticale

Herbicides — Active ingredients: 1. Pre-emergence; 2. Post-emergence (Control at the highest recommended rate). Trade names (Always check product literature for rates, timing, safe use and restrictions).

Crop	Active ingredients	Trade names	Wild oats*	Blackgrass*	Barren brome	Annual meadow-grass	Rough-stalked meadow-grass	Seedling ryegrass*	Awned canary grass	Loose silky bent	Charlock	Chickweed	Cleavers	Crane's-bill	Dead-nettle	Forget-me-not	Field pansy	Fumitory	Mayweed	Poppy	Redshank	Shepherd's purse	Speedwell
w b r d t	tri-allate 1 (2)	Avadex Excel	S	s	s	s	s	–	–	–	r	r	r	r	r	r	r	r	r	r	R	R	r
w b r d t	tralkoxydim 2	Grasp	S	S	–	s	s	S	s	s	R	R	R	R	R	R	R	R	R	R	R	R	R
w	fenoxaprop-P-ethyl 2	Cheetah Super	S	S	–	R	S	–	S	s	R	R	R	R	R	R	R	R	R	R	R	R	R
w r d t	clodinafop-propargyl 2	Topik	S	S	–	R	S	–	–	–	R	R	R	R	R	R	R	R	R	R	R	R	R
w b r t	isoproturon (IPU) 2	Various	s	S	s	s	S	S	–	S	S	S	R	R	–	–	R	–	S	S	R	S	R
most vars w b t d	chlorotoluron 1, 2	Various	s	S	s	S	S	S	–	s	–	S	s	–	r	–	–	–	S	S	R	S	–
w b r d t	pendimethalin 1, 2	Stomp	S	s	–	S	s	–	s	–	S	S	s	S	S	S	S	S	S	S	S	S	S
w b r t	IPU+diflufenican(DFF) 2	Javelin Gold	S	S	s	S	S	s	–	S	S	S	s	S	S	S	S	–	S	S	S	S	S
w	flurpyrsulfuron 2	Lexus	–	S	–	s	s	–	–	S	S	S	r	S	s	S	–	s	S	S	S	S	S
w o r t	flurpyrsulfuron + carfentrazone-ethyl 2	Lexus Class	–	S	–	s	s	s	–	S	S	S	S	S	S	S	s	S	S	s	s	S	S
w	sulfosulfuron 2	Monitor	s	s	s	S	S	–	–	S	S	S	S	–	–	S	r	S	S	r	–	S	–
w b	flufenacet + pendimethalin 1, 2	Crystal	–	s	s	s	s	–	–	s	–	S	S	–	S	S	S	S	s	S	S	S	S
w	trifluralin + clodinafop-propargyl 2	Hawk	S	S	–	S^	S	s	–	–	–	S^	S	S^	s^	S^	–	s^	–	S^	S^	S	–
w b	picolinafen + pendimethalin 2	PicoPro	–	–	–	S	s	–	–	–	S	S	S	S	S	–	S	S	R	S	–	S	S^

*Note strains of some of these grass weeds have developed resistance to some herbicides (ureas, fops and dims)
^Weed control pre-emergence of weeds
Symbols: S – susceptible, s – moderately susceptible, r – moderately resistant, R – resistant, – no information.
Always check the product label for details of timing, application rates, restrictions and safe use.

Table 3 General guide to spring broad-leaved weed control in cereals

Crop	Herbicides (Active ingredients – all post-emergence. Control at the highest recommended rates. See product literature for rates, timing, tank mixes, safe use and restrictions)	Trade names	Black bindweed	Charlock	Chickweed	Cleavers	Crane's-bill	Dead-nettle (red)	Fat hen	Forget-me-not	Fumitory	Hemp-nettle	Knotgrass	Marigold, corn	Mayweed	Nettle, small	Pansy, field	Poppy, common	Redshank	Shepherd's purse	Sow-thistle	Speedwell, common	Speedwell, ivy-leaved	Spurry, corn
w b o r	MCPA	Various	r	S	–	R	r	R	S	–	S	S	s	R	r	S	–	S	r	S	S	–	–	r
w b o	mecoprop-P	Various	r	S	S	S	r	r	s	R	S	S	r	R	s	S	R	s	s	S	s	–	r	r
w b o r d t	fluroxypyr	Various	s	R	S	S	r	s	R	S	s	r	s	R	r	R	s	R	R	R	–	r	r	–
w b o r t	ioxynil + bromoxynil (HBN)	Various	S	S	s	r	–	S	S	S	S	r	S	r	S	S	r	R	r	S	S	S	s	R
w b o d t	HBN + mecoprop-P	Swipe P	S	S	S	S	S	S	S	S	S	S	S	s	S	S	s	S	S	S	–	S	S	–
w b o	HBN + benazolin	Asset	S	S	S	S	–	S	S	S	S	–	S	S	S	S	r	S	S	S	S	S	S	r
w b o r d t	amidosulfuron	Eagle	–	S	–	S	–	–	–	S	–	–	–	–	–	–	–	–	–	S	–	–	S	–
w b o t	metsulfuron-methyl	Various	r	S	S	R	S	S	s	s	–	S	s	S	S	S	s	S	S	S	S	S	–	S
w b d t	carfentrazone-ethyl	Aurora	s	S	–	S	S	S	S	–	s	S	–	–	–	–	s	s	S	S	–	S	R	–
w b o d t	metsulfuron-methyl + carfentrazone-ethyl	Ally Express	s	S	S	S	S	S	S	S	–	S	S	S	S	S	S	S	S	S	S	S	S	S
w spring b	thifensulfuron-methyl + metsulfuron-methyl	Harmony M	S	S	S	s	S	S	s	S	S	S	S	S	S	S	s	S	S	S	S	S	s	S
w b o d t	tribenuron-methyl	Quantum	s	s	S	R	s	s	s	s	R	s	s	s	R	S	R	S	S	S	S	s	s	–
w b r d	cinodon-ethyl	Lotus	S	s	R	S	–	S	–	–	–	S	–	–	–	–	–	R	–	–	–	S	s	–

Symbols: S – susceptible, s – moderately susceptible, r – moderately resistant, R – resistant, – no information
Always check the product label for details of timing, application rates, restrictions, tank mixes and safe use.

Table 4 General table of weed susceptibility to some commonly used herbicides in some broad-leaved crops

Crop	Herbicides	Timing	Black bindweed	Charlock	Chickweed	Cleavers	Fat hen	Hemp-nettle	Knotgrass	Marigold, corn	Mayweeds	Nettle, small	Nightshade, black	Pennycress	Poppy, common	Redshank	Shepherd's purse	Speedwell	Thistle, creeping	Blackgrass	Brome barren	Meadow-grass annual	Cereal volunteers	Wild oats
Potatoes	paraquat + diquat 'PDQ'	Contact only pre/early post-em	s	S	S	r	r	S	r	r	S	r	S	S	S	S	S	S	r	S	s	S	S	S
	metribuzin 'Sencorex'	pre- and post-em	S	S	S	r	s	S	S	s	S	S	r	S	S	S	S	S	R	S	s	S	S	r
	linuron – various	pre-em	r	S	S	S	S	S	–	S	S	S	r	–	S	S	S	S	–	R	–	S	–	R
	rimsulfuron 'Titus'	post-em	S	s	S	r	r	S	r	S	S	S	s	–	S	S	S	S	–	S	–	S	S	S
Sugar beet	chloridazon – various	pre-em	r	S	S	r	r	S	r	–	S	S	r	–	S	s	s	S	–	r	–	S	–	R
	metamitron 'Goltix'	pre-or-post-em	S	S	S	R	S	S	S	S	S	S	s	S	S	S	S	S	r	R	–	S	–	R
	phenmedipham – various	post-em	r	s	S	s	S	s	s	s	r	S	r	s	s	s	s	S	r	R	R	R	R	R
	ethofumesate		s	S	S	s	S	S	S	–	S	S	S	s	S	S	S	S	R	S	s	s	s	s
Oilseed rape	metazachlor – various 'Nortron'	pre- or post-em	s	R	s	s	s	r	s	–	S	s	–	R	s	s	–	s	R	S	s	R	s	r
	benazolin+clopyralid 'Benazalox'	pre- or post-em	s	s	S	s	s	–	s	S	S	–	–	–	s	s	–	R	–	R	R	R	r	R
	propyzamide 'Kerb'	post-em	S	R	S	–	S	–	S	–	R	S	S	–	R	S	R	s	–	S	S	S	r	R
	clopyralid 'Shield'	post-em	s	–	–	–	–	–	–	S	S	–	–	–	–	r	–	–	S	R	R	R	R	R

Table 4 *(Contd)*

		Soil																		
Fodder brassicas	trifluralin – various	incorporated	S	R	S	S	S	S	R	S	R	R	–	S	R	S	R	S	–	s
	propachlor – various	pre-em	R	R	S	S	r	S	S	S	–	R	S	S	S	–	S	R	R
Peas	cyanazine – 'Fortrol'	pre or post-em	S	S	S	–	s	S	S	–	S	–	S	S	s	S	S	–	R
	pendimethalin 'Stomp'	pre-em	S	–	S	s	S	S	s	S	S	–	S	R	S	–	S	R	s
	bentazone + MCPA 'Pulsar'	post-em	S	S	S	s	S	R	S	S	–	S	S	S	R	–	R	–
			S	S	S	s	S	S	S	–	S	S	S	S	S	–	S	r
Field beans	simazine – various	pre-em	S	S	S	R	S	S	S	–	S	S	S	R	R	–	s	–	R
	trietazine + simazine 'Remtal SC'	pre-em	s	S	S	R	S	s	S	–	S	S	S	S	S	–	S	S	S
Many of the above crops (Check the label)	Graminicides e.g. cycloxydim 'Laser', fluazifop-P-butyl 'Fusilade' and propaquizafop – various	post-em	R	R	R	R	R	R	R	R	R	R	R	R	R	S	R	S	S

Symbols S – susceptible, s – moderately susceptible, r – moderately resistant, R – resistant, – no information
Always check the product label for details of timing, application rates, restrictions and safe use.

Table 5 Some common weeds and their control in grasslands

Weed	Control
Bracken	This weed is a very serious problem which is increasing, especially in hill areas. The fronds (leaves) should be cut or crushed twice a year when they are almost fully open. If possible, the field should be ploughed deep and then cropped with potatoes, rape or kale before reseeding. Rotavating to about 25 cm deep chops up and destroys the rhizomes. Very good chemical control is possible with asulam applied in the summer when the fronds are just fully expanded. Glyphosate can be used as a non-selective spray before reseeding or, selectively, with a rope-wick machine.
Buttercups	Spray with MCPA. The bulbous buttercup is the most resistant type. Improved drainage and soil fertility will lessen this problem.
Chickweed	Several chemicals will control chickweed including mecoprop-P, ethofumesate and fluroxypyr (note they kill clover). Benazolin mixtures are clover-safe.
Docks	Grazing can be helpful. Asulam gives good control when sprayed on the expanded leaves in the spring or autumn. Triclopyr, fluroxypyr, dicamba mixtures, mecoprop-P, thifensulfuron-methyl and MCPA alone or in mixtures will give good control, although treatment may need repeating. If reseeding, the old sward should be sprayed with glyphosate before ploughing.
Horsetails	If possible, drainage should be improved. Spraying with MCPA or 2,4-D will kill aerial parts only and regrowth will occur. However, if it is done two to three weeks before cutting, any hay crop should be safe for feeding.
Nettles	Spray with triclopyr alone or in mixtures; glyphosate can also be used through a rope-wick applicator.
Ragwort	Cut before the buds develop to prevent seeding. Spray with 2,4-D or MCPA in early May or in the autumn. Grazing with sheep in winter can be helpful.
Rushes	An increasing problem in many areas. If possible, drainage should be improved. Spray with MCPA or 2,4-D. The hard and jointed rush should be cut several times a year or treated with triclopyr. Apply glyphosate with a rope-wick when the rushes are growing well. Grasses and clovers should be encouraged by good management.
Sorrel	Lime should be applied if necessary. Spray with MCPA or 2,4-D.
Thistles	Spray with clopyralid or MCPA. Creeping thistle should be sprayed at the early flower bud stage. Over-grazing should be avoided. Cutting in July is very effective for control of the biennial spear thistle.
Tussock grass	Drainage should be improved. The 'tussocks' should be cut off with a flail harvester or topper. A hay or silage crop could be taken. Glyphosate can be applied with a rope-wick when the tussock grass is growing well.

Appendix 10

Map of Ontario heat units showing areas most suited to growing maize

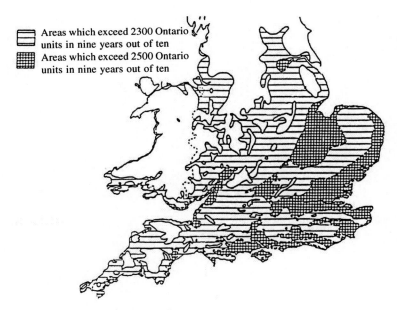

Areas which exceed 2300 Ontario units in nine years out of ten
Areas which exceed 2500 Ontario units in nine years out of ten

Source: Meteorological Office, Bracknell

Index